Supply Chain Management
for Sustainable Food Networks

Supply Chain Management for Sustainable Food Networks

Edited by

Eleftherios Iakovou
Aristotle University of Thessaloniki, Greece

Dionysis Bochtis
Aarhus University, Denmark

Dimitrios Vlachos
Aristotle University of Thessaloniki, Greece

Dimitrios Aidonis
Technological Educational Institute of Central Macedonia, Greece

Library of Congress Cataloging-in-Publication data applied for

ISBN: 9781118930755

A catalogue record for this book is available from the British Library.

Set in 10/12pt Times by SPi Global, Pondicherry, India
Printed and bound in Singapore by Markono Print Media Pte Ltd

1   2016

Contents

Notes on contributors

Chapter 1

Eleftherios Iakovou is a Full Professor and the Chairman of the Department of Mechanical Engineering of the Aristotle University of Thessaloniki, Greece. He is also the Director of the Laboratory of Advanced Supply Chain Management (LASCM) of the Department of Mechanical Engineering. Furthermore, he has been the President of the Greek Association of Supply Chain Management (GASCM) since 2005. He is the Project Coordinator for the leading European Union (EU) funded "GREEN-AgriChains" project (http://www.green-agrichains.eu). His teaching, research, and consulting interests include supply chain management, port management, intermodal logistics, inventory management, applied operations research, and innovation and strategy development. He has published extensively having more than 180 papers in top peer reviewed journals and conference proceedings and two books.

Dionysis Bochtis is a member of the Operations Management group of the Department of Engineering at Aarhus University, Denmark. His primary research is industrial engineering focused on bio-production and related supply chain systems including activities relate to fleet management (for conventional and autonomous field machinery), field robots (high level control aspects: mission planning, path planning, task allocation), biomass and agri-food supply chains management, and Decision Support Systems. He is the author of more than 200 articles in peer reviewed journals and conference proceedings including a number of invited and key-note presentations. He was the President of CIOSTA (Commission Internationale de l'Organisation Scientifique du Travail en Agriculture, founded in Paris, 1950) between 2011 and 2013.

Dimitrios Vlachos is an Associate Professor at the Department of Mechanical Engineering of Aristotle University of Thessaloniki, Greece. He has developed research activities in the fields of inventory control in supply chain management, green logistics, risk management, applied operational research, and business evaluation. He has participated in more than 45 research projects. He has authored or co-authored more than 120 papers in scientific journals and conference proceedings.

Dimitrios Aidonis is Lecturer at the Department of Logistics of the Technological Educational Institute of Central Macedonia, Greece. He received both his Diploma in Mechanical Engineering and a PhD from the Department of Mechanical Engineering of Aristotle University of Thessaloniki, Greece and his MBA from Staffordshire University, UK. His research interests focus on the areas of Applied Operations Research in Supply Chain, Reverse, and Green Logistics participating in European projects. He is the author of more than 50 papers in scientific journals and conference proceedings. He is a member of the Society of Logistics Engineers and the Council of Supply Chain Management Professionals.

Chapter 2

Spyros Fountas is an Assistant Professor at the Agricultural University of Athens, Greece. He is the Editor in Chief of *Computers and Electronics in Agriculture* and on the Editorial Board of the *Precision Agriculture* journal. He has 15 years of experience in agricultural robotics, precision agriculture technology and management and implementation of information systems within crop production. He holds an MSc from Cranfield University, UK in Information Technology, and a PhD from Copenhagen University, Denmark. He was also Visiting Scholar at Purdue University, USA. He has published 130 papers, including 45 papers in international journals, 105 in conference proceedings, and 8 book chapters.

Katerina Aggelopoulou is a graduate from the Department of Physics, University of Ioannina, Greece (BSc), and the Department of Agriculture, University of Thessaly, Greece (BSc). She has a postgraduate degree (MSc) in Agricultural Engineering & Management of Natural Resources from the University of Thessaly; her MSc thesis was entitled "Long-term effect of five soil treatment methods on soil and sugar beet crop". Her PhD thesis, also from the University of Thessaly, was on "Precision Agriculture in apples." Her research interest is in the application of Precision Agriculture in various crops and analyzing spatial data. She has participated in 5 research projects, with 6 publications in international journals, and 23 papers in international and national scientific conference proceedings.

Theofanis A. Gemtos was born in Larissa, Greece in 1947. He studied Agriculture at the University of Thessaloniki, Greece and Agricultural Engineering at the University of Reading, UK (PhD). Between 1979 and 1993, he was a Professor at the Technological Education Institution of Larissa, Greece and between 1993 and 2014, he was a Professor at the Department of Crop Production and Rural Environment in the University of Thessaly, Greece. Since 2014 he has been Professor Emeritus. He has participated in more than 25 research projects. He has published more than 47 journal papers, and presented more than 100 papers at national and international conferences and has written 4 book chapters.

Chapter 3

Sven G. Sommer is a specialist in animal waste engineering with a focus on recycling organic waste, gas emission (ammonia and greenhouse gases), separation of animal manure and biogas production. His research has a strong focus on manure chemistry and biological and chemical processes related to gas emission and facilitating manure separation.

Lorie Hamelin is an Agricultural Engineer with a PhD in Environmental Technology. She is currently an Assistant Professor at the University of Southern Denmark. Her key expertise lies in performing high quality life cycle assessments. Her research focuses on integrating agricultural, energy, and waste systems in order to optimize our use of biomass and land resources while sustainably supplying food, feed, materials, and renewable energy. She has managed 7 Danish and international projects and has acted as a Danish Technical Working Group member for the BREF (Best Available Techniques Reference Document) revision on intensive rearing of poultry and pigs, carried out by the European Integrated Pollution Prevention and Control Bureau.

Jørgen E. Olesen is a Professor (Section Manager) at the Department of Agroecology (Climate and Water Section) at Aarhus University, Denmark. His research interests include adaptation of agricultural production systems to climate change, emission of greenhouse gases from agriculture, organic and integrated crop production systems, and modeling of the soil–plant–atmosphere system and at the farm scale. He has led several interdisciplinary projects, including projects on integrated wheat production, application of remote sensing and Geographic Information Systems in agriculture, and development of a whole-farm simulation model. He has participated in more than 10 EU projects and concerted actions on the effect of climate change on agriculture and on greenhouse gas emissions from agricultural activities.

Felipe Montes is a civil and environmental engineer specializing in the development and the use of mathematical models to assess and mitigate the environmental impact of nutrient and pollutant transfer from agricultural production systems into soil, water, and the atmosphere. His research interests include the development of models for ammonia, hydrogen sulfide, volatile organics, and greenhouse gas emission from animal production systems; life cycle analysis and carbon foot print of milk production; and distributed hydrology models for sediment and nutrient load reduction in impaired agricultural watersheds.

Wei Jia graduated from China Agricultural University, Beijing, China. He focuses on manure nutrient management. His current research work is related to Comprehensive Nutrient Management Plan (CNMP) in intensive livestock production as a post-doctor at the China Academy of Agriculture Science (CAAS).

Qing Chen is working on organic waste recycling and utilization with focus on added-value fertilizer, crop nutrient management, and land application. His research strongly focuses on root zone nutrient management, nitrogen and phosphorus transformation, and utilization in soil and organic waste.

Jin M. Triolo is a specialist in environmental biotechnology with a focus on biogas production, aerobic and anaerobic waste water treatment, greenhouse gas emission, and energy-driven biorefinery systems. Her research has a strong focus on biomass characteristics and recalcitrance, integrating new biomass to bioenergy production, biochemical processes related to biogas production, and developing spectroscopic method for biomass characterization.

Chapter 4

Marcus Wagner is a Full Professor and holder of the Chair of Management, Innovation, and International Business at Augsburg University, Germany. He is also visiting Professor at the International Hellenic University of Thessaloniki, Greece. His research interests are strategic and public management, innovation, sustainability and international business, and their intersections. He has published on these topics in journals such as *Research Policy*, *Long Range Planning*, *Journal of Environmental Economics and Management*, *Journal of World Business*, *Journal of Business Research*, and *Journal of Business Venturing*.

Chapter 5

Rommert Dekker is a Full Professor in Operations Research in the Faculty of Economics at the Erasmus University Rotterdam, The Netherlands and is ranked no. 10 among all Dutch economists/researchers. His research focus is in the field of reverse logistics, sustainability, spare parts management, transportation optimization, port and container logistics, maintenance optimization and reliability. He has numerous publications in top-ranked international journals and books.

Jacqueline Bloemhof-Ruwaard is a Professor in the Operations Research and Logistics Group, Wageningen University, The Netherlands. Her main field of research is sustainable supply chain management, both in forward chains (food and other agricultural distribution networks) and closed loop supply chains (recycling and recovery of products, parts, materials, or energy). She has published on these topics in a large number of articles in ISI journals in the field of Operations Research and Environmental Science, Engineering and Environmental Sciences.

Ioannis Mallidis is a Senior Post-Doctoral Researcher at the Laboratory of Advanced Supply Chain Management (LASCM) of the Department of Mechanical Engineering, Aristotle University of Thessaloniki, Greece, and is currently employed under the

"GREEN-AgriChains" European Project. His research focus is in the field of green supply chain network design and planning and he has published in top peer reviewed transportation and Operations Research journals.

Chapter 6

Fabrizio Dabbene received his Laurea degree in 1995 and his PhD degree in 1999, both from Politecnico di Torino, Italy. He is currently Senior Researcher at the CNR-IEIIT Institute. His research interests include randomized and robust methods for systems and control, and modeling of environmental systems. He has published more than 100 research papers and 2 books, and was the recipient of the 2010 EurAgeng Outstanding Paper Award. He served as Associate Editor for *Automatica* (2008–2014) and *IEEE Transactions on Automatic Control* (2008–2012). He is a Senior Member of the IEEE, and has taken various responsibilities within the IEEE-CSS: he is an elected member of the Board of Governors (2014–2016) and currently serves as Vice President for Publications (2015).

Paolo Gay received his Laurea degree in 1994 and his PhD in Information and System Theory in 1999 from Politecnico di Torino, Italy. Since 2000 he has been with the Department of Agricultural, Forest and Food Sciences of the University of Torino, where he is currently serving as Associate Professor. His research interests are in the fields of robotics and automation for agriculture and food industries, technologies for food traceability, and supply chain optimization. He is the author of more than 120 research papers and was the recipient of the 2010 EurAgeng Outstanding Paper Award. In 2014 Elsevier awarded him the Certificate of Reviewing Excellence for the *Biosystems Engineering* journal.

Cristina Tortia has a degree in Agricultural Science from the University of Turin (1989), and a PhD in Fungi Biotechnology and Microbiology (1993), also from the University of Turin. Her research activity has involved logistics in the agro-food sector, modeling, simulation, optimization of agro-food processes, radiofrequency technologies for food traceability, treatments for soil disinfestation, microbiology and technology in oenology, dairy industry, and industrial fermentations (ethanol production from biomass). She is enrolled as a senior technical expert at DISAFA, University of Turin. She is the author of more than 70 research papers. She is a member of AIIA (Associazione Italiana di Ingegneria Agraria), CIGR, and Eurageng.

Chapter 7

Katerina Pramatari is an Associate Professor at the Department of Management Science and Technology of the Athens University of Economics and Business (AUEB), Greece and scientific coordinator of the ELTRUN-SCORE (Supply Chain & Demand Management Collaboration and Electronic Services) research group. She holds a BSc in Informatics and MSc in Information Systems from AUEB, and a PhD in Information Systems and Supply Chain Management also from AUEB. Prior to joining the academia, she worked as a systems analyst and assistant brand manager at Procter & Gamble. She has received various academic distinctions and scholarships and has published more than 100 papers in scientific journals, peer reviewed academic conference proceedings, and book chapters.

Chapter 8

Agorasti Toka is a Mechanical Engineer and Senior Post-Doctoral Researcher at the Laboratory of Advanced Supply Chain Management (LASCM) of the Department of Mechanical Engineering, Aristotle University of Thessaloniki, Greece. She has also been the General Director of the Greek Association of Supply Chain Management (GASCM) since 2012. Her teaching and research interests include biomass supply chain management, sustainable energy, and policy-making. She has published her research in top peer reviewed journals, scientific books, and refereed conference proceedings. She was also awarded "Best Young Researcher on Biomass" at World Sustainable Energy Days 2015 in Wels, Austria.

S. C. Lenny Koh is a Chair Professor in Operations Management, Founders and Directors of the Logistics and Supply Chain Management (LSCM) Research Center at the Management School, the Faculty's Centre for Energy, Environment and Sustainability (CEES), and the Advanced Resource Efficiency Centre (AREC) of The University of Sheffield, UK. With high h-index (world number 2) and high research income generation in her discipline internationally, she is also Cross-Cutting Chief of a 2022 Futures initiative for advancing supply chain resource sustainability.

Victor Guang Shi is based in the Advanced Manufacturing Research Centre with Boeing at the University of Sheffield, UK. His research interests include high value manufacturing, complex product service systems, environmental sustainability, and organizational theories.

Chapter 9

Ioannis Manikas is a Principal Lecturer of Logistics and Supply Chain Management in the Department of Systems Management and Strategy at the University of Greenwich, UK. He holds a BSc in Agriculture and Food Science from the Aristotle University of Thessaloniki, Greece and an MSc from Cranfield University, UK. He holds a PhD from the Department of Agricultural Economics of the School of Agriculture in Aristotle University of Thessaloniki, Greece. He has conducted research for projects regarding supply chain modeling, development of IT solutions for agri-food supply chain management and traceability both in Greece and the UK.

Karen Hamann (MSc Agric.) is the Managing Director of The Institute for Food Studies & Agroindustrial Development (IFAU), Denmark. From 1994 to 1997 she worked as a project coordinator in IFAU, and since 1998, she has been the Managing Director. Her research and work in IFAU have provided her with an extensive knowledge of the global agri-food system. She has continuously developed a personal network to organizations, authorities, and companies all over Europe and elsewhere. She has contributed to many research projects on food sub-sectors, food chains, and food technology and innovation. Further, she has worked with market development issues across Europe and elsewhere with special attention to agriculture, food supply and demand, and studies about trends and forecasts. She has experience in project work including management from IFAU's customized research projects and EU funded projects. She is used to working in an international forum.

Anton Sentic is a doctoral student working with the Sustainability, Technology & Innovation Research Group (STIR) at the Faculty of Business of the University of Greenwich, UK. The main focus of his work is on dynamics of sustainable transitions in energy systems as well as sustainable transitions in general. He also engages in research on the connections of sustainable development and food safety and the role of sustainability in fulfilling human needs. His previous work includes several international projects in the waste management sector and research on sustainability management in small- and medium-sized enterprises, undertaken at the University of Graz, Austria.

Chapter 10

Christos Keramydas is a Senior Researcher at the Laboratory of Quantitative Analysis, Logistics and Supply Chain Management (LASCM), Department of Mechanical Engineering, Aristotle University of Thessaloniki (AUTh), Greece. He also holds a Diploma degree in Mechanical Engineering (School of Engineering, AUTh), and an MSc degree in Statistics and Modeling (School of Sciences, Department of Mathematics, AUTh). His research interests, and his teaching experience are spread over the fields of statistics, operations research, and simulation applications, as well as in supply chain management, logistics, risk management, and food security. He has participated in several national and international research projects, and has published several papers in refereed scientific journals, and conference proceedings. Finally, he has been awarded several honors and scholarships during his career, thus far.

Konstantinos Papapanagiotou is currently working as a researcher and engineer at the Aristotle University of Thessaloniki, Greece. He received a PhD degree in Supply Chain Risk Management, a MSc degree in Information Systems & Business Administration, and a Diploma/MSc degree in Electrical Engineering & Computer Engineering from the Aristotle University of Thessaloniki. His research interests include risk management, game theory, security and operations management, as well as the development of modeling, simulation, and optimization tools. Since

2001 he has managed and participated in research and industrial projects, and published his work in scientific journals, conference proceedings, and book chapters.

Dimitrios Vlachos and **Eleftherios Iakovou** as in Chapter 1.

Chapter 11

Sirpa Kurppa is a specialist in the ecology of food production systems. She has wide-ranging science experience of more than 30 years, and has provided expert input into EU Rural Foresight studies and in work of the Committee for the Future of the Finnish Parliament, in the National Food Strategy and in the Strategy for Sustainable Consumption and Production. She was involved in preparing a proposal on green growth for the Finnish Parliament and preparing the Finnish strategy for bioeconomy. From 2013 to 2015 she was a member of the EU Bioeconomy Panel, and has been a member of the National Nutrition Council since 2014.

Preface

The agrifood industry is a sector of significant economic and political significance. It is one of the most regulated and protected sectors with significant implications for sustainability such as the fulfillment of human needs, the support of employment and economic growth, and its impact on the natural environment. According to the European Commission, the food and drink sector contributes to some 23% of global resource use, 18% of greenhouse gas emissions, and 31% of acidifying emissions.[1] Growing environmental, social and ethical concerns and increased awareness of the effects of food production and consumption on the natural environment have led to increased pressure by consumer organizations, environmental advocacy groups, policy-makers, and several consumer groups on agrifood companies to deal with social and environmental issues related to their supply chains within product lifecycles, from "farm to fork."

The agrifood industry is expected to grow in the next couple of years after a long period of recession. To that end the industry is facing new challenges that arise from:[2]

1. *New consumer trends.*

2. *The need to comply with stricter and often non-harmonized national regulatory interventions regarding product safety, quality and traceability.*

3. *Increased sources of risk throughout its supply chains.*

This book was motivated by a three-year leading research project (2012–2015) that was funded by the European Union (EU). Specifically, the Green-AgriChains Project has received funding from the EU's Seventh Framework Programme (FP7-REGPOT-2012-2013-1 under Grant Agreement No. 316167) and it involved eight leading EU universities along with four business clusters. To that effect, the gracious support of the European Commission is gratefully acknowledged.

This book intends to provide a holistic, up-to-date, interdisciplinary framework for designing and operating sustainable supply chains for agrifood products and it intends to add value to both practitioners and academics alike. The aim is to present

---

[1] http://www.euractiv.com/specialreport-prods-green-planet/cutting-food-waste-greening-diet-news-513731

[2] http://www.grant-thornton.co.uk/en/Publications/2013/Hunger-for-growth--Food-and-Beverage-looks-to-the-future/

sustainable practices that are unique for agriculture (such as organic products or precision farming), as well as practices that already have been implemented in other industrial sectors [such as transportation emissions control or corporate social responsibility (CSR)]. All book chapters include decision-making procedures and methodologies, most of which are quantitative. Even though we do discuss the most emerging state-of-the-art relevant technologies, our focus is more on the managerial dimension of the examined policies.

Chapter 1 is an introduction to *sustainable agrifood supply chain (AFSC) management*. The chapter summarizes the unique challenges for supply chain managers especially related to sustainability. These challenges are then further fine-tuned for the agrifood business. The purpose of this introductory chapter is to provide the basic managerial knowledge and motivation for the readers of the book, merging the worlds of operations management, supply chain management, and agriculture. It begins by presenting the generic system components along with the unique characteristics of AFSC networks that differentiate them from traditional supply chains. The authors then identify and discuss the most critical issues for the design and planning of AFSCs, along with the most relevant emerging technologies. They then present a critical synthesis of the related existing state-of-the-art literature efforts in order to identify major gaps, overlaps, and opportunities. These issues were further mapped accordingly on the recognized natural hierarchy of the relevant decision-making process and key findings and managerial insights are presented.

Chapter 2 discusses *knowledge-based farming*. The chapter covers the implementation of engineering management in agrifood production systems as a basis for the creation of a new generation of intelligent and sustainable processes by employing novel system approaches realized by embedded intelligent technologies for planning and controlling the use of all involved resources. It further demonstrates how robustly managed systems can address the inherent complexity and dynamic nature of bio-production systems. First, a general outline of Precision Agriculture (PA) as applied to crops is provided. Then, a general plan of its application describing the relevant data collection methods for capturing the variability of the fields and the crops is discussed. An account of the data analysis and the methods to use the data in the site specific management of the crops is provided. Several applications are presented indicating the potential of PA to lead to an optimization of the usage of resources such as fertilizers, chemicals, water, and energy leading to reduced inputs and minimizing adverse effects to the environment. In several applications the economic benefits to the farmers are also substantiated. PA can address the main components of agriculture sustainability. From an economic perspective PA can improve income to farmers, from a social perspective it can improve working conditions for farmers and the farming communities bringing the farmers to the cutting edge technological era, while from an environmental perspective the adverse effects on the environment are greatly addressed by reducing inputs and resource use.

Chapter 3 deals with *biomass from agricultural wastes*. Biomass logistics encompasses two parts, each different in scope. The first part involves the farm production of biomass and the dedicated transport system as the initial steps in the biomass supply chain. It is characterized as a low industrialized process, where planning and execution remain very much implicit and internal with only a sparse tradition for using formalized planning tools. The other part involves the biomass processing facilities comprising the specific bio-energy production/processing; it is characterized as a highly industrialized process with a long tradition of explicit formalized planning and execution tools. The overall goal of this chapter is to identify research actions for improving the overall biomass waste logistics systems by extending the methods and technologies of industrial operations and production management to include a biologically constrained production system, while taking into account the sustainability (environmental, greenhouse gas emissions, energy balance) of the entire system.

Chapter 4's theme is *maintaining sustainability.* Stakeholders' demands have been suggested to affect the environmental and social activities of firms which in turn influence various performance dimensions. This chapter contributes to the analysis of this relationship by looking at the extent to which stakeholder demands are related to the integration of management activities within the firm, and by testing the hypotheses related to moderation effects of industry, firm type, and governance structures. Using data from the manufacturing sector in the UK and Germany, it examines the way in which stakeholder pressures are associated with management integration and economic as well as environmental performance, as defined by means of six sub-dimensions. Applying structural equation modeling it is documented that stakeholder pressure inevitably leads to economic and environmental performance integration, but that important moderation effects do exist.

Chapter 5 is an amalgamation of academic research efforts, offering *a review of quantitative optimization methodologies employed for evaluating the economic and environmental impacts of implementing green supply chain management (GSCM) decisions.* More specifically, the main GSCM decisions that may affect the economic and environmental performance of the three main physical drivers of a supply chain, namely products, facilities, and transportation, are identified, along with the quantitative optimization models employed for quantifying these impacts. Finally, these decisions are mapped into strategic, tactical, and operational decision phases accordingly and a critical synthesis of the academic research efforts is provided.

Chapter 6 discusses *safety, security, and traceability.* Traceability is a tool for sharing product related information among all members in the AFSC. It helps in terms of transparency of the network, contributing toward improved production and distribution management, promoting health, safety, and quality issues of agrifood product while mitigating associated risks in the entire chain. Overall, traceability can help in terms of product differentiation and provides significant financial benefits. The first part of the chapter deals with an extensive investigation of the most effective tracking and tracing systems that are already used in AFSCs. The aim of the second part of the chapter is to present new technologies, mainly based on information technology (IT) systems, for more sophisticated traceability systems, which can ensure the quality and safety of agrifood products in the entire chain.

Chapter 7 revolves around *IT in agrifood supply chains.* Nowadays, the emerging role of Information and Communication Technology (ICT) and farming technologies is recognized as a driver for change in the agrifood sector. Many researchers stress the importance of adopting ICT and farming technologies by AFSC stakeholders, as these technologies are a major driver for innovation. The chapter intends to demonstrate the main technological trends that can be employed in the entire agrifood chain and their key role. A holistic approach is employed and an analysis of all available IT applications and techniques is conducted in all levels of the AFSC. Emphasis is given on the primary sector where a number of IT innovations has been employed (e.g., satellites, sensors for precision agriculture, etc.). IT has been a key enabler in the supply chain environment, acting as the power of process automation, the enabler of information sharing and collaboration or the supporter of management decisions and optimization logic. Especially in food supply chains, this role has been even more critical due to enhanced requirements for short life cycles and speed of response, traceability and food quality considerations, environmental constraints and sustainability. Ultimately, in the chapter, a contemporary view of this enabling role of IT in the supply chain environment is provided and a high-level IT architecture integrating the views of automation, information sharing/collaboration, and decision support is proposed. This architecture is then discussed in the context of current opportunities and challenges of food supply chains, namely radio frequency identification (RFID)-enabled supply chain management, carbon footprint monitoring and shared logistics. The chapter concludes with an overall discussion of the main barriers and drivers behind IT adoption in the supply chain and a future outlook on anticipated developments.

Chapter 8 deals with the much-debated *carbon footprint management*. Carbon footprint management has emerged as an issue of pivotal corporate importance that has led to its inclusion at both the design and management phases of contemporary AFSC networks, in which profitability and environmental impacts have to be balanced. This chapter aims at identifying the most significant carbon hot-spots that may arise across the entire AFSC, while providing sophisticated decision support management tools, both qualitative and quantitative, for "decarbonizing" the entire chain. More specifically, state-of-the-art tools for measuring the carbon footprint of agrifood products from cradle-to-grave are presented and practice-oriented low carbon interventions are proposed to aid the related decision-making process.

Chapter 9 discusses *quality management systems*. Ecolabel/Certification Quality management systems are very popular in the agrifood sector as they often demonstrate the company's ability to control food safety hazards in order to ensure that food is safe at the time of consumption. Nowadays, companies in the agrifood sector can develop and implement a number of available quality management systems (e.g., ISO22000:2005, ISO/TS 22004:2005, etc.). The main aim of the chapter is to present the key elements of the existing quality management systems available in the agrifood sector and to further investigate the employment of certain tools and techniques (e.g., trace and tracking systems) ensuring food quality and safety through the implementation of quality management systems.

Chapter 10's focus is on *risk management for agrifood supply chains*. Modern AFSCs are exposed to a wide variety of natural, technological, and humanitarian risks, such as natural disasters, adverse weather conditions (related or unrelated to global warming), biological incidents, market instability and fluctuation of food and raw materials' prices, logistical and infrastructural disruptions, public policy interventions, and institutional reforms. These risks may inhibit their normal operations and provoke deviations, disruptions, or shutdowns to the supply chain's fundamental flows. Furthermore, despite the usually low probability of the associated triggering events, these risks may have a dramatic impact on their cost, efficiency, and reliability performance. To that end, there is a need for specific and efficient pre-, as well as post-, event risk mitigation and management strategies especially in the agrifood sector that becomes even more pressing due to the direct environmental impact of the sector.

Finally, **Chapter 11** deals with *regulatory policies/trends.* There are many researchers that have addressed the significant pressures from governmental regulators as one of the most important driving factors toward the sustainability of AFSCs. Regulatory interventions force AFSC stakeholders to adopt a high level of commitment to food safety and sustainable practices in the context of their CSR activities. On the other hand, the regulatory environment in the agrifood products is indeed rather complex. In many cases, the regulatory heterogeneity (indicatively, some impressive differences on import requirements among EU countries) on agrifood trade is a major challenge that AFSC stakeholders face. This chapter aims to demonstrate and analyze the main characteristics of the regulatory policies and their impacts on the various aspects of the AFSC (including, among others, food safety, quality, and implementation costs).

This book is aimed at both practitioners and academics alike. The potential audience includes researchers, C-level executives from throughout the food and beverage industry, supply chain managers, producers/manufacturers, farm managers/ contractors, as well as stakeholders of AFSCs [producers, retailers, cooperatives, third-party logistics (3PL) companies, distributors, warehouse operators, policy-makers and other administrative and technical personnel].

The information contained in this book will be core to the interested parties that have to deal with the entire hierarchical decision-making process for the field. Specifically, this book provides essential input to policy-makers and C-level executives that deal with strategic decision-making (including the design of AFSC networks), as well as to supply chain stakeholders and farmers that have to tackle issues of competitiveness at the tactical and operational levels. Readers who will find this book a "must-have" include practitioners from different fields related to agriculture and the agrifood industry. Moreover, practitioners from the logistics and supply chain management sector can use the book as a guideline. In academia, the book can be used as a textbook in both existing and emerging Master courses in relative graduate programs including, for example, Sustainable Production, Agriculture Production Management, Operations Management, Logistics and Supply Chain Management, and Business Administration. Additionally, readers who will find this book "nice-to-have" may include researchers in fields of Operations Management, Logistics and Supply Chain Management, Business Administration, and Agriculture Sciences, as well as undergraduate students close to completing their studies, who will find it an essential aid for conducting their senior theses.

Sustainable supply chain management for agrifood companies is clearly an evolving and critical subject that has not been comprehensively addressed in the literature. While there are books that discuss the unique characteristics of AFSCs, they provide rather limited coverage on sustainability issues. Moreover, there are interesting books that address GSCM in general. We envision this new book to synthesize policies, practices, technologies and solutions offering a comprehensive, interdisciplinary, and customized paradigm. This condensed and targeted information will be of significant added value to leading executives and practitioners in the field, as well as researchers and interested academics.

Acknowledgments

This book consists of the collective work of many individuals from a wide range of organizations. It would not have been impossible to write and edit this book on sustainable food supply chains without the contributions and efforts of numerous scholars and practitioners. There are many individuals to thank, and we apologize in advance in case we have omitted anybody. First, we would like to thank the members, associates, and the EU's project officers (Mr Andrew Bianco and Mrs Carmen Madrid Gonzalez) affiliated with the Green-AgriChains project, a three-year leading research project (2012–2015) that was funded by the EU under the EU's Seventh Framework Programme (FP7-REGPOT-2012-2013-1 under Grant Agreement No. 316167). The project involved nine leading EU universities along with three business clusters. The academic partners include the Aristotle University of Thessaloniki (Greece), Brunel University (UK), INSEAD (France), Aarhus University (Denmark), International Hellenic University (Greece), University of Sheffield (UK), Alexander Technological Educational Institute of Thessaloniki (Greece), Karlsruhe Institute of Technology (Germany), and Erasmus University (Netherlands), while the business institutions include the world-class R&D logistics cluster EffizienzCluster (Germany), the Thessaloniki Innovation Zone (Greece), and the Organic Products Cluster (Greece).

Next, we would like to thank all of the authors who have contributed chapters to this book: Katerina Aggelopoulou, Jacqueline Bloemhof-Ruwaard, Fabrizio Dabbene, Rommert Dekker, Spyros Fountas, Paolo Gay, Theofanis A. Gemtos, Victor Guang Shi, Karen Hamann, Lorie Hamelin, Christos Keramydas, S.C. Lenny Koh, Sirpa Kurppa, Ioannis Mallidis, Ioannis Manikas, Felipe Montes, Jørgen E. Olesen, Konstantinos Papapanagiotou, Katerina Pramatari, Chen Qing, Anton Sentic, Sven G. Sommer, Agorasti Toka, Cristina Tortia, Jin M. Triolo, Marcus Wagner, and Jia Wei. If it was not for their diligent work, dedication, and ferocious commitment we would never have been able to have produced this book in the first place. We would also like to thank the numerous corporate organizations and the individuals who participated along with the authors in the various research projects on sustainable supply chains in the food sector. Their insights and their generosity of time allowed the authors of these chapters to conduct targeted research that formed the nexus of this book.

We would also like to thank the staff at Wiley (especially Mrs Liz Wingett) for their enthusiasm in supporting this effort from its very beginning and their impressive professionalism throughout the publication process. In addition, we would like to thank Dr Naoum Tsolakis and the doctoral student Dimitrios Tsiolias for their tireless work in formatting this book. Finally, we would like to thank our families for all of their support during the writing and editing of this book. Without their support this project would never have been completed.

1

Sustainable Agrifood Supply Chain Management

Eleftherios Iakovou,[1] Dionysis Bochtis,[2] Dimitrios Vlachos,[1] and Dimitrios Aidonis[3]

[1] *Laboratory of Quantitative Analysis, Logistics and Supply Chain Management, Department of Mechanical Engineering, Aristotle University of Thessaloniki, Thessaloniki, Greece*
[2] *Department of Engineering, Aarhus University, Inge Lehmanns Gade 10, Aarhus C, Denmark*
[3] *Department of Logistics, Technological Educational Institute of Central Macedonia, Branch of Katerini, Katerini, Greece*

1.1 Introduction – Agrifood Supply Chain Management

The agrifood sector is one of the most regulated and protected sectors worldwide, with major implications for sustainability such as the fulfillment of human needs, the support of employment and economic prosperity, the environmental impact, the tackling of poverty, and the creation of new markets (Humphrey and Memedovic, 2006). Indicatively, the European Commission (EC) is promoting significant reforms to its Common Agricultural Policy (CAP) in order to respond to the plethora of internationally emerging agrifood supply challenges (EC, 2010; Scheherazade, 2014). Growing environmental, social as well as ethical concerns, and increased awareness

Supply Chain Management for Sustainable Food Networks, First Edition. Edited by Eleftherios Iakovou, Dionysis Bochtis, Dimitrios Vlachos and Dimitrios Aidonis.

of the impact of food production and consumption on the natural environment have led to increased pressures by consumer organizations, policy-makers, and environmental advocacy groups on agrifood companies to manage social and environmental issues across their supply chains (SCs) from "farm-to-the-fork" and along products' life cycles (Courville, 2003; Weatherell and Allinson, 2003; Ilbery and Maye, 2005; Maloni and Brown, 2006; Vachon and Klassen, 2006; Welford and Frost, 2006; Matos and Hall, 2007; Grimm, Hofstetter, and Sarkis, 2014).

In this context, designing appropriate effective global strategies for handling agrifood products to fulfill consumers' demand, while responding to ever-increasing changes of lifestyle and dietary preferences, has become quite a complex and challenging task. Specifically, adverse weather conditions, volatile global food demand, alternative uses of agricultural production and fluctuating commodities' prices have led to a volatile supply of agricultural products that is expected to exceed its capacity limit in the forthcoming years. To that effect, developed countries have been increasing their agricultural production in agrifood supply chain (AFSC) operations in order to respond to the projected rise of 70% on global food demand by 2050 (FAO, 2006, 2009; Nelson *et al.*, 2010). At the same time, the value of family farms and the development of local food SCs is clearly recognized for both the developing and developed countries (FAO, 2014).

One of the most critical bottlenecks in agrifood production and distribution is the complexity and cost-efficiency of the relevant SC operations. Modern, global agrifood networks require multi-tier supply chain management (SCM) approaches due to the increased flows of goods, processes, and information both upstream and downstream the value chain. These increased requirements are related to the modern, emerging model of agrifood retailers (i.e., grocery retailers, fast-food and catering services' providers, etc.), the need for vertical and horizontal integration along the AFSCs, the plethora of differentiated product offerings, the market segmentation, the dominance of multinational enterprises in the food processing and retailing sectors, the need for limiting food waste and overexploitation of natural resources, as well as the branding of firms (van Roekel *et al.*, 2002; Chen, Chen, and Shi, 2003; Mena *et al.*, 2014).

Furthermore, SCM has been recognized as a key concept for the agrifood industry competitiveness. The rapid industrialization of agricultural production, the oligopoly in the food distribution sector, the advancement of Information and Communication Technologies (ICT) in logistics, customer concerns, and a divergence of governmental food safety regulations, the establishment of specialized food quality requirements, the emergence of modern food retailer forms, the increasing importance of vertical integration and horizontal alliances, as well as the emergence of a large number of multinational corporations, are just a few of the real-world challenges that have led to the adoption of SCM in the agrifood sector (Chen, 2006). To this end, SCM embraces the challenge to develop and deploy efficient value chains tailored to the specifications of the modern, uncertain environment, subject to the constraints of local and cross-regional conditions, with respect to logistics means and infrastructure, access to land and water resources, allocation of harvesting areas and the various processing and storing facilities, innovative and sustainable good-practice methods, regulatory and techno-economic environments, and rapid changes of food market characteristics.

In order to develop competitive and sustainable AFSCs, there are a few critical issues that have to be first recognized:

1. the unique attributes of AFSCs that differentiate them from other SC networks;

2. the decisions that should be made on the strategic, operational, and tactical levels;

3. the necessary policies to ensure sustainability of the agrifood chains; and

4. the appropriate innovative interventions, which are required to foster major advances and competitiveness within the evolving AFSC context.

Therefore, more frequent changes in AFSC designs are necessary and strategic actions should be taken to foster sustainability (Halldorsson, Kotzab, and Skjøtt-Larsen, 2009), and thus to achieve higher efficiency in logistics' operations performance and resource usage (e.g., Gold, Seuring, and Beske, 2010; Carter and Easton, 2011).

In general, an AFSC is encompassing a set of operations in a "farm-to-the-fork" sequence including farming, processing/production, testing, packaging, warehousing, transportation, distribution, and marketing (Iakovou *et al.*, 2012). These operational echelons have to be harmonized in order to support five flow types, namely:

1. physical material and product flows;

2. financial flows;

3. information flows;

4. process flows; and

5. energy and natural resources' flows.

The aforementioned operations, services, and flows are integrated into a dynamic production–supply–consumption ecosystem of research institutions, industries, producers/farmers, agricultural cooperatives, intermediaries, manufacturers/processors, transporters, traders (exporters/importers), wholesalers, retailers, and consumers (van der Vorst, 2006; Matopoulos *et al.*, 2007; Jaffee, Siegel, and Andrews, 2010). Moreover, the continuous evolution of AFSCs, and the overall complexity of the agrifood environment along with global market trends further highlight the need for integration of individual SCs into a unified AFSC concept. In such a structure, strategic relationships and collaborations among enterprises are dominant, while these organizations are further required to secure their brand identity and autonomy (Van der Vorst, da Silva, and Trienekens, 2007). A conceptual configuration of AFSCs is depicted in Figure 1.1.

The actors involved in the AFSC system can be generally partitioned into public authorities and private stakeholders. The former category includes mainly national governments and the associated ministries, administrative authorities (regional,

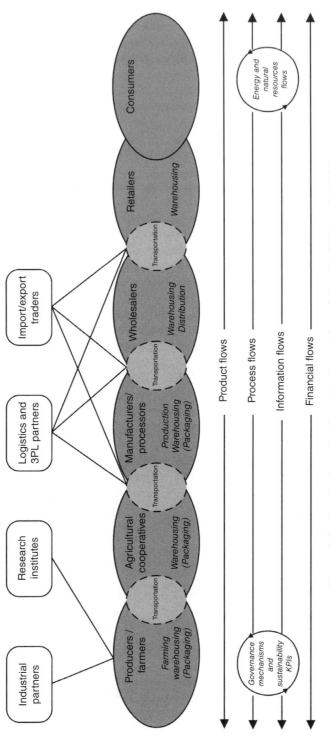

Figure 1.1   A typical agrifood supply chain. Adapted from Tsolakis et al., 2014a.

district, urban), as well as international organizations (e.g., Food and Agriculture Organization), while the latter encompasses individual farmers/growers, cooperatives, research institutes and innovation centers, chemical industries, agro-industries and processors, food traders, logistics providers, transporters, supermarket chains and food stores, as well as financial institutions (Jaffee, Siegel, and Andrews, 2010). In this context, highly concentrated agro-industrial enterprises and retailers have recently morphed into dominant players in the agrifood field, while the public sector has emerged as a key-governance actor (Bachev, 2012).

Furthermore, AFSCs exhibit a set of unique characteristics that differentiate them from classical supply networks and raise an imperative need for customized managerial capabilities. According to Van der Vorst (2000, 2006), AFSCs are characterized by:

1. the unique nature of their products as in most cases they deal with short life-cycle and perishable goods;

2. high product differentiation;

3. seasonality in harvesting and production operations;

4. variability of quality and quantity on farm inputs and processing yields;

5. specific requirements regarding transportation, storage conditions, quality and safety, and material recycling;

6. a need for complying with national/international legislation, regulations, and directives regarding food safety and public health, as well as environmental issues (e.g., carbon and water footprints);

7. a need for specialized attributes, such as traceability and visibility;

8. a need for high efficiency and productivity of expensive technical equipment, despite often lengthy production times;

9. increased complexity of operations; and

10. the presence of significant capacity constraints.

The remaining of this chapter provides an in-depth examination of AFSCs and the related decision-making across the involved operations. Specific focus is provided on the three dimensions of sustainability, that is, economic, social, and environmental (Beske, Land, and Seuring, 2014) that modern, competitive AFSCs need to accommodate.

1.2 Why Sustainable Agrifood Supply Chain Management

The world has encountered and is expected to face even greater volatility and related challenges in the future, including economic crises, social exclusion, and climate change, with direct impact upon business activities (Validi, Bhattacharya, and

Byrne, 2014; Brorström, 2015). The design and adoption of sustainability strategies throughout business operations has emerged as a meaningful intervention to accommodate such challenges (Shaw, 2013). Interestingly enough, the concept of sustainability cannot be easily defined and is, in fact, determined by academicians and decision-makers alike (Parr, 2009). Initially, researchers and practitioners were solely focused on environmental aspects to accommodate corporate needs and drive shareholder value (Caniato et al., 2011). Nonetheless, in the contemporary global contextual framework, sustainability transcends the environmental dimensions and further relates to market competition, availability of raw and virgin materials, access to energy sources and increasing global population (Bajaj, Jha, and Aggarwal, 2013). Hence, the concept of the "Triple Bottom Line" (Kleindorfer, Singhal, and van Wassenhove, 2005) or the "Three Pillars" (White and Lee, 2009) of sustainability has been introduced to highlight the need for a balanced approach to the three P's, namely people, profit, and planet. The aforementioned dimensions provide corporate growth opportunities emanating from the adoption of sustainable good practices (Byrne, Ryan, and Heavey, 2013; Sezen and Turkkantos, 2013).

The value proposition of linking research to sustainable development is strongly acknowledged. This is further affirmed in the most recent research and development policy documents of the European Union (EU). Specifically, the European Research Area (ERA) vision 2020 calls for a focus on societal needs and ambitions toward sustainable development. The three "Key Thrusts" identified by the European Technology Platform on the "Food for Life" Strategic Research Agenda 2007–2020 meet all of the criteria required to stimulate innovation, to create new markets, and to meet important social and environmental goals. These "Key Thrusts" are:

- Improving health, well-being, and longevity.

- Building consumer trust in the food chain.

- Supporting sustainable and ethical production.

While, the topic of "sustainability" is inherent to SCM (Ahi and Searcy, 2013; Pagell and Shevchenko, 2014), it is only during the last two decades that sustainability in SCM has attracted increased academic and business interest, further reflecting the fact that SC operations are a field where most organizations can and actually implement green strategies (Kewill, 2008; Seuring, 2013). Indicatively, Seuring and Muller (2008) present a comprehensive literature review of almost 200 relevant papers while further outlining the major research directions in the field. Moreover, in the work of Gupta and Palsule-Desai (2011), the existing sustainable SCM literature is classified under four broad categories related to decision-making, namely strategic considerations, decisions at functional interfaces, regulation/ government policies, and decision support tools. Similarly, Seuring (2013) reviews papers that tackle the issue of sustainable SCs with a focus on the application of quantitative models. More recently, Ahi and Searcy (2014) conducted a structured review of 445 articles and provide an analysis of 2.555 unique metrics employed in assessing green and sustainable SCM.

The issue of sustainability is even more vital for the food industry which is dominated by a growing demand for sustainably produced food as consumers today are highly cognizant of the manner in which the food is produced, processed, and distributed (Beske, Land, and Seuring, 2014). In general, AFSCs are dynamically evolving over time in order to follow the persistent changes within the broader agrifood environment and to further accommodate the continued introduction of new environmental and food safety legislation from both European and international directives (Glover *et al.*, 2014). In the forthcoming years, modern AFSCs will have to cope with a plethora of major challenges that are underway, encompassing amongst others: rapid urbanization, growth and liberalization of domestic/global factors and markets, decrease of public sector funding, dominance of global SCs, concerns for food quality and safety, changes in technology, weakness of regional rural populations to comply with the requirements posed by dominant enterprises, climate change effects on farming, and the adoption of corporate social responsibility (CSR) practices. Therefore, the recognition of the most critical issues that need to be addressed by all AFSCs' stakeholders toward an integrated decision-making process emerges as a prerequisite for designing and managing such complex, multi-tier SCs and ensuring their overall efficiency and sustainability.

Furthermore, societal stakeholders demand corporate responsibility to transcend product quality and rather extend to areas of labor standards, health and safety, environmental sustainability, non-financial accounting and reporting, procurement, supplier relations, product life cycles, and environmental practices (Bakker and Nijhof, 2002; Waddock and Bodwell, 2004; Teuscher, Grüninger, and Ferdinand, 2006). Therefore, sustainable SCM expands the concept of sustainability from a company to the SC level (Carter and Rogers, 2008) by providing companies with tools for improving their own and the sector's competitiveness, sustainability, and responsibility toward stakeholder expectations (Fritz and Schiefer, 2008). In addition, the principles of accountability, transparency, and stakeholder engagement are highly relevant to sustainable SCM (Waddock and Bodwell, 2004; Teuscher, Grüninger, and Ferdinand, 2006; Carter and Rogers, 2008). More specifically, in response to pressures for transparency and accountability, agrifood companies need to measure, benchmark, and report environmental sustainability performance of their SCs; whilst on the other hand, policy-makers need to measure the sectorial performance within the SC context for effective target setting and decision-making interventions.

Particularly, as dictated by the third "Key Thrust" that ERA articulates, food chains need to operate in a manner that exploits and optimizes the synergies among environmental protection, social fairness, and economic growth. This would ensure that the consumers' needs for transparency and for affordable food of high quality and diversity are fully met. Progress in this area is expected to have important benefits for the industry in terms of reduced uses of resources, increased efficiency, and improved governance. An overview of emerging global trends, policy developments, challenges, and prospects for European agri-futures, points to the need for novel strategic frameworks for the planning and delivery of research. Such frameworks should address the following five challenges:

- *Sustainability*: facing climate change in the knowledge-based bio-society.

- *Security*: safeguarding European food, rural, energy, biodiversity, and agri-futures.

- *Knowledge*: user-oriented knowledge development and exchange strategies.

- *Competitiveness*: positioning Europe in agrifood and other agricultural lead markets.

- *Policy and institutional*: facing policy-makers in synchronizing multi-level policies.

Addressing these challenges could usher the European agrifood sector to the knowledge-based bio-economy, while ensuring that the sector (and food retailers) remains globally competitive further addressing climate change and sustainable development concerns, such as the maintenance of biodiversity and prevention of landscape damage. Meeting these multi-faceted sustainable development challenges facing the agrifood sector worldwide, will require a major overhaul in the current agriculture research system. Recent foresight work under the aegis of Europe's Standing Committee for Agricultural Research (SCAR), has highlighted that in the emerging global scenario for European agriculture, research content needs to extend to address a diverse and often inter-related set of issues relating to sustainable development, including food safety/security (Keramydas *et al.*, 2014), environmental sustainability, biodiversity, bio-safety and bio-security, animal welfare, ethical foods, fair trade, and the future viability of rural regions. These issues cannot simply be added to the research agenda. Rather, addressing them comprehensively and holistically in agriculture research requires new methods of organizing research, in terms of priority-setting, research evaluation and selection criteria, and in bringing together new configurations of research teams, as well as managing closer interactions with the user communities and the general public in order to ensure that relevant information and knowledge is produced and the results are properly disseminated.

Furthermore, in order to unleash value, it is important to exploit the potential of utilizing agrifood waste and the associated by-product biomasses for energy recovery and nutrient recycling, to mitigate climate change and eutrophication (Kahiluoto *et al.*, 2011). To that end, biomass has emerged as a promising option, mainly due to its potential worldwide availability, its conversion efficiency, and its ability to be produced and consumed on a CO_2-neutral basis. Biomass is a versatile energy source, generating not only electricity but also heat, while it can be further used to produce biofuels (Verigna, 2006; Watanabe *et al.*, 2014; Toka *et al.*, 2014). Iakovou *et al.* (2010) provide a critical synthesis of the state-of-the-art literature on waste biomass SCM. Agrifood biomass is usually free of toxic contaminants and is determined spatially and temporally by the respective local/regional profile of the pertinent activities.

It is well documented that 31% of the greenhouse gas (GHG) emissions and more than 50% of eutrophication are related to food chains, thus highlighting the need to intervene in the AFSC to ameliorate its impact on the environment (CEC, 2006). In order to promote "green" AFSCs and elaborate agrifood biomass operations on a large scale, the application of appropriately designed innovative policies and systems is necessary (Van der Vorst, Tromp, and van der Zee, 2009; Negro, Hekkert, and Smits, 2007). Green SCM is one of the top two strategic priorities for global corporations (McKinsey, 2011). The benefits of going green are substantial, as green SCM cannot only reduce an organization's carbon footprint but it can also lead to reduced costs, improved reputation with customers, investors, and other stakeholders, thus further leading to a competitive edge in the market and increased profitability. Indicatively, a case study for the new business model for agricultural material sourcing of Nestle, a leading food company (Goldberg and Fries, 2012), summarizes a set of trends that are valid for most food companies.

Indeed, the post-2009 recession period has further underlined the need to turn the business focus, across the world, not only to profitability, but to sustainability as well. Today, one of the key priorities in corporate strategic design for an organization is to emerge as socially responsible and sustainable. Companies are structuring their sustainability reports to disclose their strategy to address the growing concerns of environmental degradation and global warming. Today, 93% of the global Fortune 250 companies release their annual sustainability report (KPMG, 2013), up from 37% in 2005 (Singh, 2010). As a focal part of sustainability initiatives, green SCM has unequivocally emerged as a key discipline that can provide competitive advantage with substantial gains for the company's bottom line. In designing green SCs, the intent is to adopt, comprehensively and across business boundaries, best practices right from product conception to the end-of-life recycling stage. In this context, green initiatives relate to both tangible and intangible corporate benefits. Sustainability reports of many companies indicate that the greening of their SCs has helped them to reduce their operating costs with increased sustainability of their business.

Additionally, modern AFSCs are exposed to a wide variety of natural, technological, and man-made risks, such as weather related risks and extreme weather events (e.g., hail storms, floods, and droughts), natural disasters (e.g., earthquakes, volcano eruptions), biological and environmental risks (e.g., livestock diseases), production risks (e.g., yield uncertainties), human resource risks (e.g., seasonal personnel unavailability), management and operational risks (e.g., forecasting errors), logistical, infrastructural, and technological risks (e.g., uncertainty of new technologies adoption), price and market risks (e.g., price volatility of inputs and outputs), financial risks (e.g., disruptions of farm business financing), policy, institutional, and regulatory risks (e.g., uncertainties of tax and fiscal policies), and political risks (e.g., political and/or social instability) (Jaffee, Siegel, and Andrews, 2010). These risks may inhibit normal operations of AFSCs and could provoke deviations, disruptions, or shutdowns to the SC's fundamental flows. Furthermore, they may have a dramatic impact on cost, efficiency, and reliability of the included activities and operations.

The associated core risk-related decisions refer to: (i) the selection of appropriate risk governance modes; and (ii) the implementation of suitable risk mitigation strategies. The first set of decisions explores the options of the market, private and public risk governance along with the relevant intervention levels. The second set refers to the nature of the applied risk mitigation policy including technology development and adoption, enterprise management practices, financial instruments, investments in infrastructure, policy and public financial support schemes, and private collective actions (OECD, 2009).

The existing research has focused only on few critical aspects of the agrifood risk management concept including cross-border transaction risks (Ameseder *et al.*, 2009), chemical and biological risks (Bachev, 2011), agricultural contracts (Ligon, 2003), catastrophic/disaster risk management (Antón, Kimura, and Martini, 2011; RPDRM, 2012), income risk management (OECD, 2000), climate risk management (Wall, Smit, and Wandel, 2004), and insurance schemes (Bielza Diaz-Caneja *et al.*, 2009).

To sum up, the nature of the overall decision-making process in sustainable AFSCs is purely dynamic, as it unfolds in real-time within an uncertain environment that changes continuously bringing new challenges and opportunities. Consequently, the decisions along with the associated implemented strategies should be continuously evaluated and reconsidered in order to ensure the long-term efficiency and sustainability of an AFSC.

1.3   Hierarchy of Decision-Making for AFSCs

Designing, managing, and operating AFSCs involves a complex and integrated decision-making process. This is even more accentuated when AFSCs deal, for example, with fresh, perishable, and seasonable products in the context of high volatility of supply and demand. In general, the design and planning of sustainable AFSCs needs to address a wide range of issues including crops planning, harvesting practices, food processing operations, marketing channels, logistics activities, vertical integration and horizontal cooperation, risk and environmental management, food safety, and sustainability assurance.

1.3.1   Strategic Level

The strategic decisions involve all stakeholders that are interested in participating in a sustainably driven SC network of agricultural goods. Thus, decisions at the strategic level of the hierarchy span the following aspects: selecting the appropriate farming technologies, SC partnership relations, design of SC networks, establishment of a performance measurement system along the AFSC, and finally, quality assurance. Below, these decisions are further discussed, while a synthesis of the relevant and up-to-date research efforts is provided.

1.3.1.1   Selection of Farming Technologies

Today's trends toward diversified crops, quality standards, increased environmental concerns, biological and weather implications, and safety regulations dictate the need for a careful selection of the farming technologies to be employed (Søgaard and

Sørensen, 2004). To this end, farming technologies range from traditional farming machinery to sophisticated information technology (IT) and precision agriculture (PA) applications; the latter are recognized as a major contributor to increased farming efficiency and environmentally sustainable farming practices (Aubert, Schroeder, and Grimaudo, 2012; Bochtis, 2013).

The main decisions involved in the selection process of the farming technologies relate to:

1. the determination of the capital requirements and expenditure on farming equipment;

2. the development of cooperative schemes in the utilization of farming machinery; and

3. the adoption of innovative farming applications.

In terms of capital expenditure and cooperative actions, the optimum solution must be investigated with relevance to the type of planting, tillage practices, harvesting methods, ownership costs, operating costs, labor costs, and timeliness costs. In terms of innovation and performance, the factors that affect the selection of farming technologies can include, indicatively, the size of the yielded production, the required quality of the agricultural products, and the volatility of weather and soil conditions.

Farming technologies ensure the uninterrupted supply of adequate goods so that a particular AFSC can respond to market demand over the strategic horizon. In the literature, there are well documented quantitative models that deal with the optimal mechanization level of farms with regard to the capital expenditure, economic efficiency, and capacity utilization (e.g., Glen, 1987; Godwin et al., 2003; Søgaard and Sørensen, 2004; Sørensen, Madsen, and Jacobsen, 2005; Pandey, Panda, and Panigrahi, 2006; Katalin et al., 2014). Moreover, many researchers stress the importance of cooperative schema in machinery utilization, especially in the case of small- and medium-scale farms, which are characterized by common agricultural factors such as the cultivated crop varieties, farm size, soil type, environmental impact, and labor employability (e.g., de Torro and Hansson, 2004; Aurbacher, Lippert, and Dabbert, 2011; Abebaw and Haile, 2012; Dai and Dong, 2014). Today, modern research deals with the incorporation of innovative approaches into applied farming technologies. Robotics and IT applications toward production automation, image analysis, and quality sensing are only a few of the radical advances that have been developed for vegetable propagation, picking, trimming and packaging, robotic milking, and livestock monitoring (Wrest Park History Contributors, 2009). Finally, the utilization of PA technologies (i.e., satellite imagery and geospatial tools that allow the selective treatment of a field as a heterogeneous entity) has emerged as a viable intervention to promote farming efficiency and foster environmental sustainability though drastic reductions in the use of contaminants (by even 90%) (e.g., Du et al., 2008; Isgin et al., 2008; Aubert, Schroeder, and Grimaudo, 2012; Busato et al., 2013; Hameed et al., 2010).

1.3.1.2 Supply Chain Partnership Relations

In terms of business relationships, AFSCs present common features and characteristics with the traditional supply networks of commercial products and services. An interesting characteristic of AFSCs is the high level of relationship complexity throughout the entire chains, as there are many stakeholders with shared, but also in some cases conflicting, goals and targets. In any case, effective business relationships contribute to the sustainability of the AFSCs by reducing environmental uncertainty, fostering the development of dynamic capabilities and resulting in higher levels of business productivity (Dyer and Singh, 1998; Fischer *et al.*, 2008; Beske, Land, and Seuring, 2014). Moreover, effective business relationships have been characterized as one of the pillars for SCs' integration (Akkermans, Bogerd, and Vos, 1999; Thakkar, Kanda, and Deshmukh, 2008) which further leads to improved inventory control management and renders SCs with increased levels of resilience (Fernández Lambert *et al.*, 2014).

The issue of business relationships has been analytically examined in the literature. Tsolakis *et al.* (2014a) identify efficient business relationships among the partners of the AFSCs as the key factor for sustaining high performance. Such relationships should be built upon certain principles such as integration, collaboration, coordination, and cooperation. Many authors highlight that it is unlikely for all partners to share equally the benefits stemming from collaboration; however, in cases where the partners share similar paradigms, there is a great possibility for success (Mungandi, Conforte, and Shadbolt, 2012). On the other hand, there are many cases where collaborating parties in AFSCs do not share balanced relations. Matopoulos *et al.* (2007) argue that the most powerful stakeholder dominates the SC by imposing its rules convincingly on the other parties. Therefore, a critical issue is the rivalry between collaborating partners in AFSCs mainly due to this asymmetry in their relationships. Conflicting objectives always affect negatively the relationship schema. Burch and Goss (1999) discuss the competitiveness among manufacturing and retail channels in specific SCs. Moreover, Bijman *et al.* (2006) present the high levels of competition and rivalry between wholesalers and retailers in the Dutch fresh vegetable SCs.

Alliance members in different chain stages (e.g., farmer–processor, processor–retailer, etc.) should invest in building successful partnerships and promoting the sustainability of their AFSCs. To that end, Fischer *et al.* (2008) analyze the factors that affect sustainability in partners' relationships in the agrifood sector in different European countries.

Collaboration and coordination among partners can further help in establishing long term and robust relationships through synergies and common activities. Effective collaboration can only be attained when all members of the SC operate under a "win-win" paradigm, working jointly under the same framework trying to achieve common goals and targets (Barratt, 2004). Through collaborative relationships, all partners can share the added value stemming from the integration of SCs while they can further improve risk management. To that effect, collaboration between farmers, processors, and retailers is pivotal for facing contemporary challenges, such as high consumers' expectations, strict legislative framework for environmental and social issues, and so on (Schiemann, 2007; Lamprinopoulou *et al.*, 2014). Mussell and Gooch (2008) present four case studies of collaboration in agrifood value chains (The Ontario Processing Tomato Industry, The Warburtons Value Chain, Perfection Fresh Australia Pty Ltd, Milk Marketing in the Upper Midwest US). In all cases, the collaborative relationships of the partners increased the level of efficiency of the chains. Additionally, Hobbs and Young (2000) present a conceptual framework for analyzing vertical SC coordination in the agrifood sector.

In the literature, key factors have been recognized that affect the quality of coordination among the partners in a specific SC. Communication, through the sharing of information between stakeholders has been recognized as a vital element for the sustainability of the business relationships in AFSCs (Reynolds, Fischer, and Hartmann, 2009; Del Borghi *et al.*, 2014). Fischer (2009) presents the results of an empirical analysis of survey data dealing with the main determinants of a relationship's sustainability in all stages of AFSCs. In the survey, 1442 partners (farmers, processors, and retailers) acting in two AFSCs (one for meat and another for cereals) from six different European countries (UK, Germany, Spain, Poland, Ireland, Finland) participated. According to the results, effective communication is the most important factor for the sustainability of the SC.

Moreover, trust has been documented as another essential factor influencing the quality and stability of business relationships in the AFSCs. According to Lindgreen (2003), trust can be considered as a complex multidimensional and dynamic concept of strategic importance in the food sector. More specifically, it is a vital indicator for sustainability in young relationships (Reynolds, Fischer, and Hartmann, 2009), where collaboration history data are missing. To that end, mistrust, for many authors is the main obstacle for implementing successful business relationships in the food industry (Kumar, Scheer, and Steenkamp, 1995; Fearne, Hughes, and Duffy, 2001). Kottila and Rönni (2008) present interesting findings of a case study with two cases in organic food chains, where development of trust among the partners is a more significant factor for success than the frequency and quality of communication.

Finally, contracting among actors can be considered as another fundamental issue for collaboration and integration of AFSCs. Ligon (2003) investigates the risk mitigation related with optimal contracts in the agricultural sector. Fischer and Hartman (2010) analyze the main characteristics of the agrifood SC that influence the selection of the optimal contract type. Da Silva (2005) proposes contract farming as a key component for the development of the agrifood systems. The appropriate regulatory environment, the minimization of contractual hold-ups, the minimization of transaction costs, and the contract design are recognized as key success factors. However, contract farming has been responsible for the emergence of certain problems throughout the AFSCs, such as concerns about unequal power relations, shifting of management decisions, and quality control. Such issues of concern are even more evident in small farmers, as agro-industrial firms tend to work with large farmers and cooperatives in order to minimize transaction costs (Sartorius and Kirsten, 2007; Mungandi, Conforte, and Shadbolt, 2012).

1.3.1.3   Design of Supply Chain Networks

The configuration of an AFSC is a vital issue for the operation and sustainable efficiency of the network in the long-term, in order to respond to increased manufacturing costs, shortened product life cycles, and the global market economies (Beamon, 1998; Farahani *et al.*, 2014; Govindan *et al.*, 2014). In this context, the core set of decisions regarding the configuration of the AFSC network includes:

1. the identification of agricultural capacity over a region, and the selection of the optimal sourcing policies;

2. the development of efficient procurement channels;

3. the allocation of processing/production facilities;

4. the allocation of intermediate warehouses;

5. the design of the transportation networks;

6. the design of the retailers' networks; and

7. the selection of markets.

Despite the significance of the aforementioned decisions and the plethora of relevant papers within the general SCM context, the agrifood literature that focuses on these issues is rather poor, probably due to difficulties generated by the structure and complexity of the relationships across an entire agrifood chain, as well as the uncertainties that characterize this type of network (Mena *et al.*, 2014; Tsolakis *et al.*, 2014a).

Taking into account that very few aspects of agrifood supply network configuration have been addressed in the literature, only a small number of papers have focused on transportation network design. More specifically, Govindan *et al.* (2014) propose a sustainable perishable food SC network design model that minimizes logistic costs and environmental impacts in terms of CO_2 emissions. Furthermore, Boudahri, Bennekrouf, and Sari (2011) propose a model for the design and optimization of the transportation network of an AFSC, tailored to the specific case of chicken meat. Additionally, Higgins *et al.* (2004) propose a framework for the integration of harvesting and transport systems for sugar production. Furthermore, Burch and Goss (1999) discuss the global sourcing issue for retail chains and its impact on the agrifood system. Finally, there is a considerable volume of research addressing SC configuration issues including methodologies and practises that could be appropriately employed in AFSC design, concerning market selection (e.g., Ulaga, Sharma, and Krishnan, 2002), plant location (e.g., Bhatnagar and Sohal, 2005), warehouse location (e.g., Demirel, Demirel, and Kahraman, 2010), and transportation network design (e.g., Akkerman, Farahani, and Grunow, 2010).

1.3.1.4   Key Performance Indicators

Real-world practice has highlighted the measurement of performance as a critical process for companies and organizations in order to improve their SC efficiency and effectiveness, and to further ensure their long-term success and profitability

(Chan, 2003; Neely, Gregory, and Platts, 2005; Aramyan *et al.*, 2007). In this context, sophisticated measurement systems have been developed for the continuous monitoring and evaluation of the SCs' performance. These performance measurement systems are even more complicated in the case of the AFSCs, due to explicit technical and managerial uniqueness (Aramyan *et al.*, 2006; Tsolakis *et al.*, 2014a). The development of measurement systems is mainly based on the selection of the Key Performance Indicators (KPIs). According to van der Vorst (2006), performance indicators in the AFSC networks can be grouped into three main levels, namely: (i) SC network level; (ii) organizational level; and (iii) process level. Aramyan *et al.* (2007) propose a conceptual performance measurement framework for AFSCs based on KPIs in four main categories: efficiency; flexibility; responsiveness; and food quality.

The latest agenda in the field of KPIs deals with the sustainability measurement and the reporting of the SCs' performance. Taticchi, Tonelli, and Pasqualino (2013) recognize transparency and communication to stakeholders, improvement of operations, and strategy alignment as the main drivers for organizations to measure the levels of sustainability in their SCs. Tsolakis *et al.* (2014b) propose a conceptual framework of financial KPIs to measure sustainability interventions in the AFSCs, while they provide a map of existing sustainability KPIs for all echelons in the AFSCs (e.g., chemical industries, farmers, wholesalers, etc.). Further, Bourlakis *et al.* (2014) propose a performance measurement framework for sustainable food SCs. Within this framework, 18 sustainable measures were identified and categorized into 5 main groups of performance elements: consumption; flexibility; responsiveness; product quality; and total SC. In addition, Yakovleva, Sarkis, and Sloan (2012) propose a four-stage methodological framework for the evaluation of the food SCs' sustainability performance. The first stage deals with the selection of the appropriate economic, environmental, and social indicators, while in the second and third stage data gathering, transformation and adjustment using Analytical Hierarchy Process are conducted. In the final stage, a sensitivity analysis is proposed, in order to obtain meaningful managerial insights. Finally, Tajbakhsh and Hassini (2014) present an envelopment analysis model for the evaluation of SC sustainability focusing on the evaluation of all operations relevant to economic, environmental, and social issues.

1.3.1.5   Quality Assurance

Over the last few years, numerous crises and incidents in the food sector [e.g., the major outbreak of Bovine Spongiform Encephalopathy (BSE), commonly known as mad cow disease, the Variant Creutzfeldt–Jakob disease (vCJD), the avian influenza, etc.] have been recorded. According to Resende-Filho and Hurley (2012), 47.8 million people in the USA (approximately 16.7% of the total population) were affected by an illness related to food in 2011. The outcome of these food crises has been the dramatically increased consumers' awareness of food safety. To that end, the implementation of food safety control systems has become an emerging issue for all stakeholders in the sector.

In terms of food management systems, there is a number of outstanding tools available, such as Hazard Analysis Critical Control Points (HACCP), Good Manufacturing

Practice (GMP), and Good Hygiene Practice (GHP) (van Schothorst, 2004; Gorris, 2005). Moreover, a plethora of well-established Quality Management Systems (QMSs) is also available, ensuring the delivery of high quality food products to end-users. Through the implementation of QMSs, companies can adopt common standards for food safety issues, product characteristics, production and business processes, hygiene levels, and so on. The implementation of QMSs schema can be either applied individually by companies or in some cases QMSs can be implemented horizontally throughout the entire SC. The horizontal implementation can guarantee the continuity of increased food safety levels, as all stakeholders employ quality assurance mechanisms and tools with common characteristics and qualifications.

ISO 22000:2005 is one of the most popular and well-established QMSs in the food sector. It is a food safety management system specifying the minimum requirements for any stakeholder in the food chain. These requirements, among others, include the ability of companies to control food safety hazards, to fulfill all applicable statutory and regulatory requirements and to communicate food safety issues to all interested parties (ISO 22000:2005, 2005).

In the same framework, the British Retail Consortium (BRC) has developed a number of BRC Standards for Food Safety, providing quality and operational criteria for suppliers, manufacturers, and global retailers in order to ensure compliance to legal and statutory requirements (BRC Global Standards, 2012). BRC standards are widely used, as there are over 21 000 certified companies in 123 countries. Indicative examples of BRC standards include issues for food safety, consumer products, packaging and materials, storage and distribution, and best practice guidelines.

Another certification scheme with characteristics similar to those of the BRC is the International Features Standards (IFS) for food. The basic objectives of IFS for food include the establishment of evaluation systems, the enhancement of transparency throughout the entire food SC and the reduction of costs and waste time for all players in the chain.

An interesting quality certification scheme, mainly focused on the primary food sector, is the Global Good Agricultural Practices (GlobalGAP). GlobalGAP has published a number of voluntary norms and standards for the certification of primary production in the food sector. Its main objective is to link farmers from developing countries to key international retailers (Asfaw, Mithöfer, and Waibel, 2009; Tipples and Whatman, 2010).

Despite the many initiatives which have been developed in the field of QMSs for the food sector, there are still specific barriers that prevent the development of these systems and tools. According to Bas, Yüksel, and Çavuooflu (2007), such barriers include the lack of knowledge and of qualification programs for food safety systems along with insufficient facilities. To that end, the contribution of several researchers (e.g., Akkerman, Farahani, and Grunow, 2010; Wever et al., 2010) who analyze the integration of QMSs in food SCs focusing mainly on the optimization of processes, economy, and governance is deemed quite valuable.

1.3.2   Tactical and Operational Levels

In this subsection, we discuss the decision-making process at the tactical and operational levels for managing Afscs. We first address the common characteristics that the Afscs display when compared with the traditional SCs and then proceed by pointing out unique and challenging issues, including the planning of harvesting and logistics operations along with transparency and traceability issues.

1.3.2.1   Harvesting Planning

The role of harvesting planning on the performance of the entire AFSC is of pivotal importance. One of the most critical issues that needs to be tackled is the extreme vulnerability of harvesting planning to disruptions, such as weather conditions and poor sunlight, plant diseases, poor soil performance, and so on (Epperson and Estes, 1999). At the same time, during the planning of agricultural operations several environmentally sustainable practices must be adopted in order to reduce GHG emissions, maintain biodiversity and foster ecological resilience (Dile *et al.*, 2013). These challenges are even more accentuated in the case of perishable goods, where time is a critical parameter that affects planning throughout all echelons of an AFSC. In this case, the trade-off between the quality of the products (time to reach the market) and the incurred costs (due to agrifood spoilage and wastage) needs further scrutiny and due diligence.

The decisions related to the harvesting operations involved in an AFSC include: (i) the scheduling of planting and harvesting; and (ii) the effective resource management among competing crops. Throughout the literature, factors such as timing of planting and harvesting, planting varieties, fertilizer utilization, water consumption, labor scheduling, and post-harvesting operations have been recognized as very important for cost minimization and maximization of yielded quality (e.g., Higgins *et al.*, 2004; Ahumada and Villalobos, 2009). In addition, several researchers have adopted the concept of Life Cycle Analysis (LCA) in order to assess the sustainable efficiency of on-farm operations (Biswas, Barton, and Carter, 2008; Meisterling, Samaras, and Schweizer, 2009).

More recently, Ahumada and Villalobos (2011) developed a comprehensive quantitative modeling approach for the complex decision-making of the harvesting and the distribution of perishable goods. Furthermore, the location of farms according to the overall AFSC planning, the matching of soil types with the desired crops, the design of crop rotations, the irrigation development and fallow systems and resource utilization balance among multiple farms are key capital-dependent decisions in order to deploy effective and sustainable AFSCs (Tan and Fong, 1988; Glen and Tipper, 2001; Rodrigues *et al.*, 2010; Schönhart, Schmid, and Schneider, 2011).

1.3.2.2   Logistics

The logistics operations in an AFSC deal with the management of the flow of goods along the entire SC in order to provide superior value to the customer at the least cost and in compliance with predetermined performance criteria and regulations.

The significance of the logistics operations upon the sustainability domain is clearly documented in the case of perishable and ready-to-eat products as agrifood products have to comply with quality specifications (Brunner, van der Horst, and Siegrist, 2010), while the sourcing and distribution of the commodities at a global scale and the increased distances between SC partners further highlight the growing awareness toward environment conservation (Soysal, Bloemhof-Ruwaard, and van der Vorst, 2014). It is no surprise that transportation is reported to be one of the main sources of CO_2 emissions (Delgado *et al.*, 1999).

The relative logistics decisions are listed below:

1. fleet management, vehicle planning, and scheduling;

2. the identification of the optimal inventory management and control systems; and

3. the selection of the appropriate packaging techniques.

Ting *et al.* (2014) propose a decision support system to assist managers in food brands to draft logistics plans in order to secure food quality and safety, while ensuring SC sustainability. In addition, Akkerman, Farahani, and Grunow (2010) provide a thorough review of agrifood distribution and logistics operations, such as unitization of goods, packaging, stacking, bundling, wrapping, unstacking, and inventory control (e.g., van Beek *et al.*, 2003).

The optimization of the transport system of AFSCs has been addressed by many researchers. For example, Higgins *et al.* (2004) propose a modeling framework to improve the efficiency of both the harvesting and transport operations while further presenting two real-world case studies encountered in the Australian sugar industry. Additionally, Higgins (2006) proposes a mixed integer programming model for scheduling road transport vehicles in sugarcane transport. A number of researchers have developed optimization models in order to solve truck scheduling problems for transporting biomass and to determine the operating parameters under various management practices in biomass logistics systems (e.g., Ravula, Grisso, and Cundiff, 2008a,b; Han and Murphy, 2012). More specifically, agricultural fleet management deals with resource allocation, scheduling, routing, and the real-time monitoring of vehicles and materials that is mostly undertaken by farmers or machine contractors. Intensive agricultural production systems involve complex planning and coordination of field operations, mainly due to uncertainties associated with yield, weather, and machine performance. The planning of such operations in general, involves four highly interconnected stages, namely harvesting, out-of-field removal of biomass, rural road and public road transportation, supported by the appropriate machinery system (harvesters, transport units, medium and high capacity transport trucks, unloading equipment) (Sørensen and Bochtis, 2010). Current scientific research has contributed to the development of models for the scheduling of field operations involving fleets of agricultural machines with off-line management systems (e.g., Higgins and Davies, 2005; Busato, Berruto, and Saunders, 2007; Berruto and Busato, 2008), with on-line planning (e.g., Bochtis and Vougioukas, 2007) or based on methods form other scientific areas (e.g., Guan *et al.*, 2008). Indicatively, Sørensen and Bochtis (2010) propose a conceptual model of fleet management in agriculture that embeds the on-line positioning of vehicles, machine monitoring/ tracking with an improved general knowledge of the production process and management, coordination of multiple machines, route, and path guidance, and so on. Jensen *et al.* (2012) present a path planning method for transporting units in agricultural operations involving in-field and inter-field transports. Vehicle routing in the agricultural sector also constitutes an interesting research field (e.g., Sigurd, Pisinger, and Sig, 2004; Zanoni and Zavanella, 2007; Ahumada and Villalobos, 2011), in food logistics applications (Tarantilis and Kiranoudis, 2004) analogous to other general commodities, or for in-field operations (Bochtis and Sørensen, 2009, 2010).

Regarding the literature of inventory management and control for AFSCs, great importance is attributed to the deterioration of products and their implications on the planning of production and distribution operations (e.g., Akkerman, Farahani, and Grunow, 2010; Bakker, Riezebos, and Teunter, 2012; Zanoni and Zavanella, 2012). Notably, Karaesmen, Scheller-Wolf, and Deniz (2011) provide a comprehensive review and classification of research efforts concerning inventory management of perishable goods, while they further highlight the need for future research in areas such as multiple-products' inventory management, inventory capacity planning, freshness, disposal and outdating, inventory issuance and demand competition, contracting and pricing. Finally, Yu, Wang, and Liang (2012) developed an integrated modeling approach for a Vendor Managed Inventory (VMI) chain and concluded that the deterioration rate of the final products can increase total inventory costs by more than 40%.

Additionally, the packaging techniques along food SCs, from raw materials to final products, are strongly connected with the delivered quality to consumers, and thus they have been thoroughly scrutinized in the literature (e.g., Appendini and Hotchkiss, 2002; Vitner, Giller, and Pat, 2006; Restuccia *et al.*, 2010). In their pioneering work, Wikström *et al.* (2014) highlight packaging design attributes that can influence the volume of food waste and which need to be considered by relevant AFSC stakeholders. Most of the existing sectorial studies focus on specific agri-product cases. For example, Sothornvit and Kiatchanapaibul (2009) determine the optimum atmospheric packaging conditions for fresh-cut asparagus so as to increase the food safety and extend the shelf-life of the product. Other indicative works are those of Hertog *et al.* (1999) and Zhang, Xiao, and Salokhe (2006). The latter, examined weight loss, respiration rate, and susceptibility to fungal contamination of fresh strawberries and managed to extend their shelf-life through testing different atmospheric treatment and packaging conditions.

Finally, the decision-making process concerning the logistics operations is closely interrelated to other key attributes such as transparency, food safety, and traceability. In this context, Van der Vorst, van Kooten, and Luning (2011) provide a holistic framework for optimizing the performance of an AFSC with regard to product quality and availability.

1.3.2.3   Food Safety Transparency

Following a number of serious food safety incidents, investors, advocates, and consumers alike, demand that companies ensure food quality in all stages of their SCs and to further disclose quality information about their products (Dai, Kong, and

Wang, 2013). Indeed, food safety is one of the most critical aspects of the AFSCs, enforcing all stakeholders to increase the level of transparency in all stages of their own SCs. Transparency refers to the shared understanding and product-related information exchange among a SC's stakeholders and can guarantee food quality and provenance to all users of food products (Hofstede *et al.*, 2004; Wognum *et al.*, 2011; Trienekens *et al.*, 2012; Tsolakis *et al.*, 2014a).

The adoption of tracking and tracing technologies is a key element for a "smart" AFSC. Innovative traceability systems at all tiers of the supply network can also improve transparency (Kassahun *et al.*, 2014). According to the European Parliament (2002) "traceability means the ability to trace and follow a food, feed, food producing animal or substance intended to be, or expected to be incorporated into a food or feed, through all stages of production, processing, and distribution"; while according to the International Standard Organization (2007) "traceability is the ability to follow the movement of a feed or food through specified stage(s) of production, processing, and distribution". Wilson and Clarke (1998) define traceability as the available information regarding the history of food production from farm to final consumer.

Leat, Marr, and Ritchie (1998) outline the increased need for traceability in food safety by providing a number of drivers, for example, identification of the source of the infected product, disease control, labeling regulations, and so on. Bosona and Gebresenbet (2013) outline the driving forces for food traceability. More specifically, they partition the driving forces into: regulatory (e.g., new food safety legislations, ownership disputes, etc.); safety and quality (e.g., tracking food safety crises, value preservation in food SCs, etc.); social (e.g., increase in consumers' awareness, changing lifestyles, etc.); economic (e.g., market share, products' prizing, etc.); and technological (advancement in technology).

Contemporary traceability systems are rather sophisticated as they are developed capitalizing on the usage of ICT instruments. The adoption of radio frequency identification (RFID) tags, barcodes, and alphanumerical codes can assist in securing visibility among the partners of the SCs by facilitating data acquisition and processing and reduce significantly management costs in the entire SC network (Gandino *et al.*, 2009; Dabbene and Gay, 2011; Zhang and Li, 2012; Grunow and Piramuthu, 2013). According to Wang and Li (2012), tracking and tracing technologies can help in the development of a product's quality assessment model and in the decision regarding appropriate pricing strategies. On the other hand, Beulens *et al.* (2005) outline that even if innovative tracking and tracing systems can be easily installed and implemented by each player in the SC, the most important element still remains the coordination at a physical unit's level.

To this end, the establishment of appropriate channels for exchanging information and data and the promotion of the required mechanisms for collaboration and coordination are essential in order to overcome certain difficulties due to the dynamic nature and the high levels of complexity in the structure of modern AFSCs. Finally, Trienekens *et al.* (2012) present a comprehensive framework for transparency analysis in food SCs by identifying the necessary governance mechanisms adapted to different stakeholders' demands for transparency in all echelons of the SCs.

1.4   Emerging Trends and Technologies in Primary Production

On a global scale, GHG emissions from agriculture account for almost 14% of total emissions. Agriculture production is the most important source of nitrous oxide (N_2O) from organic and mineral nitrogen fertilizers, and methane (CH_4) from livestock digestion processes and stored animal manure. At the EU-27 level, emissions from agriculture account for 9.2% of total emissions (corresponding to 462 Mt of CO_2 equivalent in absolute numbers). However, this figure does not include agriculture-related emissions such as the emissions from agricultural land use (57 Mt CO_2 in EU-27 accounting for approximately 1% of the total emissions of all sectors), from fossil fuel use in agricultural buildings and agricultural machinery for field operations, which account for around 1% of CO_2 emissions of all sectors [following the reporting scheme of the United Nations Framework Convention on Climate Change (UNFCCC) these emissions are accounted in the "energy" inventory], and emissions from the manufacturing of fertilizers and animal feed.

Finally, it is worth noting that although agricultural emissions of N_2O and CH_4 rose globally by approximately 17% in the period 1990–2007, mainly due to the increased production in developing countries, during the same period in the EU-27, agricultural emissions declined by approximately 20% mainly due to reductions in livestock numbers and the improved fertilizer applications. Additional reductions in N_2O and CH_4 emissions could be achieved by various farm management practices including, among others, the overall reduction of external inputs (e.g., by employing precision agriculture principles and ICT tools), and the implementation of alternative tillage systems. These issues are further discussed in the following sections.

1.4.1   Alternative Production Systems

Current intensive tillage production systems highly influence soil structure decreasing the soil organic matter leading to significant GHG emissions due to the loss of CO_2 from arable soil. The introduction of less intensive methods in terms of soil preparation (indicated as conservation agriculture systems) and agricultural vehicle traffic (indicated as controlled traffic systems), is expected to keep reducing the agricultural impact on the global CO_2 balance (Chatskikh *et al.*, 2008). Conservation agriculture systems include reduced and zero tillage systems, and direct seeding combined with a varied crop rotation which eliminates disease and pest complications. According to FAO (2001) conservation agriculture conserves and improves arable soil conditions, while conserving water and biological resources thus enhancing and sustaining farm production. It maintains either a permanent or a semi-permanent organic soil cover (e.g., dead mulch) which protects the arable soil from the negative effects of sun, rain and wind, allowing micro-organisms living in the soil and fauna to further preserve nutrient balancing since all inherent natural processes are not disturbed by the mechanical tillage intervention.

Conventional in-field traffic systems can cause a trafficked area of 80–100% of the total field area, while in conservation tillage systems the trafficked area is reduced to 30–60% (Tullberg, Yule, and McGarry, 2007). Sørensen et al. (2014) studied the environmental effects of the implementation of reduced soil tillage and no soil tillage systems. They documented that the average of the total GHG emissions per kilogram of product for the conventional soil tillage, the reduced soil tillage, and the no soil tillage scenarios amounted to 915, 817, and 855 g CO_2/kg, respectively. The reductions in CO_2 emissions occurred in conservation systems when compared with the conventional system mainly stem from the reduced CO_2 emissions from carbon mineralization. Furthermore, when considering the operational cost benefits in conservation production systems in conjunction with the above mentioned environmental benefits, it becomes clear that conservation systems provide an overall advantage compared with conventional methods. However, for a comprehensive evaluation, the increased demands for management aimed at sustaining yields should also be an integral part under a systems approach; otherwise, the environmental benefits will be compromised.

In-field traffic, on the other hand, is a main concern in terms of soil sustainability and energy consumption. Controlled-traffic farming (CTF) is a traffic system for agricultural vehicle for their in-field activities which diversifies the cropped area and the trafficked area by creating permanent parallel field-work tracks (Chamen et al., 2003). CTF reduces the trafficked area of a field area (in the range of 20% of the total field) even more compared with various conservation tillage systems. Various studies demonstrate that the implementation of CTF is able to reduce the effects of arable crop production systems on environmental impacts, such as climate change, acidification, eutrophication, non-renewable resources depletion, human-toxicity, eco-toxicity, and furthermore, on soil erosion and land use. Based on a comprehensive review conducted by Gasso et al. (2013), a state-of-the-art analysis on the environmental impacts of CTF compared with the conventional traffic systems demonstrated that CTF is able to reduce:

- soil fluxes of N_2O in the range of 21–45%;
- water runoff in the range of 27–42%;
- in-field operations direct emissions up to 23%;
- indirect impacts associated with fertilizers up to 26%;
- indirect impacts associated with pesticides up to 26%;
- indirect impacts associated with seeds up to 36%; and
- indirect impacts associated with fuels up to 23%.

From an operations execution point of view, advanced navigation aiding and auto-steering systems for agricultural machinery ensure accurate driving on predetermined tracks making the implantation of CTF feasible. However, modifications are needed so that the wheel distance widths of the implemented machinery are able to match the permanent tracks offset, allowing the tires to run exclusively on the permanent tracks. This compatibility between the machinery and the spatial configuration of the permanent tracks remains a major impediment to a wide adoption of the CTF; this hindrance can be addressed only with the active engagement of the agricultural machinery industry (Tullberg, 2010).

1.4.2   Innovative Technologies

Advanced engineering and systems engineering approaches in bio-production systems provide great potential for supporting producers to amend environmental impacts in various ways. Selected examples of the implementation of these technologies are listed in the following paragraphs.

1.4.2.1   Satellite-Based Navigation

Global Positioning System (GPS) based navigation-aiding systems and auto-steering systems for agricultural vehicles can reduce the overlapping application of fertilizers and pesticides. Specifically, continuous recording of the field areas where material is applied drives the automatic turning on or off sections of the sprayer preventing double coverage of previously sprayed field areas. The potential savings using automatic section control have been reported to be up to 25% (Stombaugh, Zandonadi, and Dillon, 2009). In general, these systems have provided a number of tangible benefits including the elimination of overlaps and underlaps (untreated areas) leading to savings in input materials, fuels, operational time, and operational cost, reduced operator fatigue, reduced soil compaction, and improved crop establishment. Especially, the latter is a crucial KPI for an effective implementation of the precision agriculture principles as it reduces the spatial uncertainty inherent in crop production systems. Finally, the usage of GPS-based navigation technologies for agricultural machinery is a prerequisite for the utilization of CTF.

1.4.2.2   Satellite-Based Monitoring

Satellite imagery is a powerful tool for crop production which can provide micro-variations in a dynamic and comprehensive manner on crop productivity parameters, such as spatial and structural distribution of soil properties, growing status, moisture, and water content. In contrast to proximal sensing, remote sensing applications in agriculture are based typically, on the reflecting electromagnetic radiation of soil and plant material. These satellite monitoring technologies are replacing the intensive and costly process of laboratory analyzed soil and crop samples. A typical cost in the USA of satellite imagery services is less than US$15 per hectare for multiple readings per year providing a potential increase to the yield of as much as 10% (The Economist, 2009).

The spatial resolution of satellite imagery has improved from 80 m, at the time of the first application in agriculture (Bhatti, Mulla, and Frazier, 1991) (with Landsat), to sub-meter resolution in modern applications (with GeoEye and WorldView).

Furthermore, the visit frequency has improved from 18 (with Landsat) to 1.1 days (with WorldView-2) (Mulla, 2013). The added value of satellite-based monitoring has been proven for the level of large-scale applications, for example, for monitoring areas in relation to EU directives and policies (Alexandridis, Zalidis, and Silleos, 2008). However, modern agricultural production management systems, such as precision agriculture, require spatial information of a higher accuracy in order to support reliable decision-making. To this end, integrated frameworks have been proposed which combine satellite, aerial [i.e., based on unmanned aerial vehicles (UAVs)], and ground (i.e., mobile vehicles and static stations) sensing providing multi-sources and multi-scales monitoring approaches (Shi *et al.*, 2014). These approaches appear to be extremely valuable in the case of small-holder agricultural production systems and, in general, to geographical areas with fragmented agricultural land.

1.4.2.3 Robotics

For over six decades, robots have been playing a leading and often innovative role in increasing the efficiency and reducing the cost within industrial production. In the case of agricultural production, their usage is expected to highly improve sustainability. This conjecture stems from the hypothesis that the current large (in terms of power and size) machinery systems, developed under the economies of scale paradigm, can be replaced by multiple-unit robotic systems consisting of lighter and more autonomous units. However, the challenge is that in contrast to the floor production, where tasks and the environment are predefined, intelligent robotic systems have to be developed to be able to cope flexibly with outdoor, non-structured (i.e., arable farming), or in the best case semi-structured (e.g., orchard farming), environments where agricultural production takes place.

A targeted area for the use of field robots is in pesticide application. Pesticide usage represents a substantial chemical load for the environment with a high risk of undesirable side effects on human health. There is a significant potential for reducing pesticide by implementing patch spraying based on the combination of machine vision and subsequent image analysis techniques combined with precision spraying systems carried out by conventional machinery of small field robots (Bochtis *et al.*, 2011). A state-of-the-art case of robotic variable rate application has been recently presented (Pérez-Ruiz *et al.*, 2015), where based on field trials it was documented that the estimated cost reduction for site-specific flame weeding was approximately 28 €/ha when compared with a conventional system (from 52 to 24 €/ha).

In addition to the ground unmanned vehicles, UAVs appear to have great potential. The use of UAVs is the new trend for small-scale monitoring operations with a current global unmanned aerial systems market revenue of 5400 M€ and this is expected to grow up to 6350 M€ by 2018 (MarketsandMarkets, 2013). Agricultural production belongs to an area that is likely to be able to considerably expand the use of UAVs to high rates, as it involves flying solely on unpopulated areas where restrictions dealing with built-up locations are non-existent (Kuchler, 2014). For agricultural production applications, UAVs offer a complementary solution for crop management and monitoring combined with satellite and ground monitoring layers. Furthermore, the use of

UAVs in agricultural production provides a fast deployment monitoring system at low cost, with the ability to deliver high image resolution suitable for small-scale investigations, and able to overcome the difficulty of repeated measurements during the crop (a barrier inherent in full implementation of satellite-based monitoring; Colomina and Molina, 2014). Finally, regarding small-farm-based production systems, the benefits obtained by the employment of UAVs for monitoring small productive areas have still to be proven (Lelong *et al.*, 2008).

1.5   Conclusions

SCM is widely accepted as an area of critical importance for the agrifood sector. SC stakeholders involved in both the design and the execution of AFSCs are called to address systemically an array of complex and often interwoven decisions spanning all levels of the natural hierarchical decision-making process. To that effect, this chapter captures comprehensively and in a novel interdisciplinary framework, both the associated challenges and the complexity of the decision-making process for the design and planning of AFSCs.

We began by presenting the generic system components along with the unique characteristics of AFSC networks that differentiate them from traditional SCs. We proceeded by identifying and discussing the most critical issues for the design and planning of AFSCs, along with the most relevant emerging technologies, as well as by presenting a critical synthesis of the related existing state-of-the-art literature efforts in order to identify major gaps, overlaps, and opportunities. These issues were further mapped accordingly on the recognized natural hierarchy of the relevant decision-making process.

Our critical analysis reveals the following key findings:

- Even though SCs of the agrifood sector have been addressed by the research community, there is a lack of integrated systemic approaches that could support effectively the design and planning of such networks.

- There is a need for the development of appropriate channels for exchanging information and data alongside the promotion of the required mechanisms for collaboration and coordination within modern AFSCs in order to address various challenges stemming from the dynamic nature and the inherent high levels of complexity of these SCs.

- The decision-making process concerning the logistics operations should be closely interrelated to other key attributes such as transparency, food safety, and traceability.

- The integration of QMSs in the AFSCs focusing on the optimization of processes, the economy, and governance is a critical aspect for ensuring a sustainability-driven flow of information, processes, and materials.

- More integrated and sophisticated measurement systems have to be developed and standardized for the continuous monitoring and evaluation of the AFSCs' performance in terms of sustainability aspects.

- Even though in the general SCM literature there is a significant volume of relevant research, a number of core customized decisions regarding the configuration of AFSC networks are still lacking. Targeted research actions have to overcome the difficulties imposed by the structure and complexity of the relationships across an entire agrifood chain toward the development of dedicated decision-making approaches for this type of network.

- The implementation of advanced engineering and systems engineering approaches (such as satellite-based navigation, remote sensing and monitoring, and robotic systems) in primary production provides great potential to amend environmental impacts in both large-scale and small-holder agricultural production systems. In parallel, a widespread adoption of less intensive methods in terms of soil preparation and in-field traffic, are expected to reduce the agricultural impact on global CO_2 balance and prevent soil degradation as a "growth medium."

We envision that the presented decision-making framework, along with the respective critical synthesis, which merge the worlds of operations management, SCM, and agriculture could provide a platform of great value for researchers and practitioners alike to build upon, in their evolving efforts toward the scientific development and management of highly competitive and sustainable AFSCs.

Acknowledgments

This study has been conducted in the context of the GREEN-AgriChains project that is funded from the European Community's Seventh Framework Programme (FP7-REGPOT-2012-2013-1) under Grant Agreement No. 316167. This chapter reflects only the authors' views; the EU is not liable for any use that may be made of the information contained herein.

References

Abebaw, D. and Haile, M.G., 2012. The impact of cooperatives on agricultural technology adoption: empirical evidence from Ethiopia. *Food Policy*, **38**, 82–91.

Ahi, P. and Searcy, C., 2013. A comparative literature analysis of definitions for green and sustainable supply chain management. *Journal of Cleaner Production*, **52**, 329–341.

Ahi, P. and Searcy, C., 2014. An analysis of metrics used to measure performance in green and sustainable supply chains. *Journal of Cleaner Production*, **86**, 360–377.

Ahumada, O. and Villalobos, J.R., 2009. Application of planning models in the agrifood supply chain: a review. *European Journal of Operational Research*, **195**, 1–20.

Ahumada, O. and Villalobos, J.R., 2011. Operational model for planning the harvest and distribution of perishable agricultural products. *International Journal of Production Economics*, **133**, 677–687.

Akkerman, R., Farahani, P., and Grunow, M., 2010. Quality, safety and sustainability in food distribution: a review of quantitative operations management approaches and challenges. *OR Spectrum*, **32**, 863–904.

Akkermans, H., Bogerd, P., and Vos, B., 1999. Virtuous and vicious cycles on the road towards international supply chain management. *International Journal of Operations & Production Management*, **19**, 565–581.

Alexandridis, T.K., Zalidis, G.C., and Silleos N.G., 2008. Mapping irrigated area in Mediterranean basins using low cost satellite earth observation. *Computers and Electronics in Agriculture*, **64**(2), 93–103.

Ameseder, C., Canavari, M., Cantore, N., Deiters, J., Fritz, M., Haas, R., Matopoulos, A., Meixner, O., and Vlachopoulou, M., 2009. Perceived risks in cross-border transactions in agrifood chains. 113th EAAE Seminar: A Resilient European Food Industry and Food Chain in a Challenging World, Chania, Crete, Greece, September 3–6, 2009.

Antón, J., Kimura, S., and Martini, R., 2011. *Risk Management in Agriculture in Canada.* OECD Food, Agriculture and Fisheries Working Papers, No. 40. Paris: OECD.

Appendini, P. and Hotchkiss, J.H., 2002. Review of antimicrobial food packaging. *Innovative Food Science & Emerging Technologies*, **3**, 113–126.

Aramyan L., Ondersteijn C., van Kooten O., and Lansink, A.O., 2006. *Performance Indicators in Agrifood Production Chains.* Quantifying the Agrifood Supply Chain. Dordrecht: Springer, pp. 47–64.

Aramyan, L.H., Oude Lansink, A.G.J.M., Van der Vorst, J.G.A.J., and van Kooten, O., 2007. Performance measurement in agrifood supply chains: a case study. *Supply Chain Management: An International Journal*, **12**, 304–315.

Asfaw, S., Mithöfer, D., and Waibel, H., 2009. EU food safety standards, pesticide use and farm-level productivity: the case of high-value crops in Kenya. *Journal of Agricultural Economics*, **60**, 645–667.

Aubert, B.A., Schroeder, A., and Grimaudo, J., 2012. IT as enabler of sustainable farming: an empirical analysis of farmer's adoption decision of precision agriculture technology. *Decision Support Systems*, **54**, 510–520.

Aurbacher, J., Lippert, C., and Dabbert, S., 2011. Imperfect markets for used machinery, asynchronous replacement times, and heterogeneity in cost as path-dependent barriers to cooperation between farmers. *Biosystems Engineering*, **108**, 144–153.

Bachev, H., 2011. Management of Chemical and Biological Risks in Agrifood Chain. Munich Personal RePEc Archive, MPRA Paper No. 30905. Institute of Agricultural Economics, Sofia.

Bachev, H., 2012. Risk Management in Agrifood Chain. Munich Personal RePEc Archive, MPRA Paper No. 39594. Institute of Agricultural Economics, Sofia.

Bajaj, S., Jha, P.C., and Aggarwal, K.K., 2013. Single-source, single-destination, multi product EOQ model with quantity discount incorporating partial/full truckload policy. *International Journal of Business Performance and Supply Chain Modelling*, **5**, 198–220.

Bakker, F.D. and Nijhof, A., 2002. Responsible chain management: a capability assessment framework. *Business Strategy and the Environment*, **11**, 63–75.

Bakker, M., Riezebos, J., and Teunter, R.H., 2012. Review of inventory systems with deterioration since 2001. *European Journal of Operational Research*, **221**, 275–284.

Barratt, M., 2004. Understanding the meaning of collaboration, *Supply Chain Management: An International Journal*, **9**, 30–42.

Bas, M., Yüksel, M., and Çavuooflu, T., 2007. Difficulties and barriers for the implementing of HACCP and food safety systems in food businesses in Turkey. *Food Control*, **18**, 124–130.

Beamon, B.M., 1998. Supply chain design and analysis: models and methods. *International Journal of Production Economics*, **55**, 281–294.

Berruto, R. and Busato, P., 2008. System approach to biomass harvest operations: simulation modeling and linear programming for logistic design. ASABE Annual International Meeting, Rhode Island, Paper No. 084565.

Beske, P., Land, A., and Seuring, S., 2014. Sustainable supply chain management practices and dynamic capabilities in the food industry: a critical analysis of the literature. *International Journal of Production Economics*, **152**, 131–143.

Beulens, A.J.M., Broens, D.F., Folstar, P., and Hofstede, G.J., 2005. Food safety and transparency in food chains and networks: relationships and challenges. *Food Control*, **16**, 481–486.

Bhatnagar, R. and Sohal, A., 2005. Supply chain competitiveness: measuring the impact of location factors, uncertainly and manufacturing practices. *Technovation*, **25**, 443–456.

Bhatti, A.U., Mulla, D.J., and Frazier, B.E., 1991. Estimation of soil properties and wheat yields on complex eroded hills using geostatistics and thematic mapper images. *Remote Sensing of Environment*, **37**, 181–191.

Bielza Diaz-Caneja, B.M., Conte, C.G., Gallego Pinilla, F.J., Stroblmair, J., Catenaro, R., and Dittmann, C., 2009. *Risk Management in Agricultural Insurance Schemes in Europe*. Ispra: The Institute of the Protection and Security of the Citizen.

Bijman, J., Omta, S.W.F., Trienekens, J.H., Wijnands, J., and Wubben E., 2006. Management and organization in international agrifood chains and networks (eds J. Bijman, S.W.F. Omta, J.H. Trienekens, J.H.M. Wijnands, and E.F.M. Wubben) *International Agrifood Chains and Networks-Management and Organisation*. Wageningen: Wageningen Academic Publishers, pp. 15–28.

Biswas, W.K., Barton, L., and Carter, D., 2008. Global warming potential of wheat production in Western Australia: a life cycle assessment. *Water and Environment Journal*, **22**, 206–216.

Bochtis, D., 2013. Satellite based technologies as key enablers for sustainable ICT-based agricultural production systems. *Procedia Technology*, **8**, 4–8.

Bochtis, D.D. and Sørensen, C.G., 2009. The vehicle routing problem in field logistics part I. *Biosystems Engineering*, **104**, 447–457.

Bochtis, D.D. and Sørensen, C.G., 2010. The vehicle routing problem in field logistics part II. *Biosystems Engineering*, **105**, 180–188.

Bochtis, D.D., Sørensen, C.G., Jørgensen, R.N., Nørremark, M., Hameed I.A., and Swain, K.C., 2011. Robotic weed monitoring. *Acta Agriculturae Scandinavica, Section B: Plant Soil Science* **61**, 202–208.

Bochtis, D.D. and Vougioukas, S.G., 2007. Agricultural machine allocation based on simulation. Proceedings of the Second IFAC International Conference on Modeling and Design of Control Systems in Agriculture, Osijek, Croatia, pp. 147–152.

Bosona T. and Gebresenbet G., 2013. Food traceability as an integral part of logistics management in food and agricultural supply chain. *Food Control*, **33**, 32–48.

Boudahri, F., Bennekrouf, M., and Sari, Z., 2011. Optimization and design of the transportation network of agrifoods supply chain: application chicken meat. *International Journal of Advanced Engineering Sciences and Technologies*, **11**, 213–220.

Bourlakis, M., Maglaras, G., Aktas, E., Gallear, D., and Fotopoulos, C., 2014. Firm size and sustainable performance in food supply chains: insights from Greek SMEs. *International Journal of Production Economics*, **152**, 112–130.

BRC Global Standards, 2012. British Retail Consortium, http://www.brcglobalstandards.com/ (accessed July 8, 2015).

Brorström, S., 2015. Strategizing sustainability: the case of River City, Gothenburg. *Cities*, **42**, 25–30.

Brunner, T.A., van der Horst, K., and Siegrist, M., 2010. Convenience food products. Drivers for consumption. *Appetite*, **55**, 498–506.

Burch, D. and Goss, J., 1999. Global sourcing and retail chains: shifting relationships of production in Australian agrifoods. *Rural Sociology*, **64**, 334–350.

Busato, P., Berruto, R., and Saunders, C., 2007. Modeling of grain harvesting: interaction between working pattern and field bin locations. *Agricultural Engineering International: The CIGR Ejournal*, **IX**, http://www.cigrjournal.org/index.php/Ejounral/issue/view/29 (accessed July 8, 2015).

Busato, P., Sørensen, C.G., Pavlou, D., Bochtis, D.D., Berruto, R., and Orfanou, A., 2013. DSS tool for the implementation and operation of an umbilical system applying organic fertilizer. *Biosystems Engineering*, **114**, 9–20.

Byrne, P.J., Ryan, P., and Heavey, C., 2013. Sustainable logistics: a literature review and exploratory study of Irish based manufacturing organizations. *International Journal of Engineering and Technology Innovation*, **3**, 200–213.

Caniato, F., Caridi, M., Crippa, L., and Moretto, A., 2011. Environmental sustainability in fashion supply chains: an exploratory case based research. *International Journal of Production Economics*, **135**, 659–670.

Carter, C.R. and Easton, P.L., 2011. Sustainable supply chain management: evolution and future directions. *International Journal of Physical Distribution & Logistics Management*, **41**, 46–62.

Carter, C.R. and Rogers, D.S., 2008. A framework of sustainable supply chain management: moving towards new theory. *International Journal of Physical Distribution & Logistics Management*, **38**, 360–387.

CEC, 2006. Environmental Impact of Products of Products (EIPRO). Analysis of Consumption of the EU-25. Technical Report EUR 22284, http://ec.europa.eu/environment/ipp/pdf/eipro_report.pdf (accessed July 8, 2015).

Chamen, T., Alakukku, L., Pires, S., Sommer, C., Spoor, G., Tijink, F., and Weisskopf, P., 2003. Prevention strategies for field traffic-induced subsoil compaction: a review. Part 2: equipment and field practices. *Soil and Tillage Research*, **73**, 161–174.

Chan, F.T.S., 2003. Performance measurement in a supply chain. *The International Journal of Advanced Manufacturing Technology*, **21**, 534–548.

Chatskikh, D., Olesen, J.R.E., Hansen, E.M., Elsgaard, L., and Petersen, B.R.M., 2008. Effects of reduced tillage on net greenhouse gas fluxes from loamy sand soil under winter crops in Denmark. *Agriculture, Ecosystems & Environment*, **128**, 117–126.

Chen, K., 2006. Agrifood supply chain management: opportunities, issues, and guidelines. International Conference on Livestock Services, Beijing, People's Republic of China, April 16–22, 2006.

Chen, K.Z., Chen, Y., and Shi, M., 2003. Globalization, pesticide regulation, and supply chain development: a case of Chinese vegetable export to Japan. FAO Scientific Workshop. Globalization, Urbanization and the Food Systems of Developing Countries: Assessing the Impacts on Poverty, Food and Nutrition Security, Rome, Italy, October 8–10, 2003.

Colomina, I. and Molina, P., 2014. Unmanned aerial systems for photogrammetry and remote sensing: a review. *ISPRS Journal of Photogrammetry and Remote Sensing*, **92**, 79–97.

Courville, S., 2003. Use of indicators to compare supply chains in the coffee industry. *Greener Management International*, **43**, 94–105.

Dabbene, F. and Gay, P., 2011. Food traceability systems: performance evaluation and optimization. *Computers and Electronics in Agriculture*, **75**, 136–146.

Dai, J. and Dong, H., 2014. Intensive cotton farming technologies in China: achievements, challenges and countermeasures. *Field Crops Research*, **155**, 99–110.

Dai, Y., Kong, D., and Wang, M., 2013. Investor reactions to food safety incidents: evidence from the Chinese milk industry. *Food Policy*, **43**, 23–31.

Da Silva, C.A.B., 2005. *The Growing Role of Contract Farming in Agrifood Systems Development: Drivers, Theory and Practice*. Rome: Food and Agriculture Organization of the United Nations.

De Torro, A. and Hansson, P.-A., 2004. Machinery co-operatives – A case study in Sweden. *Biosystems Engineering*, **87**, 13–25.

Del Borghi, A., Gallo, M., Strazza, C., and Del Borghi, M., 2014. An evaluation of environmental sustainability in the food industry through Life Cycle Assessment: the case study of tomato products supply chain. *Journal of Cleaner Production*, **78**, 121–130.

Delgado, C., Rosegrant, M., Steinfeld, H., Ehui, S., and Courbois, C., 1999. Live Stock to 2020: The Next Food Revolution. Technical Report, Food, Agriculture, and the Environment Discussion Paper 28, International Food Policy Research Institute, Washington, DC.

Demirel, T., Demirel, N.C., and Kahraman, C., 2010. Multi-criteria warehouse location selection using choquet integral. *Expert Systems with Applications*, **37**, 3943–3952.

Dile, Y., Karlberg, L., Temesgen, M., and Rockström, J., 2013. The role of water harvesting to achieve sustainable agricultural intensification and resilience against water related shocks in sub-Saharan Africa. *Agriculture, Ecosystems & Environment*, **181**, 69–79.

Du, Q., Chang, N.-B., Yang, C., and Srilakshmi, K.R., 2008. Combination of multispectral remote sensing, variable rate technology and environmental modeling for citrus pest management. *Journal of Environmental Management*, **86**, 14–26.

Dyer, J. and Singh, H., 1998. The relational view: cooperative strategy and sources of interorganisational competitive advantage. *Academy of Management Review*, **23**, 660–679.

EC, 2010. *The Common Agricultural Policy after 2013: Summary Report*. Brussels: Department of Agriculture and Rural Development.

Epperson, J.E. and Estes, E.A., 1999. Fruit and vegetable supply-chain management, innovations, and competitiveness: cooperative regional research project S-222. *Journal of Food Distribution*, **30**, 38–43.

EU Regulation 178/2002, 2002. Regulation (EC) No. 178/2002 of the European Parliament and of the Council of 28 January 2002 Laying Down the General Principles and Requirements of Food Law, Establishing the European Food Safety Authority and Laying Down Procedures in matters of Food Safety.

FAO, 2001. Conventional ploughing erodes the soil – zero-tillage is an environmentally-friendly alternative. *FAO International Conference on Conservation Agriculture*, Madrid, Spain, October 1–5, 2001. Rome: Food and Agriculture Organization of the United Nations.

FAO, 2006. *World Agriculture towards 2030/2050: Interim Report*. Rome: Food and Agriculture Organization of the United Nations.

FAO, 2009. *How to feed the World in 2050*. Rome: Food and Agriculture Organization of the United Nations.

FAO, 2014. *The State of Food and Agriculture*. Rome: Food and Agriculture Organization of the United Nations.

Farahani, R.Z., Rezapour, S., Drezner, T., and Fallah, S., 2014. Competitive supply chain network design: an overview of classifications, models, solution techniques and applications. *Omega*, **45**, 92–118.

Fearne, A., Hughes, D., and Duffy, R., 2001 Concepts of collaboration – supply chain management in a global food industry (eds J.F. Eastham, L. Sharples, and S.D. Ball) *Food and Drink Supply Chain Management – Issues for the Hospitality and Retail Sectors*. Oxford: Butterworth-Heinemann, pp. 55–89.

Fernández Lambert, G., Aguilar Lasserre, A.A., Miranda Ackerman, M., Moras Sánchez, C.G., Ixmatlahua Rivera, B.O., and Azzaro-Pantel, C., 2014. An expert system for predicting orchard yield and fruit quality and its impact on the Persian lime supply chain. *Engineering Applications of Artificial Intelligence*, **33**, 21–30.

Fischer, C., 2009. Managing sustainable agrifood chain relationships – factors affecting relationship quality and stability dimensions. *19th Annual World Forum and Symposium*, Budapest, Hungary, June 20–23, 2009. Washington ,DC: International Food and Agribusiness Management Association.

Fischer, C. and Hartman, M., 2010. *Agrifood Chain Relationships*. London: CAB International.

Fischer, C., Hartmann, M., Reynolds, N., Leat, P., Revoredo-Giha, C., Henchion, M., and Gracia, A., 2008. Agri-food chain relationships in Europe – empirical evidence and implications for sector competitiveness. 12th Congress of the European Association of Agricultural Economists, Gent, Belgium, August 26–29, 2008.

Fritz, M. and Schiefer, G., 2008. Food chain management for sustainable food system development: a European research agenda. *Agribusiness*, **24**, 440–452.

Gandino, F., Montrucchio, B., Rebaudengo, M., and Sanchez, E.R., 2009. On improving automation by integrating RFID in the traceability management of the agrifood sector. *IEEE Transactions on Industrial Electronics*, **56**, 2357–2365.

Gasso, V., Sørensen, C.A.G., Oudshoorn, F.W., and Green, O., 2013. Controlled traffic farming: a review of the environmental impacts. *European Journal of Agronomy*, **48**, 66–73.

Glen, J.J., 1987. Mathematical models in farm planning: a survey. *Operations Research*, **35**, 641–666.

Glen, J.J. and Tipper, R., 2001. A mathematical programming model for improvement planning in a semi-subsistence farm. *Agricultural Systems*, **70**, 295–317.

Glover, J.L., Champion, D., Daniels, K.J., and Dainty, A.J.D., 2014. An Institutional Theory perspective on sustainable practices across the dairy supply chain. *International Journal of Production Economics*, **152**, 102–111.

Godwin, R.J., Richards, T.E., Wood, G.A., Welsh, J.P., and Knight, S.M., 2003. An economic analysis of the potential for precision farming in UK cereal production. *Biosystems Engineering*, **84**, 533–545.

Gold, S., Seuring, S., and Beske, P., 2010. Sustainable supply chain management and inter-organizational resources: a literature review. *Corporate Social Responsibility and Environmental Management*, **17**, 230–245.

Goldberg, R.A. and Fries, L.A., 2012. *Nestlé: Agricultural Material Sourcing Within the Concept of Creating Shared Value (CSV)*. Harvard Business School Case 913-406, December 2012. Boston, MA: Harvard Business School.

Gorris, L., 2005. Food safety objective: an integral part of food chain management. *Food Control*, **16**, 801–809.

Govindan, K., Jafarian, A., Khodaverdi, R., and Devika, K., 2014. Two-echelon multiple-vehicle location – routing problem with time windows for optimization of sustainable supply chain network of perishable food. *International Journal of Production Economics*, **152**, 9–28.

Grimm, J., Hofstetter, J., and Sarkis, J., 2014. Critical factors for sub-supplier management: a sustainable food supply chains perspective. *International Journal of Production Economics*, **152**, 159–173.

Grunow, M. and Piramuthu, S., 2013. RFID in highly perishable food supply chains – remaining shelf life to supplant expiry date? *International Journal of Production Economics*, **146**, 717–727.

Guan, S., Nakamura, M., Shikanai, T., and Okazaki, T., 2008. Hybrid Petri nets modelling for farm work flow. *Computers and Electronics in Agriculture*, **62**, 149–158.

Gupta, S. and Palsule-Desai, O., 2011. Sustainable supply chain management: review and research opportunities. *IIMB Management Review*, **23**, 234–245.

Halldorsson, A., Kotzab, H., and Skjøtt-Larsen, T., 2009. Supply chain management on the crossroad to sustainability: a blessing or a curse? *Logistics Research*, **1**, 83–94.

Hameed, I.A., Bochtis, D.D., Sørensen, C.G., and Nøremark, M., 2010. Automated generation of guidance lines for operational field planning. *Biosystems Engineering*, **107**, 294–306.

Han, S. and Murphy, G.E., 2012. Solving a woody biomass truck scheduling problem for a transport company in Western Oregon, USA. *Biomass and Bioenergy*, **44**, 47–55.

Hertog, M.L.A.T.M., Boerrigter, H.A.M., van den Boogaard, G.J.P.M., Tijskens, L.M.M., and van Schaik, A.C.R., 1999. Predicting keeping quality of strawberries (cv. 'Elsanta') packed under modified atmospheres: an integrated model approach. *Postharvest Biology and Technology*, **15**, 1–12.

Higgins, A., 2006. Scheduling of road vehicles in sugarcane transport: a case study at an Australian sugar mill. *European Journal of Operational Research*, **170**, 987–1000.

Higgins, A., Antony, G., Sandell, G., Davies, I., Prestwidge, D., and Andrew, B., 2004. A framework for integrating a complex harvesting and transport system for sugar production. *Agricultural Systems*, **82**, 99–115.

Higgins, A. and Davies, I., 2005. A simulation model for capacity planning in sugarcane transport. *Computers and Electronics in Agriculture*, **47**, 85–102.

Hobbs, J.E. and Young, L.M., 2000. Closer vertical co-ordination in agrifood supply chains: a conceptual framework and some preliminary evidence. *Supply Chain Management: An International Journal*, **5**, 131–143.

Hofstede, G.J., Spaans, H., Schepers, H., Trienekens, J.H., and Beulens, A.J.M., 2004. *Hide or Confide: the Dilemma of Transparency*. Hilversum: Reed Business Information.

Humphrey, J. and Memedovic, O., 2006. *Global Value Chains in the Agrifood Sector*. Vienna: United Nations Industrial Development Organization.

Iakovou, E., Karagiannidis, A., Vlachos, D., Toka, A., and Malamakis, A., 2010. Waste biomass-to-energy supply chain management: a critical synthesis. *Waste Management*, **30**, 1860–1870.

Iakovou, E., Vlachos, D., Achillas, C., and Anastasiadis, F., 2012. A Methodological Framework for the Design of Green Supply Chains for the Agrifood Sector. Working Paper.

Ilbery, B. and Maye D., 2005. Food supply chains and sustainability: evidence from specialist food producers in the Scottish/English borders. *Land Use Policy*, **22**, 331–344.

International Standard Organization ISO 22000:2005, 2005. *Food Safety Management Systems – Requirements for Any Organization in the Food Chain*. ISO, Geneva.

International Standard Organization ISO 22005:2007, 2007. *Traceability in the Feed and Food Chain — General Principles and Basic Requirements for System Design and Implementation*. ISO, Geneva.

Isgin, T., Bilgic, A., Forster, D.L., and Batte, M., 2008. Using count data models to determine the factors affecting farmers' quantity decisions of precision farming technology adoption. *Computers and Electronics in Agriculture*, **62**, 231–242.

Jaffee, S., Siegel, P., and Andrews, C., 2010. Rapid Agricultural Supply Chain Risk Assessment: A Conceptual Framework. Agriculture and Rural Development Discussion Paper 47. The World Bank , Washington, DC.

Jensen, M.A.F., Bochtis, D., Sørensen, C.G., Blas, M.R., and Lykkegaard, K.L., 2012. In-field and inter-field path planning for agricultural transport units. *Computers & Industrial Engineering*, **63**, 1054–1061.

Kahiluoto, H., Kuisma, M., Havukainen, J., Luoranen, M., Karttunen, P., Lehtonen, E., and Horttanainen, M., 2011. Potential of agrifood wastes in mitigation of climate change and eutrophication – two case regions. *Biomass and Bioenergy*, **35**, 1983–1994.

Karaesmen, I.Z., Scheller-Wolf, A., and Deniz, B., 2011 Planning production and inventories in the extended enterprise (eds K.G. Kempf, P. Keskinocak, and R. Uzsoy) *Managing Perishable and Aging Inventories: Review and Future Research Directions*. New York: Springer, pp. 393–436.

Kassahun, A., Hartog, R.J.M., Sadowski, T., Scholten, H., Bartram, T., Wolfert, S., and Beulens, A.J.M., 2014. Enabling chain-wide transparency in meat supply chains based on the EPCIS global standard and cloud-based services. *Computers and Electronics in Agriculture*, **109**, 179–190.

Katalin, T.-G., Rahoveanu, T., Magdalena, M., and István, T., 2014. Sustainable new agricultural technology – economic aspects of precision crop protection. *Procedia Economics and Finance*, **8**, 729–736.

Keramydas, C., Tsolakis, N., Vlachos, D., and Iakovou, E., 2014. A system dynamics approach towards food security in agrifood supply networks: a critical taxonomy of modern challenges in a sustainability context. *MIBES Transactions*, **8**, 68–83.

Kewill, 2008. Logistics and Transport Industry Environmental Survey. Report Code TIEL0807WP, Transport Intelligence, Brinkworth.

Kleindorfer, P., Singhal, K., and van Wassenhove, L., 2005. Sustainable operations management. *Production and Operations Management*, **14**, 482–492.

Kottila, M.R. and Rönni, P., 2008. Collaboration and trust in two organic food chains. *British Food Journal*, **110**, 376–394.

KPMG, 2013. KPMG International Survey of Corporate Responsibility Reporting 2013. KPMG, http://www.kpmg.com/Global/en/IssuesAndInsights/ArticlesPublications/corporate-responsibility/Documents/kpmg-survey-of-corporate-responsibility-reporting-2013.pdf (accessed January 28, 2015).

Kuchler, H., 2014. Drones at Work: Farmers Take Flight Into the Future. Financial Times (June 24, 2014), http://www.ft.com/intl/cms/s/0/2ac84532-f317-11e3-a3f8-00144feabdc0.html#axzz3Oz2pprMD (accessed June 27, 2015).

Kumar, N., Scheer, L., and Steenkamp, J., 1995. The effects of perceived interdependence on dealer attitudes. *Journal of Marketing Research*, **32**, 348–356.

Lamprinopoulou, C., Renwick, A., Klerkx, L., Hermans, F., and Roep, D., 2014. Application of an integrated systemic framework for analysing agricultural innovation systems and informing innovation policies: comparing the Dutch and Scottish agrifood sectors. *Agricultural Systems*, **129**, 40–54.

Leat, P., Marr, P., and Ritchie, C., 1998. Quality assurance and traceability – the Scottish agri-food industry's quest for competitive advantage. *Supply Chain Management: An International Journal*, **3**, 115–117.

Lelong, C.C.D., Burger, P., Jubelin, G., Roux, B., Labbe, S., and Baret, F., 2008. Assessment of unmanned aerial vehicles imagery for quantitative monitoring of wheat crop in small plots. *Sensors*, **8**, 3557–3585.

Ligon, E., 2003. Optimal risk in agricultural contracts. *Agricultural Systems*, **75**, 265–276.

Lindgreen, A., 2003. Trust as a valuable strategic variable in the food industry: different types of trust and their implementation. *British Food Journal*, **105**, 310–327.

Maloni, J.M. and Brown, M.E., 2006. Corporate social responsibility in the supply chain: an application in the food industry. *Journal of Business Ethics*, **68**, 35–52.

MarketsandMarkets, 2013. Unmanned Aerial Vehicle Market (2013–2018). Technical Report, MarketsandMarkets, Dallas, TX.

Matopoulos, A., Vlachopoulou, M., Manthou V., and Manos, B., 2007. A conceptual framework for supply chain collaboration: empirical evidence from the agrifood industry. *Supply Chain Management: An International Journal*, **12**, 177–186.

Matos, S. and Hall, J., 2007. Integrating sustainable development in the supply chain: the case of life cycle assessment in oil and gas and agricultural biotechnology. *Journal of Operations Management*, **25**, 1083–1102.

McKinsey, 2011. *The Business of Sustainability: McKinsey Global Survey Results*. New York: McKinsey & Company.

Meisterling, K., Samaras, C., and Schweizer, V., 2009. Decisions to reduce greenhouse gases from agriculture and product transport: LCA case study of organic and conventional wheat original research article. *Journal of Cleaner Production*, **17**, 222–230.

Mena, C., Terry, L., Williams, A., and Ellram, L., 2014. Causes of waste across multi-tier supply networks: cases in the UK food sector. *International Journal of Production Economics*, **152**, 144–158.

Mulla, D.J., 2013. Twenty-five years of remote sensing in precision agriculture: key advances and remaining knowledge gaps. *Biosystems Engineering*, **114**, 358–371.

Mungandi, S., Conforte, D., and Shadbolt, N.M., 2012. Integration of smallholders in modern agrifood chains: lessons from the KASCOL model in Zambia. *International Food and Agribusiness Management Review*, **15**, 155–176.

Mussell, A. and Gooch, M., 2008. *Case Studies on Agrifood Value Chain Collaboration*. Guelph: George Morris Centre and Value Chain Management Centre.

Neely, A., Gregory, M., and Platts, K., 2005. Performance measurement system design: a literature review and research agenda. *International Journal of Operations & Production Management*, **25**, 1228–1263.

Negro, O.S., Hekkert, M.P., and Smits, R.E., 2007. Explaining the failure of the Dutch innovation system for biomass digestion – a functional analysis. *Energy Policy*, **35**, 925–938.

Nelson, G., Rosegrant, M., Palazzo, A., Gray, I., Ingersoll, C., Robertson, R., Tokgoz, S., Zhu, T., Sulser, T., Ringler, C., Msangi, S., and You, L., 2010. *Food Security, Farming and*

Climate Change to 2050: Scenarios, Results, Policy Options, 1st edn. Washington, DC: International Food Policy Research Institute.

Organization for Economic Co-operation and Development, 2000. *Income Risk Management in Agriculture*. Paris: OECD.

Organization for Economic Co-operation and Development, 2009. *Managing Risk in Agriculture: A Holistic Approach*. Paris: OECD.

Pagell, M. and Shevchenko, A., 2014. Why research in sustainable supply chain management should have no future. *Journal of Supply Chain Management*, **50**, 44–55.

Pandey, P.K., Panda, S.N., and Panigrahi, B., 2006. Sizing on-farm reservoirs for crop-fish integration in rainfed farming systems in Eastern India. *Biosystems Engineering*, **93**, 475–489.

Parr, A., 2009. *Hijacking Sustainability*. Cambridge, MA: MIT Press.

Pérez-Ruiz, M., Gonzalez-de-Santos, P., Ribeiro, A., Fernandez-Quintanilla, C., Peruzzi, A., Vieri, M., Tomic, S., and Agüera J., 2015. Highlights and preliminary results for autonomous crop protection. *Computers and Electronics in Agriculture*, **110**, 150–161.

Ravula, P.P., Grisso, R.D., and Cundiff, J.S., 2008a. Comparison between two policy strategies for scheduling trucks in a biomass logistic system. *Bioresource Technology*, **99**, 5710–5721.

Ravula, P.P., Grisso, R.D., and Cundiff, J.S., 2008b. Cotton logistics as a model for a biomass transportation system. *Biomass and Bioenergy*, **32**, 314–325.

Resende-Filho, M.A. and Hurley, T.M., 2012. Information asymmetry and traceability incentives for food safety. *International Journal of Production Economics*, **139**, 596–603.

Restuccia, D., Spizzirri, U.G., Parisi, O.I., Cirillo, G., Curcio, M., Iemma, F., Puoci, F., Vinci, G., and Picci, N., 2010. New EU regulation aspects and global market of active and intelligent packaging for food industry applications. *Food Control*, **21**, 1425–1435.

Reynolds N., Fischer, C., and Hartmann, M., 2009. Determinants of sustainable business relationships in selected German agrifood chains. *British Food Journal*, **111**, 776–793.

Rodrigues, G.S., Rodrigues, I.A., Buschinelli, C.C.A., and Barros, I., 2010. Integrated farm sustainability assessment for the environmental management of rural activities. *Environmental Impact Assessment Review*, **30**, 229–239.

RPDRM, 2012. *Disaster Risk Management in Food and Agriculture*. Rome: Food and Agriculture Organization of the United Nations.

Sartorius, K. and Kirsten, J., 2007. A framework to facilitate institutional arrangements for smallholder supply in developing countries: an agribusiness perspective. *Food Policy*, **32**, 640–655.

Scheherazade, D., 2014. Climate Change Raises Risk to Food Supplies. Financial Times (April 11, 2014), http://www.ft.com/cms/s/2/fbcfbefc-a516-11e3-8988-00144feab7de.html#axzz3EguMUX7N (accessed June 27, 2015).

Schiemann, M., 2007. Inter-enterprise Relations in Selected Economic Activities. Statistics in Focus–Industry, Trade and Services, 57/2007. Eurostat, Luxembourg, http://ec.europa.eu/eurostat/documents/3433488/5295229/KS-SF-07-057-EN.PDF/83c7fc58-79f5-4bdc-ac41-b5f1a5efdbf5 (accessed July 8, 2015).

Schönhart, M., Schmid, E., and Schneider, U.A., 2011. CropRota – A crop rotation model to support integrated land use assessments. *European Journal of Agronomy*, **34**, 263–277.

van Schothorst, M., 2004. *A Simple Guide to Understanding and Applying the Hazard Analysis Critical Control Point Concept*, 3rd edn. Brussels: International Life Sciences Institute Europe.

Seuring, S., 2013. A review of modeling approaches for sustainable supply chain management. *Decision Support Systems*, **54**, 1513–1520.

Seuring, S. and Muller, M., 2008. From a literature review to a conceptual framework for sustainable supply chain management. *Journal of Cleaner Production*, **16**, 1699–1710.

Sezen, B. and Turkkantos, S., 2013. The effects of relationship quality and lean applications on buyer–seller relationships. *International Journal of Business Performance and Supply Chain Modelling*, **5**, 378–400.

Shaw, K., 2013. Docklands dreamings: illusions of sustainability in the Melbourne docks redevelopment. *Urban Studies*, **50**, 2158–2177.

Shi, Y., Ji, S., Shao, X., Tang, H., Wu, W., Yang, P., Zhang, Y., and Ryosuke, S., 2014. Framework of SAGI agriculture remote sensing and its perspectives in supporting national food security. *Journal of Integrative Agriculture*, **13**, 1443–1450.

Sigurd, M., Pisinger, D., and Sig, M., 2004. Scheduling transportation of live animals to avoid the spread of diseases. *Transportation Science*, **38**, 197–209.

Singh, A., 2010. Integrated Reporting: Too Many Stakeholders, Too Much Data? Forbes (June 9, 2010), http://www.forbes.com/sites/csr/2010/06/09/integrated-reporting-too-many-stakeholders-too-much-data/ (accessed June 27, 2015).

Søgaard, H.T. and Sørensen, C.G., 2004. A model for optimal selection of machinery sizes within the farm machinery system. *Biosystems Engineering*, **89**, 13–28.

Sørensen, C.G. and Bochtis, D.D., 2010. Conceptual model of fleet management in agriculture. *Biosystems Engineering*, **105**, 41–50.

Sørensen, C., Halberg, N., Oudshoorn, F., Petersen, B., and Dalgaard, R., 2014. Energy inputs and GHG emissions of tillage systems. *Biosystems Engineering*, **120**, 2–14.

Sørensen, C.G., Madsen, N.A., and Jacobsen, B.H., 2005. Organic farming scenarios: operational analysis and costs of implementing innovative technologies. *Biosystems Engineering*, **91**, 127–137.

Sothornvit, R. and Kiatchanapaibul, P., 2009. Quality and shelf-life of washed fresh-cut asparagus in modified atmosphere packaging. *LWT – Food Science and Technology*, **42**, 1484–1490.

Soysal, M., Bloemhof-Ruwaard, J.M., and van der Vorst, J.G.A.J., 2014. Modelling food logistics networks with emission considerations: the case of an international beef supply chain. *International Journal of Production Economics*, **152**, 57–70.

Stombaugh, T.S., Zandonadi, R.S., and Dillon C.R., 2009. Assessing the potential of automatic section control (eds E.J. Van Henten, D. Goense, and C. Lokhorst) *Proceedings of the Joint International Agricultural Conference (JIAC), Precision Agriculture 09*. Wageningen: Wageningen Academic Publishers, pp. 759–766.

Tajbakhsh, A. and Hassini, E., 2014. A data envelopment analysis approach to evaluate sustainability in supply chain networks. *Journal of Cleaner Production*. DOI: 10.1016/j.jclepro.2014.07.054.

Tan, L.P. and Fong, C.O., 1988. Determination of the crop mix of a rubber and oil palm plantation – a programming approach. *European Journal of Operations Research*, **34**, 362–371.

Tarantilis, C.D and Kiranoudis, C.T., 2004. Operational research and food logistics. *Journal of Food Engineering*, **70**, 253–255.

Taticchi, P., Tonelli, F., and Pasqualino, R., 2013. Performance measurement of sustainable supply chains. A literature review and a research agenda. *International Journal of Productivity and Performance Management*, **62**, 782–804.

Teuscher, P., Grüninger, B., and Ferdinand, N., 2006. Risk management in sustainable supply chain management (SSCM): lessons learnt from the case of GMO-free soybeans. *Corporate Social Responsibility and Environmental Management*, **13**, 1–10.

Thakkar, J., Kanda, A., and Deshmukh, S.G., 2008. Supply chain management in SMEs: development of constructs and propositions. *Asia Pacific Journal of Marketing and Logistics*, **20**, 97–131.

The Economist, 2009. Agriculture and Satellites. Harvest Moon: Artificial Satellites are Helping Farmers Boost Crop Yields. The Economist (November 5, 2009), http://www.economist.com/node/14793411 (accessed June 27, 2015).

Ting, S.L., Tse, Y.K., Ho, G.T.S., Chung, S.H., and Pang, G., 2014. Mining logistics data to assure the quality in a sustainable food supply chain: a case in the red wine industry. *International Journal of Production Economics*, **152**, 200–209.

Tipples, R. and Whatman, R., 2010. Employment standards in world food production – The place of GLOBALGAP supply contracts and indirect legislation. *New Zealand Journal of Employment Relations*, **35**, 1–5.

Toka, A., Iakovou, E., Vlachos, D., Tsolakis, N., and Grigoriadou, A.-L., 2014. Managing the diffusion of biomass in the residential energy sector: an illustrative real-world case study. *Applied Energy*, **129**, 59–69.

Trienekens, J.H., Wognum, P.M., Beulens, A.J.M., and van der Vorst, J.G.A.J., 2012. Transparency in complex dynamic food supply chains. *Advanced Engineering Informatics*, **26**, 55–65.

Tsolakis, N., Anastasiadis, F., Iakovou, E., and Vlachos, D., 2014a. Sustainable supply chain management and firm financial performance: a methodological framework for the agrifood sector. 2nd International Conference on Contemporary Marketing Issues, Athens, Greece, June 18–20, 2014.

Tsolakis, N., Keramydas, C., Toka, A., Aidonis, D., and Iakovou, E., 2014b. Agrifood supply chain management: a comprehensive hierarchical decision-making framework and a critical taxonomy. *Biosystems Engineering*, **120**, 47–64.

Tullberg, J., 2010. Tillage, traffic and sustainability—A challenge for ISTRO. *Soil & Tillage Research*, **111**, 26–32.

Tullberg, J.N., Yule, D.F., and McGarry, D., 2007. Controlled traffic farming – from research to adoption in Australia. *Soil & Tillage Research*, **97**, 272–281.

Ulaga, W., Sharma, A., and Krishnan, R., 2002. Plant location and place marketing: understanding the process from the business customer's perspective. *Industrial Marketing Management*, **31**, 393–401.

Vachon, S. and Klassen, R.D., 2006. Extending green practices across the supply chain: the impact of upstream and downstream integration. *International Journal of Operations & Production Management*, **26**, 795–821.

Validi, S., Bhattacharya, A., and Byrne, P.J., 2014. A case analysis of a sustainable food supply chain distribution system – A multi-objective approach. *International Journal of Production Economics*, **152**, 71–87.

Van Beek, P., Koelemeijer, K., van Zuilichem, D.J., Reinders, M.P., and Meffert, H.F.T., 2003. Transport logistics of food (eds L. Trugo and P.M. Finglas) *Encyclopedia of Food Sciences and Nutrition*, 2nd edn. Waltham, MA: Academic Press, pp. 5835–5851.

Van der Vorst, J.G.A.J., 2000. *Effective food supply chains-generating, modelling and evaluating supply chain scenarios*. PhD thesis. Wageningen University.

Van der Vorst, J.G.A.J., 2006. Quantifying the agrifood supply chain (eds C.J.M. Ondersteijn, J.H.M. Wijnands, R.B.M. Huirne, and O. Van Kooten) *Performance Measurement in Agrifood Supply-Chain Networks*. Dordrecht: Springer, pp. 13–24.

Van der Vorst, J.G.A.J., van Kooten, O., and Luning, P.A., 2011. Towards a diagnostic instrument to identify improvement opportunities for quality controlled logistics in agrifood supply chain networks. *International Journal on Food System Dynamics*, **2**, 94–105.

Van der Vorst, J.G.A.J., da Silva, C.A., and Trienekens, J.H., 2007. *Agro-Industrial Supply Chain Management: Concepts and Applications. Agricultural Management, Marketing and Finance Occasional Paper*. Rome: Food and Agriculture Organization of the United Nations.

Van der Vorst, J.G.A.J., Tromp, S.-O., and van der Zee, D.-J., 2009. Simulation modeling for food supply chain redesign: integrated decision making on product quality, sustainability and logistics. *International Journal of Production Research*, **47**, 6611–6631.

Van Roekel, J., Kopicki, R., Broekmans, C., and Boselie, D., 2002. *Building Agri Supply Chains: Issues and Guidelines*. Washington, DC: The World Bank.

Verigna, H.J. 2006. *Advanced Techniques for Generation of Energy from Biomass and Waste*. Petten: ECN.

Vitner, G., Giller, A., and Pat, L., 2006. A proposed method for the packaging of plant cuttings to reduce overfilling. *Biosystems Engineering*, **93**, 353–358.

Waddock, S. and Bodwell, C., 2004. Managing responsibility: what can be learned from the quality movement? *California Management Review*, **47**, 25–37.

Wall, E., Smit, B., and Wandel, J., 2004. *Canadian Agrifood Sector Adaptation to Risks and Opportunities from Climate Change*. Ontario: Canadian Climate Impacts and Adaptation Research Network for Agriculture.

Wang, X. and Li, D., 2012. A dynamic product quality evaluation based pricing model for perishable food supply chains. *Omega*, **40**, 906–917.

Watanabe, H., Li, D., Nakagawa, Y., Tomishige, K., Kaya, K., and Watanabe, M.M., 2014. Characterization of oil-extracted residue biomass of *Botryococcus braunii* as a biofuel feedstock and its pyrolytic behaviour. *Applied Energy*, **132**, 475–484.

Weatherell, A. and Allinson, J., 2003. In search of the concerned consumer: UK public perceptions of food, farming and buying local. *Journal of Rural Studies*, **19**, 233–244.

Welford, R. and Frost, S., 2006. Corporate social responsibility in Asian supply chains. *Corporate Social Responsibility and Environmental Management*, **13**, 166–176.

Wever, M., Wognum, N., Trienekens, J., and Omta, O., 2010. Alignment between chain quality management and chain governance in EU pork supply chains: a transaction-cost-economics perspective. *Meat Science*, **84**, 228–237.

White, L. and Lee, G. J., 2009. Operational research and sustainable development: tackling the social dimension. *European Journal of Operational Research*, **193**, 683–692.

Wikström, F., William, H., Verghese, K., and Clune, S., 2014. The influence of packaging attributes on consumer behaviour in food-packaging life cycle assessment studies – a neglected topic. *Journal of Cleaner Production*, **73**, 100–108.

Wilson, T.P. and Clarke, W.R., 1998. Food safety and traceability in the agricultural supply chain: using the internet to deliver traceability. *Supply Chain Management: An International Journal*, **3**, 127–133.

Wognum, P.M., Bremmers, H., Trienekens, J.H., van der Vorst J.G.A.J., and Bloemhof, J.M., 2011. Systems for sustainability and transparency of food supply chains – Current status and challenges. *Advanced Engineering Informatics*, **25**, 25–76.

Wrest Park History Contributors, 2009. Information technology and control. *Biosystems Engineering*, **103**(Suppl. 1), 142–151.

Yakovleva, N., Sarkis, J., and Sloan, T., 2012. Sustainable benchmarking of supply chains: the case of the food industry. *International Journal of Production Research*, **50**, 1297–1317.

Yu, Y., Wang, Z., and Liang, L., 2012. A vendor managed inventory supply chain with deteriorating raw materials and products. *International Journal of Production Economics*, **136**, 266–274.

Zanoni, S. and Zavanella, L., 2007. Single-vendor single-buyer with integrated transport-inventory system: models and heuristics in the case of perishable goods. *Computers & Industrial Engineering*, **52**, 107–123.

Zanoni, S. and Zavanella, L., 2012. Chilled or frozen? Decision strategies for sustainable food supply chains. *International Journal of Production Economics*, **140**, 731–736.

Zhang, M. and Li, P., 2012. RFID application strategy in agrifood supply chain based on safety and benefit analysis. *Physics Procedia*, **25**, 636–642.

Zhang, M., Xiao, G., and Salokhe, V.M., 2006. Preservation of strawberries by modified atmosphere packages with other treatments. *Packaging Technology and Science*, **19**, 183–191.

2

Precision Agriculture: Crop Management for Improved Productivity and Reduced Environmental Impact or Improved Sustainability

**Spyros Fountas,[1] Katerina Aggelopoulou,[2]
and Theofanis A. Gemtos[2]**

[1]*Department of Agricultural Engineering, Agricultural University of Athens,
Iera Odos 75, Athens, Greece*
[2]*Laboratory of Farm Mechanisation, Department of Agriculture, Crop Production
and Rural Environment, University of Thessaly, Fytoko Street, Volos, Greece*

2.1 Introduction

Precision Agriculture (PA) can be defined as the management of spatial and temporal variability in fields using Information and Communications Technologies (ICT). PA is the art and science of utilizing advanced technologies for enhancing crop production while minimizing potential environmental pollution (Khosla and Shaver, 2001).

Supply Chain Management for Sustainable Food Networks, First Edition. Edited by Eleftherios Iakovou,
Dionysis Bochtis, Dimitrios Vlachos and Dimitrios Aidonis.
© 2016 John Wiley & Sons, Ltd. Published 2016 by John Wiley & Sons, Ltd.

PA can assist crop producers, as it permits the use of precise and optimized inputs leading to reduced costs and environmental impact, and it can be utilized in a traceability system that could record the activities at a site-specific level (Fountas et al., 2011).

PA has a rather short history. Its application started about 25 years ago when Global Positioning System (GPS) and new sensor technologies were made available. GPS was available for civilian use by the end of the 1980s. Its accuracy improved when selective availability was removed in 2000 (Heraud and Lange, 2009). The initial applications were mainly for arable crops. Harvesting was mechanized and sensors were placed on the machines to map yield variability. In the early 1990s the first applications were in cereals using impact or γ-ray grain flow sensors (Godwin et al., 2003), while applications in high value crops (fruits and vegetables) were started by the end of the 1990s.

PA is not a new idea. A few decades ago farms were small and the farmer had to walk all over his fields several times each year. The farmer was able to observe all variations within the fields and take appropriate management decisions for each part. The farmer was able to add more seeds in parts where emergence was low or add more fertilizer where growth was lower or the plants were yellow. This knowledge depended on the farmer's memory combined with direct observation. One problem was that in most cases the farmer's decisions were influenced more by the memory of recent years' results that were more influenced by weather or other factors not present in the following years. This connection and knowledge of the fields were reduced with farm mechanization and the increase in farm size. The larger the field, the lower the farmer's knowledge of the field's variability. Gradually the average rule was used to manage the fields. The underlying assumption was that the field was homogeneous and the same management for all parts was justified. When the first yield monitors were developed and yield maps were created, it was proved that yield and soil properties varied greatly within even small fields. This fact marked the development of PA.

PA is a cyclic system of data collection, data analysis, use of the results for crop management decisions, evaluation of the decisions at the end of the cropping system and the cycle continues for subsequent years. Figure 2.1 presents this cycle. The first task before applying PA management is to establish soil and crop variability. A homogeneous soil planted with a homogeneous genetic material has very limited benefits from applying PA. Therefore, data collection is the first stage of the system, followed by data analysis and the application of the system. Each year data are stored in a database (library) and used as historical data for future decisions. The system can be divided into data collection, data analysis, managerial decisions, and applications and evaluation.

The present chapter aims at giving an account on the application of PA over the last 25 years, on the methods used, the results obtained, the adoption of the technology and the effects to crop management, to the environment and the sustainability of agricultural systems.

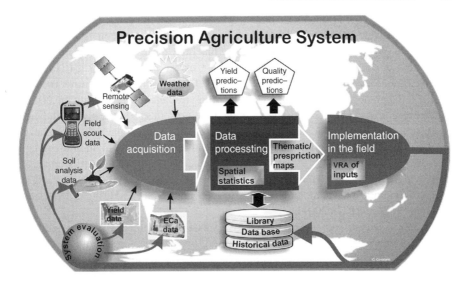

Figure 2.1   A generalized precision agriculture system. Adapted from Tagarakis, 2014.

2.2   How Precision Agriculture Is Applied

2.2.1   Data Collection

Many types of data can be collected during the growing season. Yield spatial distribution data, soil data (physical and chemical properties, topography), remote sensing data, data collected by crop scouting (crop growth, diseases, pests, weeds that currently cannot be detected by sensors), as well as weather data can be collected for every field at site-specific level to assist the farm manager in crop management. All data have to be geo-referenced using GPS technology and introduced to a Geographic Information System (GIS) database. GPS technology has different levels of accuracy. Simple GPS offers a few meters accuracy, Differential Global Positioning System (DGPS) sub meter accuracy while Real Time Kinematic Global Positioning System (RTK-GPS) 1–2 cm accuracy (Heraud and Lange, 2009). For most applications DGPS accuracy seems to be sufficient as RTK systems are too expensive for farm use. Recently, RTK-GPS central systems have been installed in agricultural regions around the world, which can be accessed by farmers at low cost by paying a subscription fee. This will enhance high accuracy GPS use.

2.2.1.1   Yield Mapping

Yield mapping can be carried out easily in mechanized crops with yield monitors attached to the harvesting machines. The first applications of yield mapping were in combine harvesters for cereals using γ-ray sensors (Godwin *et al.*, 2003). Later, sensors based on seed impact on a plate (AgLeader Technology, 2014) and volumetric applications were developed and used. A yield monitor system consists of a sensor that measures the crop flow in the harvesting machine, a sensor that measures quality properties of the crop, a GPS receiver and a Central Processing Unit (CPU) that receives the collected data and stores them for future use. Yield monitors measure yield on the go at frequent time intervals, storing yield data together with the coordinates every few square meters. The yield data are presented "on the go" on the yield monitor in the cabin of the harvesting machine. Several sensors have also been developed for harvesting machines for crops other than cereals, such as cotton using light sensors (Tomasson *et al.*, 1999), processing tomatoes using loading cells under the conveying chains of the machines (Pelletier and Upadyaya, 1999), hay producing crops (Wild and Auernhammer, 1999; Kromer, Shmittmann, and Osman, 1999), and peanuts (Vellidis *et al.*, 2001).

In vines, sensors were developed relatively early for the mechanical harvesting of grapes for wine making. They were first applied for the 1999 vintage in Australia and in the USA (Arnó *et al.*, 2009). Loading cells that weighed the crop passing on a conveying belt or an array of ultrasonic beam sensors mounted over the grape discharge chute were used to estimate the volume, and hence tonnage, of fruit harvested (Bramley and Hamilton, 2004). In Florida citrus plantations, Schueller *et al.* (1999) used a system to weigh the palette bins where the oranges were collected. In Greece, Aggelopoulou *et al.* (2011a) mapped the yield in apple orchards. The apples were handpicked and placed in about 20 kg capacity bins along the rows of the palmette formed trees (Figure 2.2). Each bin was weighed and geo-referenced using a GPS receiver. The bins corresponding to groups of 5 or 10 trees were grouped to represent their yield. A similar approach was used by Tagarakis *et al.* (2014) for yield mapping of handpicked vines. Yield spatial variability was evident in all applications, even in orchards or vineyards of 1 ha.

Fountas *et al.* (2011) measured the yield variation in olive tree orchards. Olives, in conventional orchards, were picked by hitting the fruit branches with sticks. The olives fell onto plastic sheets placed underneath each tree. They were then collected and put into bags, and each bag was weighed and geo-referenced using GPS. Ampatzidis *et al.* (2009) have mapped the yield of a peach orchard. They used radio frequency identification (RFID) tags on the bins. A weighing machine was combined with an RFID reader and a GPS device to record the weight and the place of each bin. The data collected were used to produce yield maps of the orchard. Konopatzki *et al.* (2009) have mapped the yield of a pear orchard. They measured the yield of each tree (harvested in three passes of the workers) and also found variability in the yield. Qiao *et al.* (2005) developed a mobile automatic fruit grading robot. It was taken to a plant and a worker picked peppers and placed them on the machine for grading. The machine located the plant, weighed the fruits of each plant, and analyzed the quality. Yield and quality maps showed spatial variability even in the very small plot of the experiment.

In all the above studies, it has been noted that yield spatial variability is a fact even in the small fields that are arable or have fruit and vegetables. The variability in most cases is also high enough to justify investment in PA technology.

Figure 2.2   Data collection for yield mapping in an apple orchard in Greece.

2.2.1.2   Quality Mapping

In most crops the quantity is one component of the production of a field. The quality of the product is a second component and for many crops quality is also very important. Especially in fruits and vegetables high quality secure a premium price. But in other crops like durum wheat for pasta making high protein content also receives premium price. Sensors for cereal moisture content were developed from the early stages of yield mapping. Systems using grain permittivity were developed and used successfully. Light spectrum sensors were developed for some of the grain or seed properties and are in commercial use to determine, for example, the protein content of cereal seeds or the oil content of oily seeds (Zeltex ACUHARVEST, http://www.zeltex.com/accuharvest.html). Several laboratories are working to develop sensors to measure the quality of products (i.e., NIRS Forage and Feed Testing Consortium in the USA, http://nirsconsortium.org; University of Padua, Italy). Many studies rely on manual sampling and analysis of the samples in the laboratory, such as for determining cotton lint quality (Gemtos, Markinos, and Nassiou, 2005).

According to Kondo and Ting (1998), for fruit crops, quality commonly includes outer parameters (size, color, shape, surface texture, and mass), inner parameters (sweetness, acidity, or inner diseases), and freshness. Given the high cost of hand picking of most table horticultural crops in many cases lower yields with better quality can be more profitable for the farmer. Aggelopoulou *et al.* (2010) analyzed the spatial variability of yield, soil, and the quality of apples. They measured several parameters of the quality such as color, sugars, malic acid, pH, and flesh firmness. The variability existed even in small orchards. The fruit quality (sugar content and flesh firmness) negatively correlated with the yield.

Extensive work on grape quality has been carried out. Grape samples were analyzed to assess the variability of the quality to produce high quality wine. Using remote sensing a high correlation was found between the vegetation index [e.g., Normalized Difference Vegetation Index (NDVI)] maps near veraison (beginning of maturity) and the grape quality maps. Based on that, the production was separated of the two zones of the field which produced a different quality of wines. The dense vegetation part gave lower quality with lighter color (Bramley, Pearse, and Chamberlain, 2003).It was found also that the dense vegetation part produced more (about double) than the lower. However, it was not always true that low yielding parts produced high quality (Bramley and Hamilton, 2004). Bramley (2005) presented the results of grape quality analysis in two commercial vineyards. The variability of the parameters of the quality was there although this variation was much lower than the yield's variation. The zones formed by the quality parameters were not always similar to the yield zones. It seems that the factors affecting quality are more complex than the factors affecting yield. The spatial variability of the quality characteristics was relatively low. He concluded that it is difficult to define zones of certain quality characteristics as the wine industry is requiring. Additionally, the cost of sample collection and analysis is high and only on-the-go sensors could offer the opportunity to separate grape qualities. Best, León, and Claret (2005) measured an index m^2 leaf/kg fruit in vines. They found that the quality of grapes (Brix, color factors) was lower when the index was larger (higher vigor of the plants). Sethuramasamyraja, Sachidhanantham, and Wample (2010) used a hand-held near-infrared (NIR) spectrometer to analyze anthocyanin variability in two vineyards for 2 years in California, USA. The vines were divided into two management zones based on threshold values suggested by the vineries. A harvester with two stores (gondolas) was developed and used. Based on management zone boundaries, the different quality grapes were directed to the appropriate store. The two quality lots were used separately to produce wine. Expert panels testing the wines verified the different quality and proved the usefulness of the method.

2.2.1.3 Soil Sampling and Analysis

Soil is the substrate where crops are grown. It affects several parameters of crop growth, the final yield, and its quality. Most of the cropping activities are also affecting soil through tillage, compaction fertilization, and so on. Soils were analyzed for their physical and chemical properties from the beginning of PA. Initially grid sampling was used. The idea was to mark the field by normal lines with a certain distance between them and produce small parcels from where samples were taken. The size of the parcels differs depending on the purpose of the study. In research projects smaller parcels are commonly used (less than 0.1 ha) but for commercial applications larger parcels of 0.4 ha are the usual size. Samples taken from the parcel (from different parts of the parcel) are mixed, homogenized, and then analyzed for their properties such as texture, nutrient element content, cation-exchange capacity, pH, organic matter, and so on. Soil maps are produced for each property and can be used to define fertilization. Fountas, Bartzanas, and Bochtis (2011) using grid sampling and analysis of an olive orchard defined the soil maps (Figure 2.3) and the amount of phosphorus and potassium fertilization for each tree.

Aggelopoulou *et al.* (2011a) also analyzed soils in a dense grid. They found that correlations between soil nutrients and yield were not consistent. They suggested taking into account the apple yield and the nutrients removed to produce prescription maps for fertilizer application. Best, León, and Claret (2005) found also low correlation coefficients between soil properties and yield parameters even by sampling 10 samples per hectare. They suggested that better correlation exists for yield parameters and Apparent Electrical Conductivity (ECa) maps. Soil sampling and analysis is a labor intensive and costly activity. For research purposes this can be justified but in most commercial applications it is not acceptable. A second possibility is to define management zones with another measurement such as yield mapping or ECa mapping and direct the soil sampling to the zones. This greatly reduces the number of samples and the cost and offers a good picture of the field for crop management. Tagarakis (2014) applied directed sampling in a vineyard based on ECa, elevation maps and the delineation of management zones by the farmer. Nine samples were sufficient to characterize the soil.

A third possibility is to develop sensors that can measure soil properties on the go. This is a fast and low cost method. Several methods to assess soil parameters have been developed or are under development. The soil sensors were based on properties such as electrical and electromagnetic, optical and radiometric, mechanical, acoustic, pneumatic, and electrochemical (Adumchuk *et al.*, 2004). Electrical resistivity and electromagnetic (EM) induction was used to assess the soil ECa. The ECa measures conductance through not only the soil solution, but also through the solid soil particles and via exchangeable cations that exist at the solid–liquid interface of clay minerals (Corwin and Lesch, 2003). This property is directly connected to soil properties such as texture, water content, organic matter, salinity, ions in the soil, and temperature. If we exclude saline soils from the measurements and take measurements near field capacity most measured conductivity variability is due to soil texture. Electric resistivity instruments use flat, vertical disks to apply a voltage, and measure the soil resistance by measuring the current in other similar disks. The distance between the disks defines the depth of the measurement. In EM induction sensors coils are used to induce and measure the electricity. An EM transmitter coil located at one end of the instrument induces circular eddy-current loops in the soil. The magnitude of these loops is directly proportional to the electrical conductivity of the soil in the vicinity of that loop. A second coil measures the produced current which is the result of soil properties (e.g., clay content, water content, organic matter, ions). Instrument orientation and distance from the soil define the depth of measurements.

The two instruments were used in many applications of PA combined with GPS. They provide a fast and relatively cheap way to produce maps which present the variability of the field and they are correlated to yield. Many researchers have

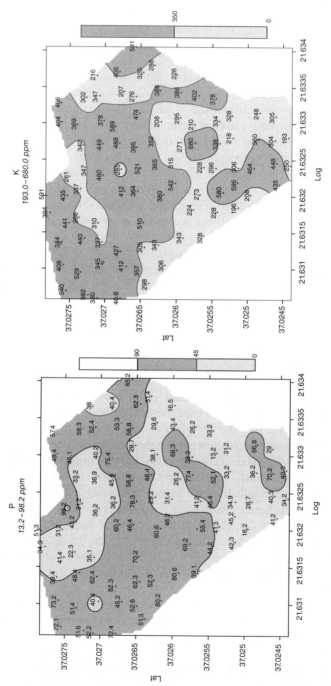

Figure 2.3 Phosphorus (P) and potassium (K) maps of an olive tree orchard (Fountas et al., 2011).

reported this connection (Kitchen *et al.*, 2005). Soil texture is a basic factor of soil variability and influences several soil and crop parameters. When EC is applied in the application of PA in the field, it can permit a directed soil sampling through EC zoning. In many cases it is directly connected to the yield and product quality. Heavier or lighter soils react differently to weather conditions, and require different water, fertilizer, and herbicides applications. The GPS readings when they are relatively accurate can also offer elevation maps. Elevation maps can help in the farm management of fields with inclinations. ECa was also correlated to the water holding capacity of the soil and was used for variable rate irrigation (Hedley and Yule, 2009).

Soil color without vegetation offers an indication of its texture and soil organic matter. Adumchuk *et al.* (2004) named such images "bare soil images." Early laboratory studies showed a correlation of soil organic matter with both visible and NIR reflectance. Mechanical sensors have been used to assess soil compaction using instrumented tines (Andrade *et al.*, 2002) or automatic penetrometers. They gave good results but they have to pass through the soil to assess the compaction. Acoustic sensors during soil braking by a tine were also tested.

Electromechanical sensors have been developed. One with commercial application can map pH. A tool is lowered into the soil when the instrument is moved in the field and a sample is extracted before the tool is returned to its initial position above the soil. The sample is analyzed by either an ion-selective electrode (glass or polymer membrane), or an ion-selective field effect transistor (ISFET) (Adumchuk *et al.*, 2004). The electrodes can measure pH, K^+, and NO_3^- but the time needed for measuring ions is long and not suitable for on-the-go measurements. The only commercial application is for pH measurements. The pH sensor is combined with an electromagnetic resistance (ECa) instrument and measures both on the go.

Sensors are under development that can assess some soil properties such as organic matter and nutrient content using the properties of light when reflected or passing through the soil. Proximal soil sensors have been developed that can provide high resolution data on spatial variation in soil properties (Stenberg *et al.*, 2010), which enables the management of land on a field and sub-field scale. Sensors based on visible and infrared radiation analysis have been developed and placed on mobile platforms. The sensors were placed at the back of a sub-soiler shank and measured the reflected light from the soil. A fiber type, visible–NIR spectrophotometer with a measurement range of 305–2200 nm was used. A good correlation was claimed between measured reflected wave lengths and soil properties such as soil texture, soil organic matter, soil water content, pH, and phosphorus but low correlation with potassium (Shaddad, 2014).

2.2.1.4 Remote Sensing

Remote sensing is defined as a group of techniques that can collect field data without being in contact with the object (plant, soil, etc.). An electromagnetic wave when falling on an object it can pass through can be reflected or absorbed. Useful information can be derived from the plants by measuring these effects. It is a useful technology for PA as it can give data for parameters of the field relatively easily. All visible objects in the field (soil, plants) can be remotely sensed, where the sun light is reflected differently from them. Sunlight is an electromagnetic wave that is formed by a spectrum of wavelengths. The sunlight is formed by ultraviolet wavelengths, visible light, and infrared wavelengths. The green plants absorb the red and blue wavelengths and reflect the green and the infrared. By measuring the reflected wavelengths with a multispectral camera the vigor of the plants (that makes them greener) can be measured. Green plants can also be seen that can have a problem such as a disease, a nutrient deficiency, or water logging, and so on. The soil can be correlated by the color with the soil organic matter, moisture, and so on. Light reflectance (sun or some artificial) has been used in PA in the form of vegetation indices. The most commonly used of them is the NDVI. NDVI is an expression of the vigor of the plants and has been correlated with crop yield and quality. Several other indices can be calculated and used offering good agreement with certain characteristics of the crop.

The measurements of plant reflectance can be carried out by satellites, airplanes, or ground instruments. Satellites can give images of large areas at relatively low cost but they cannot work when clouds are covering the Earth. Aeroplanes or helicopters do not have the cloud problem but they are more expensive. Ground sensors work well but require more labor. Ground sensors usually use an artificial light that makes measurements independent of sunlight and so measurements can be carried out even during the night. In several PA studies crop reflectance was used as an early measurement of the crop growth, with crop vigor reflecting the nitrogen availability and the health status of the plants, and for prediction of yield and product quality. In the most used application, NDVI was used to regulate nitrogen application. The hypothesis is that greener plants (higher NDVI) have more available nitrogen and require less fertilizer application compared with less green plants (lower NDVI). Sensors developed by YARA use artificial light, measure NDVI on the go and adjust nitrogen applications for crops such as cereal, rapeseed, or potatoes. Several applications in the same line in different crops offer nitrogen fertilizer savings, improved yields, and better product quality (Lan *et al.*, 2008).

Bramley, Pearse, and Chamberlain (2003) used the NDVI of vines at veraison as an indication of grape quality and used it to separate the product into high and low wine quality producing plots. The idea was successful and gave good results and profit for the farmer. For vines, high vegetation at the end of the growing season indicates high yield which in most cases but not all is followed by low quality. Best, León, and Claret (2005) in Chile found good agreement between NDVI and yield and quality characteristics of a vineyard (correlation coefficient $r^2 > 0.7$). They found also high correlation between Leaf Area Index (LAI) and NDVI ($r^2 > 0.75$). Hall *et al.* (2010) have studied the correlations between spectral images and the properties of the grapes and yield. They have estimated canopy area, canopy density, the total soluble solids, yield, berry size, and anthocyanins. Canopy area and canopy density were consistently significantly correlated with fruit anthocyanin and phenolic content, berry size, and yield. However, the total soluble solids correlations were not consistent.

Any object which has a temperature above absolute zero emits electromagnetic radiation. This is used in thermal cameras to detect differences in temperature in plants. Thermal cameras have been used in PA to assess the water status of crops and regulate irrigation (Alchanatis *et al.*, 2010). Another property of plants or a product is the absorption of electromagnetic waves when they pass through the plants or product. Every object has a characteristic absorption of specific wavelengths and this can be used to find its quality characteristics. Sensors for assessing the protein or oil content of seeds are already in commercial use (see Section 2.2.1.2). Chlorophyll fluorescence can depict the photosynthetic state in green leaves. Fluorescence sensors measure the absorption of specific wavelengths followed by the dissipation of the absorbed energy by light emission at longer wavelengths (Corpa et al., 2003). Fluorescence sensing technology can be used to detect the status of plant nitrogen. It also gives information on the status of the plant chlorophyll (Tremblay, Wang, and Cerovic, 2012). A commercial, fluorescence-based optical sensor (FORCE-A, Orsay, France) was successfully used for monitoring grape anthocyanin but other new sensors can also assess plant chlorophyll status for fertilizer applications.

2.2.1.5   Field Scouting

Field scouting is a part of each management system that currently cannot be avoided. The farmer has to go to the field to verify the indications given by the different instruments used. In many cases, measurements of emergence rates, growth of the plants measured by their height or the canopy of the trees or trunk size of the trees provide useful information to apply PA. Some of them can be measured by instruments but some still have to be measured by human labor. Farmers, even in large farms, have a good knowledge of their farm. In many cases, at the beginning of the application of PA it is useful to ask the farmer to draw a map of the field showing the characteristics of each part. In many cases the farmer's opinion does not differ much from the management zones defined using data from sensors.

2.2.2   Data Analysis and Management Zone Delineation

All data collected have to be analyzed and interpreted if a meaning is to be drawn from them. There are too many data and appropriate methods have to be used or developed for the analysis. Simple exploratory (descriptive) statics can give a first idea of the values, their spread, the range, and the distribution. Geostatistics, based on what is called, "the theory of regionalized variables," is basically a probabilistic method of spatial interpolation. Final construction of the map corresponding to parcel level is made possible, based on estimation of the error at non-sampled points, using the spatial variability structure of the sampled data (variogram) and an interpolation method (kriging). This type of information, which can be obtained for different properties and for successive years, opens new and interesting possibilities in agronomic crop analysis and management (Arnó et al., 2009). Given the spatial dependence of the values, interpolation between sampling points can be carried out using geostatical methods such as kriging. Maps covering the whole field can be produced and indicate the variability of the properties. There are several methods for data analysis although there is not a clear method to compare the produced maps. An optical impression is still used for the comparison of the maps. Correlations between parts of the field with different parameters can be carried out to assess their relationships.

Kitchen *et al.* (2005) tried to delineate productivity management zones based on ECa, elevation and yield maps using MZA software. They used a pixel agreement between zones to compare the zones based on different parameters. Tagarakis *et al.* (2013) used the same approach to compare maps of PA in a vineyard project. Taylor, McBratney, and Whealan (2007) presented a protocol for data analysis and management zone delineation using available free software. This protocol could help farmers in the better use of the data collected through PA technologies. Soft computing techniques have been employed to define correlation between the properties measured and permit a forecast of the results (Papageorgiou, Markinos, and Gemtos, 2011; Papageorgiou *et al.*, 2013). Neural networks, fuzzy logic, and fuzzy cognitive maps have been used recently to analyze data and explain yield variation. Aggelopoulou *et al.* (2013) delineated management zones in apples based on yield, soil, and quality data using a multivariate approach. Data fusion from different sensors was proposed as a method to analyze data and provide useful correlations for management zone delineation or for on-the-go variation of inputs.

The analysis of the data aims at defining parts of the field with common characteristics that can be managed separately. These parts are the management zones. The term management zone implies a part of the field with similar characteristics that can be managed in a common way. Management zone delineation should form homogeneous parts of the field where inputs or other practices can be applied in the same way. The management zones should be large enough to permit VRA (Variable Rate Application) of inputs but small enough to be homogeneous.

2.2.3   Variable Rate Application Technology

VRA technology is the major target for PA. All information gathered should result in a better management of the formed zones. Variable rate means that the appropriate rates of inputs will be applied at the appropriate time and precisely, leading either to reduced inputs, costs, and adverse environmental effects or improved yields and quality. Two methods are used to apply variable rate. The first, called "map based," is based on historical data (previous or present year). Process control technologies allow information drawn from the GIS (prescription maps) to control processes such as fertilizer application, seeding rates, and herbicide selection and application rate, thus providing for the proper management of the inputs. The second, called "sensor based," uses sensors that can adjust the application rates on the go. The sensors detect some characteristics of the crop or soil and adjust the application equipment. VRA can be applied to all inputs like fertilizer application, spraying for pests, water application and also for practices such as pruning or even separate harvesting of the zones (Auernhammer, 2001). Both systems have advantages and disadvantages. The on-the-go sensors are more acceptable to farmers as they are simple to use and facilitate their work. Probably using a mixture of both will offer the most advantages in the future.

Variable fertilizer applications in vineyard management and other practices, such as foliar nutrient programs and drip irrigation, could help to minimize variability in vine growth as well as fruit quality (Sethuramasamyraja, Sachidhanantham, and Wample, 2010). Davenport *et al.* (2003) applied variable rate fertilizer in a vineyard for 4 years. They have analyzed the nutrient content of the soil. They concluded that nitrogen and potassium applications benefited the field as they reduced the coefficient of variation (CV) of the nutrient content but not the phosphorus application where the CV remained high.

Based on management zone delineation and historical data, prescription maps can be produced defining the specific requirements of each zone. The prescription map is imported to the controller of the application machine and changes the adjustment (the amount of the input applied per unit of area as prescribed) as the machine moves through the field. Several machines have been produced to adjust according to prescription maps the seeding, fertilizer, manure, water rate, or have areas where a pesticide can be applied or not. Obviously a lot of data have to be collected and properly analyzed to make the application effective. In tree crops where temporal variability is lower, this application is more feasible than in arable crops.

Prescription maps can be produced based on several characteristics of the field or the crop.

In the case of the orchard in Figure 2.3 (Fountas *et al.*, 2011), the farmer applied the fertilizer by hand for each tree. He was able to use the map with the two zones and apply one or two portions of fertilizer for the defined trees. For apples, Aggelopoulou *et al.* (2011a) used the soil analysis data and the nutrient removal from the soil by the crop to prepare prescription maps for fertilizer application (Figure 2.4). Prescription maps can be based on characteristics measured during the growing

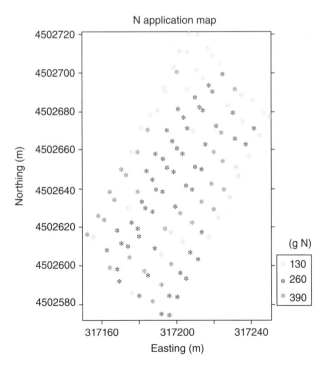

Figure 2.4   Prescription map for nitrogen (N) application per group of trees (Aggelopoulou et al., 2011a).

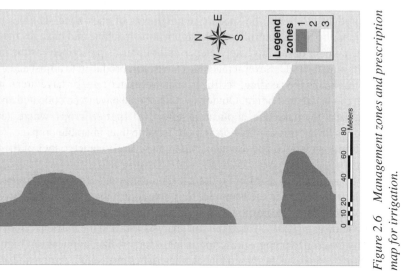

Figure 2.6   Management zones and prescription map for irrigation.

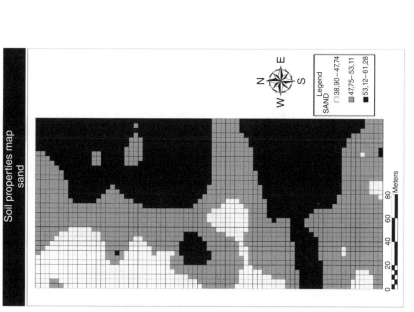

Figure 2.5   Soil clay content map.

season. Aggelopoulou *et al.* (2011b) found high correlation between flowers and yield distribution. This can be used to manage the inputs of the crop as low yielding part requirements are different to high yielding early in the season (in spring).

Several on-the-go sensors have been presented and used. The most well-known is the sensor that detects light reflectance from the crop. Using NDVI, the sensor detects the vigor of the crop. Usually crops with sufficient nitrogen supply are greener than plants with lower nitrogen. This characteristic was used to adjust nitrogen rates for the field in crops such as cereals. The most well-known sensor by YARA (N-Sensor, http://www.yara.co.uk/crop-nutrition/Tools-and-Services/n-sensor/) is used in many applications of nitrogen fertilizer. Other manufacturers have produced similar sensors (i.e., TOPCONCropSpec, http://ag.topconpositioning.com/ag-products/x20-application-kits/cropspec).

In tree crops several characteristics can be used to directly adjust inputs. Tree canopy volume, density, and height can be measured electronically (Giles, Delwiche, and Dodd, 1988). In citrus orchards in Florida, tree canopy measured by ultrasonic or laser sensors was correlated to yield. This property was used to adjust the variable chemical application (Zaman, Schuman, and Miller, 2005, Zaman *et al.*, 2006). During spraying, sensors detecting missing trees can stop spraying. This automates the spraying, stopping at the headlands and facilitates the operator's work. Other sensors detect the tree density and height (using laser scanners, ultrasonic or photoelectric sensors) (Giles, Delwiche, and Dodd, 1988; Tumbo *et al.*, 2002) and adjust the spraying direction of nozzles to reduce out-of-target spraying. New nozzles were developed to change the output. These are pulse width modulation nozzles that use fast reaction solenoids to open or close the flow several times per second varying the discharge. One other development changes the active ingredient solution by introducing it at different rates in the distribution tubes of the sprayer (after the pump) (Ess and Morgan, 2003). Gil *et al.* (2007) tested a VRA sprayer in vines. The sprayer had nozzles in three groups of five in each part of the row. Ultrasonic sensors sensed the canopy width and adjusted the sprayer. A saving of 58.8% was achieved with the same coverage of the canopy by the two sprayers (conventional and experimental) with the variable rate sprayer having better depositions inside the canopy.

Variable rate irrigation is of great importance due to the shortage of water reserves and the importance of irrigated crops in many parts of the world. Variable rate irrigation has attracted the interest of researchers. Applications in central pivot systems based on prescription maps proved that considerable savings in water and energy can be achieved. Prescription maps can be based on soil properties, crop conditions, and the real conditions of the field. In parts of the field without plants the water application is stopped. In a feasibility study of fields in Greece and Turkey based on soil variability savings of up to 7% of water and energy can be achieved (Gemtos *et al.*, 2010). Based on a soil texture map (Figure 2.5), three management zones were delineated using FUZME software (Figure 2.6) in a cotton field for variable rate irrigation.

Using the Food and Agriculture Organization CROPWAT model for cotton water application a range of water savings of between 2.5% and 7.2% were achieved. In orchards, irrigation systems have to be designed from the outset to achieve variable rate irrigation. Knowing the soil variability it is possible to develop more than one network to apply different water depths or frequency of application. The zone separation criteria can be soil texture and soil elevation (Tagarakis, 2014).

In the last few years wireless systems of sensors have been developed to measure soil water content during the growing season. The sensors are installed in the management zones and can give information to the farmer so that the farmer decides the irrigation or directly to the controllers of automatic irrigation systems that can define proper application levels.

Several direct sensing systems have been used for weed control. Some herbicides are sensitive to soil organic matter. Soil organic matter detection has been used to automatically adjust the herbicide application rate. Increased efficiency was reported (Grisso *et al.*, 2011). A second line of action is the detection of green plants and to use herbicides only when weeds are present. The system is to be used between the rows of vegetables or other crops. A sensor detects the green color of the plants from the soil and applies the herbicide (such as glyphosate) only when green plants are detected. More than 30% herbicide savings were reported. Weed recognition systems can also be used and drops of herbicides are applied only on the weeds. These systems work also on the crop row. High herbicide savings are reported. A third line of action is the use of mechanical weed control by avoiding the crop plants. The system detects the useful plants in two ways. One way is to detect the seed placement in the field using a RTK-GPS and then produce maps with the plant places. The second way is to use a camera in front of the machine to detect weeds and crop plants and direct a tool only to the weeds. Several tools have been developed. The most successful commercially is a horizontal disk system that has one sector removed. The machine vision system or the plant map or both detect the crop plants and adjust the disk rotation in such a way as to avoid damaging them (Dedousis and Godwin, 2008).

2.2.4   Auto Guidance Systems and Other Applications

PA is not only site specific management. Most of the technologies used in PA can be used in several applications improving farm management. The use of GPS technology can offer guidance systems to the tractors that help them to follow desired paths in the field. Especially RTK-GPS offers high accuracy. This can help to avoid double passing or missing strips in the field when chemicals are applied leading to savings in material and reduction of the effects to the environment. This can lead to more accurate tree planting or controlled traffic in fields reducing the compaction problem of the soils. The addition of GPS and other sensors to the tractor (using ISOBUS standardization) can offer a full record of the farm machinery movements as well as fuel and energy consumption. Recording of farm machinery activities (with inputs from the farmer) can lead to a Farm Management Information System that can cover administration requirements for certification of production systems (such as integrated crop production management systems) or EU cross compliance (Sorensen *et al.*, 2010). Keeping records on inputs and yields means the first step of a traceability system is formed as required by the consumers. PA can assist in the development of Certified Integrated Crop Production systems. Setting targets to reduce fertilizer inputs can be achieved by redistributing the fertilizers within the field without reducing yields.

Knowing the machinery movements, better use or better itineraries can be estimated that can improve efficiency. This can save time and fuel but also reduce soil compaction. The development of autonomous vehicles can lead to improved mechanization systems with fleets of small sized tractors working 24 h a day and doing all farming activities accurately (Blackmore *et al.*, 2007, 2009).

2.3   Decision Support Systems for the Farmer

A decision support system (DSS) is a computer-based system that supports business decisions. In agriculture it refers to the decision taken by the farmer for the management of the farm. PA is directly connected to decision-making by the farmer. It is true that in this respect research is not currently successful. The lack of functional tools for decision-making explains to a certain extent the difficulty faced so far for a rapid and widespread adoption of PA. This is a fact recognized by researchers in the field. Arnó *et al.* (2009) pointed out that the development of a DSS in PA is still a pending assignment. Kitchen *et al.* (2005) indicated that more precise crop models working in PA can help in the development of a successful DSS. Many efforts have been made to capture the decision-making process for farmers using PA starting from data collection in the field, capturing external data, and processing the data to derive useful decisions (Fountas *et al.*, 2006).

2.4   Profitability and Adoption of Precision Farming

The adoption of a new technology by farmers is a difficult procedure. Farmers are generally of a more conservative nature. The evolution of agriculture in many parts of the world has resulted in elderly farmers and usually with a lower education level. This makes changes and adoption of new technologies even more difficult. Different surveys indicate a lower use of computers and the Internet by farmers. In many places the infrastructure for commutations is inferior in rural areas. Kutter *et al.* (2011) defined the farmers' adoption of PA as the combined utilization of several site-specific technologies using GPS such as auto guidance and variable rate technology (VRT) of inputs and/or yield mapping on a farm. This definition does not imply that these practices have to be carried out by farm staff but can be offered by a third party as well.

The farmers to adopt a new system have to recognize, research, and implement these technologies and management practices at an on-farm production level (Koch and Khosla, 2003). Kutter *et al.* (2011) pointed out that farmers will adopt PA when they are convinced that they will have an economic benefit, it offers advantages over traditional methods, and it is less complicated. This is not clear. Additionally farmers usually like to observe an application and see the benefits before adopting any innovative technology. Research has shown that large farms adopt PA more than small ones. The same applies to young farmers. Ehsani, Sankaran, and Dima (2010) reported the results of a meeting with stakeholders in Florida, USA. They presented a summary of the requirements of the farmers for new technologies in agriculture. They expect them to be proven and robust, cost effective and when new equipment is employed for it to be reliable and well supported for servicing and repair. They expect to find sensors for disease recognition and early warning and to help them to follow regulations. Early and accurate yield predictions are important. For autonomous vehicles they require reliability and safety, and to have the option of manual driving when a problem occurs. Moreover, Lawson *et al.* (2011) carried out a wide survey across four European nations recording farmers' attitudes toward PA and information systems and they recorded the basic incentives that farmers had using the advanced systems.

Adoption is wider in the USA. In a 2013 survey conducted by Erickson, Widmar, and Holland (2013) the answers from dealers in the USA indicated that the best sellers for PA practices were the GPS-based guidance systems, which were used by about 85% of the farmers, while 40% used satellite/aerial imagery, and only 13% soil sensors (ECa or pH). GPS-enabled sprayers' boom with section control was used by 53%. VRA for nutrient application was offered by 70% of the farm dealers. Only 15% responded that they did not offer PA applications. These results gave an indication of the interest for PA applications. Additionally, in a Florida survey of farmers, 17.5% used sensor-based VRA and 16.1% soil variability mapping and GPS boundary mapping. Zarco-Tejada, Hubbard, and Loudjani (2014) claimed that similar figures are indicative for the EU as well. However, other studies in Europe have indicated a slower adoption rate. A survey in the UK for the application of PA (Department for Environment, Food and Rural Affairs, 2013) showed an increase between 2009 and 2012 for GPS receivers from 14 to 22%, for soil mapping from 14 to 20%, for VRA from 13 to 16%, and for yield mapping from 7 to 11%. The difference in adoption rate in the EU is wider in the South than in the North. A significant number of small farms in Europe feel that PA is difficult to apply to their farms. It is suggested that cooperative use of equipment or through contractors can help in this instance. Even though PA has been adopted in large farms in Northern Europe, USA, and Latin America, the application of PA in areas in the world where small farms prevail is still a big challenge and has to be explored both for its economic and environmental benefits.

Most research is pointing toward the belief that PA will be adopted by farmers if it offers economic advantages over conventional methods and is simple and easily applied. The economic returns of PA have been studied. It is clear that PA requires some new equipment (yield sensors, installation of equipment, ECa sensors, VRA equipment, computers, etc.) that has to be depreciated. Depreciation time has to be short as is the case of most electronic devices. Additional costs for training to produce maps and interpret the results are also required. Variable costs are the annual data analysis and interpretation. All these costs should be covered by the benefits from the application. In many cases improved yields and reduced costs are the benefits and can be directly estimated. In many cases such as the reduction in chemical, water, or energy use, apart from the direct reduction in costs, it is difficult to translate the additional benefits to the environment into monetary units. In high value crops, quality improvement can be of great interest. Bramley, Pearse, and Chamberlain (2003) in a separate harvest of two parts of a field found that the high quality grapes produced wine of high price ($30/bottle) while the low quality grapes gave low priced wine ($19/bottle). They comment that if the grapes were harvested all in bulk they would produce a low quality wine. The profit based on the gross price of wine was around $30 000/ha. An estimation of the application cost was $11/t of harvested fruit which is negligible compared with the profit.

2.5   Precision Agriculture and Sustainability

Sustainability is a term used for production systems that are environmentally friendly. The UN Brundtland Commission defined the term as the development able to ensure that it meets the needs of the present without compromising the ability of future generations to meet their own needs (World Commission on Environment and Development, 1987). The American Society of Agronomy defined sustainable agriculture as "one that, over the long term, enhances environmental quality and the resource base in which agriculture depends; provides for basic human food and fiber needs; is economically viable; and enhances the quality of life for farmers and the society as whole" (American Society of Agronomy, American Society of Agronomy, ASA, 1989). Sustainability is described as the intersection of economy, society, and ecology. The definitions indicate that sustainable agriculture has: (i) to be productive to cover the increasing human population with high quality food (food security and safety) and raw material, even lately energy; (ii) to secure profit for the farmers and maintain their welfare but at the same time has to make an optimum use of resources and save them for the next generation; and (iii) to reduce the adverse effects of agriculture on the environment. Resources such as soil, water, energy, and biodiversity have to be used for the present production but maintained for the next generations.

Bongiovanni and Lowenberg-Deboer (2004) reviewed the sustainability effects of PA. Several literature references indicate fertilizer input reduction and the effects on the environment. Variable rate fertilizer applications have attracted the interest of the scientific community. Nitrogen is the input with the higher energy input into the system but also causes pollution. In rain fed crops, nitrogen fertilizers account for 34% of the energy input (about the same as tillage, 39%) and account for 29% in irrigated crops (with irrigation accounting for 48% in sunflower) (Gemtos et al., 2013). Several studies indicate fertilizer saving with increased or unaffected yields and improved profit for farmers and effects on the environment. Lan et al. (2008) studied variable rate fertilizer (nitrogen, phosphorus, potassium) on maize crop. Yield analysis showed that variable rate fertilizer increased yield by 11 and 33% for the 2 years of the experiment while the amount of applied fertilizer was decreased, 32 and 29%, respectively. Morari *et al.* (2013) applied variable rate nitrogen application in a Durum Wheat field in the Veneto area of Italy. They applied nitrogen based on NDVI sensors and achieved improved grain quality and reduced nitrogen inputs. Vatsanidou *et al.* (2014) applied nitrogen with variable rate based on the replacement of the nutrients removed by the previous year's crop. They achieved a 43% reduction in the applied rate without affecting the year's yield. In a study in apples in Greece, Liakos (2013) applied homogeneous and variable rate (based on the nutrients removed by yield) fertilization in alternate rows of the orchard for 2 years. He found considerable reduction in the nitrogen inputs with a small decrease in the yield but the profit for the farmer increased. He also found an improvement in the quality of the apples.

Several examples of input savings have been given in the presentation of PA technologies. Tagarakis (2014) in a 1 ha vineyard split the drip irrigation network into two based on soil texture and elevation and achieved up to 20% water saving. It is clear that PA can offer considerable help in developing a sustainable agriculture assisting farmers in their decision-making during crop growing. New sensors able to detect any irregular reaction of the crops or the soil will improve productivity, resource use, profitability, and reduce effects on the environment.

2.6   Conclusions

PA is a crop management system that adapts inputs to the requirements of each part of the field. It assesses at the beginning the variability of the field and the crop using several technologies and sensors and then applies inputs to meet the crop requirements. Variable rate input application is the technology that offers the opportunity to adjust inputs to requirements leading to reduced inputs and/or increased yields, improved resource use, and reduced adverse effects on the environment. Additionally PA offers improved profitability and productivity of the farms. These are the components that lead to improved sustainability of agriculture. The adoption is, however, still not as anticipated especially in many regions where small farms exist, and their potential benefits should be further explored.

References

Adumchuk, V.I., Hummel, J.W., Morgan, M.T., and Upadhyaya, S.K., 2004. On-the-go soil sensors for precision agriculture. *Computers and Electronics in Agriculture*, **44**, 71–91.

Aggelopoulou, K., Bochtis, D., Fountas, S., Swain, K.C., Gemtos, T., and Nanos, G., 2011a. Yield prediction in apples based on image processing. *Journal of Precision Agriculture*, **12**, 448–456.

Aggelopoulou, K., Castrignanò, A., Gemtos, T., and De Benedetto, D., 2013. Delineation of management zones in an apple orchard in Greece using a multivariate approach. *Computers and Electronics in Agriculture*, **90**, 119–130.

Aggelopoulou, K., Fountas, S., Pateras, D., Nanos, G., and Gemtos, T., 2011b. Soil spatial variability and site-specific fertilization maps in an apple orchard. *Precision Agriculture*, **12**, 118–129.

Aggelopoulou, K., Wulfsohn, D., Fountas, S., Nanos, G., Gemtos, T., and Blackmore, S., 2010. Spatial variability of yield and quality in an apple orchard. *Precision Agriculture*, **11**, 538–556.

AgLeader Technology, 2014. http://www.agleader.com (accessed July 14, 2015).

Alchanatis, V., Cohen, Y., Cohen, S., Moller, M., Sprinstin, M., Meron, M., and Sela, E., 2010. Evaluation of different approaches for estimating and mapping crop water status in cotton with thermal imaging. *Precision Agriculture*, **11**, 27–41.

American Society of Agronomy (ASA). 1989. Decision Reached on Sustainable Agriculture. Agronomy News (January issue), ASA, Madison, WI, p. 15.

Ampatzidis, Y.G., Vougioukas, S.G., Bochtis, D.D., and Tsatsarelis, C.A., 2009. A yield mapping system for hand-harvested fruits based on RFID and GPS location technologies: field testing. *Precision Agriculture*, **10**, 63–72.

Andrade, P., Upadhyaya, S.K., Jenkins, B.M., and Filho, A.G.S., 2002. Evaluation of UC Davis compaction profile sensor. ASAE (American Society of Agricultural and Biological Engineers) 2002 ASAE Annual Meeting, ASAE, St. Joseph, MI, Paper No. 02–1185.

Arnó J., Martínez-Casasnovas, J.A., Ribes-Dasi, M., and Rosell, J.R., 2009. Review: precision viticulture. Research topics, challenges and opportunities in site-specific vineyard management. *Spanish Journal of Agricultural Research*, **7**, 779–790.

Auernhammer, H., 2001. Precision farming – the environmental challenge. *Computers and Electronics in Agriculture*, **30**, 31–43.

Best S., León, L., and Claret, M., 2005. Use of precision viticulture tools to optimize the harvest of high quality grapes. *FRUTIC 05, 7th Fruit Nut and Vegetable Production Engineering Symposium*. Montpellier, France, September 12–16, 2005. Versailles: INRA, pp. 249–258.

Blackmore, B.S., Fountas, S., Gemtos, T.A., and Griepentrog, H.W., 2009. A specification for an autonomous crop production mechanization system. *Acta Horticulturae (ISHS)*, **824**, 201–216.

Blackmore, B.S., Griepentrog, H.W., Fountas, S., and Gemtos, T., 2007. Specifications for an autonomous crop mechanization system. *Agricultural Engineering International: The CIGR Ejournal*, **9**, Manuscript PM 06 032.

Bongiovanni, R. and Lowenberg-Deboer, J., 2004. Precision agriculture and sustainability. *Precision Agriculture*, **5**, 359–387.

Bramley, R.G.V., 2005. Understanding variability in winegrape production systems 2. Within vineyard variation in quality over several vintages. *Australian Journal of Grape and Wine Research*, **11**, 33–42.

Bramley, R. and Hamilton, R.P., 2004. Understanding variability in winegrape production 1. Within vineyard variation in yield over several vintages systems. *Australian Journal of Grape and Wine Research*, **10**, 32–45.

Bramley R., Pearse, B., and Chamberlain, P., 2003. Being profitable precisely – a case study of precision viticulture from Margaret River. *Australian & New Zealand Grapegrower & Winemaker* **473a**(Annual Technical Issue), 84–87.

Corpa, L.A., McMurtrey, J.E., Middletonc, E.M., Mulchid, C.L., Chappellea, E.W., and Daughtry, C.S.T., 2003. Fluorescence sensing systems: in vivo detection of biophysical variations in field corn due to nitrogen supply. *Remote Sensing of Environment*, **86**, 470–479.

Corwin, D.L. and Lesch, S.M., 2003. Application of soil electrical conductivity to precision agriculture: theory, principles, and guidelines. *Agronomy Journal*, **95**, 455–471.

Davenport, J.R., Marden, J.M., Mills, L., and Hattendorf, M.J., 2003. Response of Concord grape to variable rate nutrient management. *American Journal of Enology and Viticulture*, **54**, 286–293.

Dedousis, A.P. and Godwin, R.J., 2008. The rotating disc-hoe-an overview of the system for mechanical weed control. *ASABE (American Society of Agricultural and Biological Engineers) Annual International Meeting*. Providence, RI, June 29–July 2, 2008. St. Joseph, MI: ASABE, pp. 451–463.

Department for Environment, Food and Rural Affairs, 2013. Farm Practices Survey Autumn 2012. London: DEFRA. https://www.gov.uk/government/uploads/system/uploads/attachment_data/file/181719/defra-stats-foodfarm-environ-fps-statsrelease-autumn2012 edition-130328.pdf (accessed June 22, 2015).

Ehsani, P., Sankaran, S., and Dima, C., 2010. Grower Expectations of New Technologies for Applications in Precision Horticulture. Gainesville, FL: Agricultural and Biological Engineering Department, Florida Cooperative Extension Service, Institute of Food and Agricultural Sciences, University of Florida. http://edis.ifas.ufl.edu/pdffiles/AE/AE46700.pdf (accessed June 22, 2015).

Erickson, B., Widmar, D., and Holland, J., 2013. Survey: An Inside Look at Precision Agriculture in 2013. Croplife. http://www.croplife.com/equipment/precision-ag/survey-an-inside-look-at-precision-agriculture-in-2013 (accessed 22 June 2015).

Ess, D.R. and Morgan, M.T., 2003. *The Precision Farming Guide for Agriculturalists*, 3rd edn. Moline, IL: Deere and Company.

Fountas, S., Aggelopoulou, K., Bouloulis, C., Nanos, G.D., Wulfsohn, D., Gemtos, T.A., and Galanis, M., 2011. Site-specific management in an olive tree plantation. *Precision Agriculture*, **12**, 179–195.

Fountas, S., Bartzanas, T., and Bochtis, D., 2011. Emerging footprint technologies in agriculture, from field to farm gate. In: M. Bourlakis, I. Vlachos and V. Zeimpekis (eds). *Intelligent Agrifood Chains and Networks*. Oxford: Wiley-Blackwell, pp. 67–85.

Fountas, S., Wulfsohn, D., Blackmore, S., Jacobsen, H.L., and Pedersen, S.M., 2006. A model of decision making and information flows for information-intensive agriculture. *Agricultural Systems*, **87**, 192–210.

Gemtos, T.A., Akdemir, B., Turker, U., and Mitev, G. (2010) A Feasibility Study for Variable Rate Irrigation in the Black Sea region: Economical and Environmental Benefits. Final Report, Black Sea Economic Cooperation, Istanbul.

Gemtos, T.A., Cavalaris, C., Karamoutis, C., Tagarakis, A., and Fountas, S., 2013. Energy analysis of three energy crops in Greece. *CIGR Journal*, **15**, 52–66.

Gemtos, T.A., Markinos, A., and Nassiou, T., 2005. Cotton lint quality spatial variability and correlation with soil properties and yield. 5th European Conference on Precision Agriculture, Uppsala, Sweden, 2005, pp. 361–368.

Gil, E., Escola, A., Rosell, J.R., Planas, S., and Val, L., 2007. Variable rate application of plant protection products in vineyard using ultrasonic sensors. *Crop Protection*, **26**, 1287–1297.

Giles, D.K., Delwiche, M. J., and Dodd, R.B., 1988. Electronic measurement of tree canopy volume. *Transactions of the ASAE*, **31**, 264–272.

Godwin R.J., Wood, G.A., Taylor, J.C., Knight, S.M., and Welsh, J.P., 2003. Precision farming of cereal crops: a review of a six year experiment to develop management guidelines. *Biosystems Engineering*, **84**, 375–391.

Grisso, R., Alley, M., Thomason, W., Holshouser, D., and Roberson, G.T., 2011. Precision Farming Tools: Variable-Rate Application. Blacksburg, VA: Virginia Cooperative Extension, Virginia Tech, Virginia State University. Publication 442–505, http://pubs.ext.vt.edu/442/442-505/442-505_PDF.pdf (accessed June 22, 2015).

Hall A., Lamb, D.W., Holzapfel, B.P., and Louis, J.P., 2010. Within-season temporal variation in correlations between vineyard canopy and winegrape composition and yield. *Precision Agriculture*, **12**, 103–117.

Hedley, C.B. and Yule, Z.I.J., 2009. Soil water status mapping and two variable rate irrigation scenarios. *Precision Agriculture*, **10**, 342–355.

Heraud, J.A. and Lange, A.F., 2009. *Agricultural automatic vehicle guidance from horses to GPS: how we got here, and where we are going.* 2009 Agricultural Equipment Technology Conference, Louisville, KY, February 9–12, 2009.

Khosla, R. and Shaver, T., 2001. Zoning in on nitrogen needs. *Colorado State University Agronomy Newsletter*, **21**, 24–26.

Kitchen, N.R., Sudduth, K.A., Myers, D.B., Drummond, S.T., and Hong, S.Y., 2005. Delineating productivity zones on claypan soil fields apparent soil electrical conductivity. *Computers and Electronics in Agriculture*, **46**, 285–308.

Koch, Br. and Khosla, R., 2003. The role of precision agriculture in cropping systems. *Journal of Crop Production*, **9**, 361–381.

Kondo, N. and Ting, K.C., 1998. *Robotics for Bioproduction Systems.* St Joseph, MI: American Society of Agricultural Engineers.

Konopatzki, M.R.S., Souza, E.G., Nóbrega, L.H.P., Uribe-Opazo, M.A., Suszek, G., Rodrigues, S., and de Oliveira, E.F., 2009. Pear tree yield mapping. *ISHS Acta Horticulturae*, **824**, 303–312.

Kromer, K.H., Shmittmann, O., and Osman A.M., 1999. *Crop Yield Monitoring on Forage Harvesters.* St Joseph, MI: American Society of Agricultural Engineers, Publication no. 99–1051.

Kutter T., Tiemann, S., Siebert, R., and Fountas, S., 2011. The role of communication and co-operation in the adoption of precision farming. *Precision Agriculture*, **12**, 2–17.

Lan, Y., Zhang, S., Li, W., Hoffmann, W.C., and Ma, C., 2008. Variable rate fertilization for maize and its effects based on the site-specific soil fertility and yield. *Agricultural Engineering International: The CIGR Ejournal*, 10.

Lawson, L.G., Pedersen, S.M., Sorensen, C.G., Pesonen, L., Fountas, S., Werner, A., Oudshoorn, F.W., Herold, L., Chatzinikos, T., Kirketerp, I.M., and Blackmore, S., 2011. A four nation survey of farm information management and advanced farming systems: a descriptive analysis of survey responses. *Computers and Electronics in Agriculture*, **77**, 7–20.

Liakos, V., 2013. Precision agriculture application in apples orchards. PhD thesis. University of Thessaly, Greece.

Morari, F., Loddo, S., Berzahgi, P., Ferdito, J.C., Berti, A., Sartori, L., Visioli, G., Marmiroli, M., Pragolo, D., and Mosca G., 2013. Understanding the effects of site-specific fertilisation on yield and protein content of durum wheat. In: J.V. Stafford (ed.). *European Conference on Precision Agriculture '13.* Wageningen: Wageningen Academic Publishers, pp. 321–327.

Papageorgiou, E.I., Aggelopoulou, K.D., Gemtos, T.A., and Nanos, G.D., 2013. Yield prediction in apples using Fuzzy Cognitive Map learning approach. *Computers and Electronics in Agriculture*, **91**, 19–29.

Papageorgiou, E.I., Markinos, T.A., and Gemtos, T., 2011. Fuzzy cognitive map based approach for predicting yield in cotton crop production as a basis for decision support system in precision agriculture application. *Applied Soft Computing*, **11**, 3643–3657.

Pelletier, G. and Upadyaya, K.S., 1999. Development of a tomato load/yield monitor. *Computers and Electronics in Agriculture*, **23**, 103–107.

Qiao, J., Sasao, A., Shibusawa, S., Kondo, N., and Morimoto, E., 2005. Mapping yield and quality using the mobile fruit grading robot. *Biosystems Engineering*, **90**, 135–142.

Schueller, J.K., Whitney, J.D., Wheaton, T.A., Miller, W.M., and Turner, A.E., 1999. Low-cost automatic yield mapping in hand-harvested citrus. *Computers and Electronics in Agriculture*, **23**, 145–153.

Sethuramasamyraja B., Sachidhanantham, S., and Wample, R., 2010. Geospatial modeling of wine grape quality indicators (Anthocyanin) for development of differential wine grape harvesting technology. *International Journal of Geomatics and Geosciences*, **1**, 372–385.

Shaddad, S.M. (2014) Proximal soil sensors and geostatistical tools in precision agriculture applications. PhD thesis. University of Sassari, Italy.

Sorensen, S., Fountas, S., Nash, E., Pesonen, L., Bochtis, D., Pedersen, S., Basso, B., and Blackmore, S., 2010. Conceptual model of a future farm management information system. *Computers and Electronics in Agriculture*, **72**, 37–47.

Stenberg, B., Viscarra Rossel, R.A., Mouazen, M.A., and Wetterlind, J. 2010. Visible and near infrared spectroscopy in soil science. *Advances in Agronomy*, **107**, 163–215.

Tagarakis, A., 2014. Vineyard management using ICT in a precision agriculture application. PhD thesis. University of Thessaly, Greece.

Tagarakis, A., Koundouras, S., Papageorgiou, E.I., Dikopoulou, Z., and Fountas, S., 2014. A fuzzy inference system to model grape quality in vineyards. *Precision Agriculture*, **15**, 555–578.

Tagarakis, A., Liakos, V., Fountas, S., Koundouras, S., and Gemtos, T., 2013. Management zones delineation using fuzzy clustering techniques in grapevines. *Precision Agriculture*, **14**, 18–39.

Taylor, J.A., McBratney, A.B., and Whealan, B.M., 2007. Establishing management classes for broadacre agricultural production. *Agronomy Journal*, **99**, 1366–1376.

Tomasson, J.A., Penington, D.A., Pringle, H.C., Colombus, E.P., Tomson, S.J., and Byler, R.K., 1999. Cotton mass flow measurements: experiments with two optical devices. *Applied Engineering in Agriculture*, **15**, 11–17.

Tremblay, N., Wang, Z., and Cerovic, Z.G., 2012. Sensing crop nitrogen status with fluorescence indicators: a review. *Agronomy for Sustainable Development*, **32**, 451–464.

Tumbo, S.D., Salyani, M., Whitney, J.D., Wheaton, T.A., and Miller, W.M., 2002. Investigation of laser and ultrasonic ranging sensors for measurements of citrus canopy volume. *Applied Engineering in Agriculture*, **18**, 367–372.

Vatsanidou, A., Fountas, S., Nanos, G., and Gemtos, T., 2014. Variable rate application of nitrogen fertilizer in a commercial pear orchard. *Fork to Farm: the International Journal of the American Farm School of Thessaloniki*, **1**, 1–8.

Vellidis, G., Perry, C.D., Durrence, J.S., Thomas, D.L., Hill, D.L., Kwien, R.W., and Rains, G., 2001. Field testing the peanut yield monitoring. In: P.C. Robert, R.H. Rust, and W.E. Larson (eds). *American Society of Agronomy Proceedings of the Third International Conference on Precision Agriculture*. Minneapolis, MN: ASA, pp. 835–844.

Wild, K. and Auernhammer, H., 1999. A weighing system for local yield monitoring of forage crops in round balers. *Computers and Electronics in Agriculture*, **23**, 119–132.

World Commission on Environment and Development, 1987. Report of the World Commission on Environment and Development: Our Common Future.

Zaman, Q.U., Schumann, A.W., and Hostler, K.H., 2006. Estimation of citrus fruit yield using ultrasonically-sensed tree size. *Applied Engineering in Agriculture*, **22**, 39–44.

Zaman, Q.U., Schuman, A.W., and Miller, W.M., 2005. Variable rate nitrogen application in Florida citrus based on ultrasonically-sensed tree size. *Applied Engineering in Agriculture*, **21**, 331–335.

Zarco-Tejada, P.J., Hubbard, N., and Loudjani, P., 2014. Precision Agriculture: An Opportunity for EU Farmers – Potential Support with the CAP 2014–2020. *Brussels: Policy Department B: Structural and Cohesion Policies*, European Parliament.

3

Agricultural Waste Biomass

Sven G. Sommer,[1] Lorie Hamelin,[1] Jørgen E. Olesen,[2] Felipe Montes,[3] Wei Jia,[4] Qing Chen,[4] and Jin M. Triolo[1]

[1] *Faculty of Engineering Institute of Chemical Engineering, Biotechnology and Environmental Technology, University of Southern Denmark, Campusvej 55, Odense M, Denmark*

[2] *Department of Agroecology, Aarhus University, Blichers Allé 20, Tjele, Denmark*

[3] *Production Systems and Modeling, Plant Science Department, The Pennsylvania State University, 245 Agricultural Science and Industries Building, University Park, USA*

[4] *College of Resources and Environmental Sciences, China Agricultural University, No. 2 Yuanmingyuan Xilu, Beijing, China*

3.1 Introduction

Organic residues and waste from farming have been used for a wide range of purposes including heating, house construction, and fertilization. The residues from crop production include all the potentially harvestable above-ground biomass after the main edible or marketable part of the crop has been harvested, such as straw, tops, leaves, bark, stalks, stems, stubbles, and husks. Such residues may also be used for livestock feed, in particular for ruminants, or for bedding materials in livestock systems. Residues from livestock production are mainly animal excreta mixed with straw, wood chips, sand, or other materials strewn on the floor where the animals are confined (Sommer *et al.*, 2013). The organic waste originating from industries

Supply Chain Management for Sustainable Food Networks, First Edition. Edited by Eleftherios Iakovou, Dionysis Bochtis, Dimitrios Vlachos and Dimitrios Aidonis.

processing agricultural products may be bagasse or vinasses from bioethanol and sugar production, blood and intestines, and a range of plant residues from food production.

Globally, large amounts of agricultural waste biomasses are produced and used. For example, a large fraction of the plant biomass is left in the field where it contributes to maintain soil fertility and improve the soil structure by bonding mineral particles into micro- and macro-aggregates (Johnston, Poulton, and Coleman, 2009). Organic agricultural waste is also a potential source of energy and using, for example, animal manure for biogas is recognized as a cost-effective mitigation technology for greenhouse gas (GHG) emissions in agriculture, although the inappropriate use of agricultural residues may also increase GHG emissions (Bruun *et al.*, 2014; Hamelin, Naroznova, and Wenzel, 2014). Furthermore, organic waste can be used in biorefineries and be a feedstock for material production of lactic and/or organic acids, amino acids and fibers for biofuels, insulation/building materials, and paper or polymers (Ecker *et al.*, 2012).

When managing agricultural residues or waste products one must bear in mind that they contain significant amounts of carbon (C) and plant nutrients, these being precursors of GHGs, ammonia (NH_3), pollution of surface waters, offensive odor, and particles (Steinfeld *et al.*, 2006; Sommer *et al.*, 2009a; Sutton *et al.*, 2011). The management of agricultural residues will strongly affect the magnitude of such emission flows. In addition to the pollution resulting from excessive nutrient use and gaseous emissions, the risk of disease spreading (Vänneras, 2013), as well as heavy metals and biogenic contamination of soil (Syakti *et al.*, 2013), should be taken into consideration when developing sustainable strategies for recycling and using agricultural waste.

Plant residues and livestock waste being left over throughout the food supply chain should be considered a raw material that can be used as a plant nutrient fertilizer, for soil amelioration, production of bio-energy carriers, and bio-products. The intention of this chapter is to present an overview of the magnitude of the global agricultural waste production, and present concrete examples of how it is recycled in selected regions. Physiochemical characteristics of the wastes are given and we present the potential use for energy production and bio-product manufacturing. The environmental risks related to recycling and use of waste biomass are briefly presented.

Throughout this chapter, various terms are used when referring to animal excreta, reflecting the various forms of management for this. It is assumed that the reader is familiar with these terms; these will therefore not be defined herein, but a glossary with additional explanations for each of these terms can be found in Pain and Menzi (2011).

3.2   Amount of Biomass

Nearly a third of our planet's land surface is currently used for agriculture (FAO, 2014a). As a result, a considerable amount of waste from the agricultural sector, worldwide, is produced every year.

Here, we present global agricultural waste production statistics and examples of waste production and use for USA, China, and Denmark, countries with very different policies regarding the handling of agricultural waste products.

3.2.1   Global Production of Agricultural Residues

Concerned with supplying food, feed, fiber, and energy needs of future populations, a number of studies have attempted to quantify the global biomass energy potential by 2050 (Dornburg *et al.*, 2010). These studies, involving a number of assumptions about, for example, future crop yields and global diets, typically recognize two main categories of biomass: dedicated crops; and an aggregated residual biomass consisting of organic wastes combined with residues from forestry and agriculture. Although the ranges presented for the former varies greatly (narrowed to 100–400 EJ year^{-1} in Dornburg *et al.*, 2010), there is a general agreement that the overall residues potential is around 100 EJ year^{-1}; considering that an average gross calorific value (i.e., higher heating value) of 18.5 MJ per kilogram of dry matter (DM), this corresponds to about 5.4 Pg DM year^{-1}. These future potentials are, of course, of utmost necessity in the perspective of long-term strategic decision making. In a shorter-time perspective, quantifying today's residue potential is also a valuable exercise, as it provides a clear picture of the type and amount of residual agricultural biomass that is available under the current conditions, and in which proportions.

One noteworthy attempt to do so is the study of Krausmann *et al.* (2008), where the global unused human extraction of biomass for the year 2000 was quantified to 0.9 Pg DM, that is, 17 EJ year^{-1}. Although the study of Krausmann *et al.* (2008) is a landmark, its focus is on the overall biomass flows rather than agricultural residues flows only. As a result, it does not present information on the available biomass from livestock manure, and the information on crop residues it presents is aggregated and not detailed at crop level. An estimate of the amount of residues produced (in terms of DM) from the world major crops and animal production systems is given in Tables 3.1 and 3.2. The values presented should be considered upper estimates of the potentials; they do not consider the amount of crop or animal residues that is already used, nor the amounts that could not technically be harvested/collected. This especially applies for manure, where the amounts excreted on pasture are often not seen as potentials (for this reason, an estimate of the portion excreted on pasture is presented in Table 3.2). For crop residues, no more than 80% may realistically be harvested in industrialized countries (Birkmose, Hjort-Gregersen, and Stefanek, 2013). It should also be highlighted that the values presented in Table 3.2 are for manure as excreted, that is, they do not reflect the input (e.g., through bedding) or losses (e.g., through emissions) of DM. The bedding should therefore be considered part of the crop residues when estimating available biomasses.

As shown in Table 3.1, the greatest amount of DM from crop residues is supplied from the cultivation of cereals, particularly maize (Asia, North America), rice (Asia) and wheat (Asia, Europe). Non-dairy cattle (Asia, Latin America) similarly provide, by far, the greatest amount of DM from animal manure, representing 40% of all the DM produced from animal manures worldwide (Tables 3.2 and 3.3). Considering

Table 3.1 Estimated global crop residue flows by world region for year 2012.[a]

	DM[b] (%)	HI[c]	Africa	Asia	Latin America	Europe	North America	Oceania	World
Cereals (Tg DM)									
Barley	85	1.5	7.7	24.6	9.0	101	16.4	11.1	**170**
Maize	86	1.0	60.0	248	114	81	246	0.58	**750**
Maize, green	31	1.0	0.54	0.37	0.42	0.24	1.3	0.12	**3.0**
Millet	90	1.5	22	18	0.026	0.83	0.095	0.061	**41**
Oats	85	1.0	0.23	0.84	1.3	11.4	3.1	1.1	**18**
Rice	86	1.5	35	841	35	5.6	12	1.2	**929**
Rye	85	1.5	0.12	1.5	0.064	16	0.67	0.052	**19**
Sorghum	85	1.5	31	12	20.0	1.0	8.2	2.9	**75**
Wheat	87	1.5	32	397	25	250	113	39	**856**
Others	85	1.5	1.4	0.36	0	7.3	0	0	**7.0**
Pulses (Tg DM)									
Beans	88	1.0	5.8	11	5.0	1.0	1.5	0.42	**25**
Bambara beans	87	1.0	0.14	0	0	0	0	0	**0.14**
Chick peas	89	1.0	0.62	8.5	0.29	0.081	0.27	0.60	**10.3**
Lentils	88	1.0	0.17	1.8	0.015	0.076	1.5	0.41	**4.0**
Lupins	93	1.0	0.018	0.0[d]	0.049	0.22	0	0.91	**1.2**
Peas, dry	87	1.0	0.52	1.7	0.16	2.9	2.9	0.31	**8.5**
Pulses	90	1.0	1.1	1.6	0.030	1.1	0	0.85	**4.6**
Others	89	1.0	5.7	3.5	0.20	0.33	0.022	0.018	**9.7**

Oil crops (Tg DM)

Linseed	92	1	0.12	0.60	0.036	0.53	0.58	0.0086	1.9
Rapeseed	92	1.5	0.28	31	0.52	31	23	4.7	90
Seed cotton	92	1.5	6.6	72	9.2	1.4	12	4.0	106
Sesame	97	2	3.3	4.2	0.34	0.0028	0	0	7.8
Soybeans	87	1	1.7	24	105	4.8	76	0.075	210
Sunflower	92	1	2.0	5.3	3.6	22	1.2	0.043	34
Palm fruit	53	1.3	13	152	9.8	0	0	1.6	176

Sugar crops (Tg DM)

Sugar beet	22	0.25	0.60	2.0	0.10	10.5	1.8	0	15
Sugar cane	30	0.25	7.1	55	71	0[d]	2.1	2.1	138

Tubers and potatoes (Tg DM)

Root and tubers	24	0.25	1.7	11	1.1	7.1	0	0.1	22
Potato	24	0.25	0.41	0.088	0	0.0016	1.4	0.025	0.58
Sweet potato	24	0.25	1.1	4.9	0.12	0.0031	0.073	0.046	6.2
Others	24	0.25	13	5.0	2.1	0[d]	0[d]	0.062	20.1

[a] Estimated as primary production × harvest index (HI), for each crop type. Primary harvest retrieved from FAO (2014b). Inconsistencies in the numbers are due to rounding. The number of significant digits (maximum 3) does not reflect precision and was selected only in order to avoid ambiguity.

[b] Retrieved from Møller et al. (2000) and Feedipedia (2014), when not available in the former. The value for palm fruit is from Sulaiman et al. (2011).

[c] Harvest index (no unit): ratio between the above-ground residues and the primary harvest; retrieved from Lal (2005).

[d] Not a zero value, but close to negligible.

Table 3.2 Global estimated excreted DM flows from manure by world region, for year 2012.[a]

	kg VS kg⁻¹ DM[b]	kg DM kg⁻¹ FM[b]	Africa	Asia	Latin America	Europe	North America	Oceania	World, total	World, % on pasture[c]
Cattle (Tg manure, dry basis)										
Cattle, dairy	0.80	0.13	45	129	57	83	25	11	**350**	41
Cattle, non-dairy	0.90	0.10	124	312	365	90	90	40	**1020**	48
Pigs (Tg manure, dry basis)										
Pigs, breeding	0.80	0.10	0.46	7.9	1.2	4.0	1.8	0.12	**16**	4
Pigs, all others	0.80	0.10	4.1	71	11	23	8.8	0.61	**118**	7
Poultry and birds (Tg manure, dry basis)										
Chickens[d]	0.80	0.26	16	108	29	9.5	13	0.59	**176**	28
Ducks	0.80	0.26	0.23	11	0.16	0.80	0.078	0.013	**12**	55
Geese and guinea fowls	0.80	0.26	0.23	3.1	0.0039	0.088	0.0015	0[e]	**3.4**	—
Turkey	0.80	0.26	0.19	0.11	0.62	3.7	8.3	0.044	**13**	12
Sheep and goat (Tg manure, dry basis)										
Goats	0.80	0.32	55	95	5.3	2.3	0.40	0.54	**159**	100
Sheep	0.80	0.28	47	77	12	23	1.1	19	**179**	100
Others (Tg manure, dry basis)										
Horses	0.84	0.15	4.5	11	16	5.4	9.9	0.37	**47**	100
Mules	0.80	0.15[e]	0.42	1.4	2.5	0.10	0.0017	0	**4.5**	71%
Rabbits and hares	0.86	0.35	0.58	22	12	5.1	0	0	**39**	—

Asses	0.80	0.30[e]	8.4	7.3	2.9	0.23	0.022	0.0039	**19**	—
Buffalos	0.80	0.25	5.6	309	2.3	0.66	0	0.00037	**317**	—
Llamas	0.80	0.30[e]	0	0	9.9	0	0	0	**9.9**	—
Camels	0.80	0.30[e]	26	4.6	0	0.0085	0	0	**30**	—
Pigeon and other birds	0.80	0.26[e]	0.12	0.15	0	0.013	0	0	**0.28**	—
Rodents, others	0.80	0.30[e]	0	0	1.2	0	0	0	**1.2**	—
VS (kg head^{-1} day^{-1})[f]									
Cattle, dairy	—	—	1.90	2.70	2.90	4.80	5.40	3.50	—	—
Cattle, non-dairy	—	—	1.50	1.85	2.50	2.65	2.40	3.00	—	—
Pigs, breeding	—	—	0.30	0.30	0.30	0.48	0.50	0.50	—	—
Pigs, market	—	—	0.30	0.30	0.30	0.30	0.27	0.28	—	—
Chickens[d]	—	—	0.02	0.02	0.02	0.01	0.01	0.01	—	—
Ducks	—	—	0.02	0.02	0.02	0.02	0.02	0.02	—	—
Turkey	—	—	0.02	0.02	0.02	0.07	0.07	0.07	—	—
Goats	—	—	0.35	0.35	0.35	0.30	0.30	0.30	—	—
Sheep	—	—	0.32	0.32	0.32	0.40	0.40	0.40	—	—
Horses	—	—	1.72	1.72	1.72	2.13	2.13	2.13	—	—
Mules	—	—	0.94	0.94	0.94	0.94	0.94	0.94	—	—

(continued overleaf)

Table 3.2 (Continued)

	kg VS kg⁻¹ DM^b	kg DM kg⁻¹ FM^b	Africa	Asia	Latin America	Europe	North America	Oceania	World, total	World, % on pasture^c
Rabbits and hares	—	—	0.10	0.10	0.10	0.10	0.10	0.10	—	—
Buffalos	—	—	3.10	3.50	3.90	3.90	3.90	3.90	—	—
Camels	—	—	2.49	2.49	2.49	2.49	2.49	2.49	—	—
Fur-bearing (rodents)	—	—	0.14	0.14	0.14	0.14	0.14	0.14	—	—

^a Estimated as VS excreted per head per day × (VS content of manure)⁻¹ × live population (head), for all animal categories available in the FAO datasets. Live population retrieved from FAO (2014c,d). Inconsistencies in the numbers are due to rounding. The number of significant digits (maximum 3) does not reflect precision and was selected only in order to avoid ambiguity.

^b For freshly excreted manure. The VS values were retrieved from the compilation presented in Hamelin (2013; Table 11). If not available in Hamelin (2013), a default value of 0.80 kg VS kg⁻¹ DM was applied. The value for rabbits was retrieved from United States Department of Agriculture and Natural Resources Conservation Service (2008). The DM values are shown for enabling an estimation of the manure flows on a wet basis, although these were not used in the calculation of the figures presented here. DM values were retrieved from ASAE (American Society of Agricultural Engineers, ASAE, 2010) unless otherwise indicated; values for goat and sheep were taken from ASAE (American Society of Agricultural Engineers, ASAE, 2003), and those of rabbits and buffalos from Nguyen, Le, and Dinh (2000).

^c Estimated based on the proportion of nitrogen, as a percentage of DM (ASAE, American Society of Agricultural Engineers, ASAE, 2010) and on the amount of nitrogen left on pastures (FAO, 2014e) for each manure type. This is a very rough estimation. No estimation could be made in cells left blank, due to a lack of reliable data.

^d For the FAO category "chickens," the values used for VS, DM as well as for kg VS per head per day are those for "broilers." However, the values for layers were always similar, if not equal.

^e Estimate from the authors.

^f From IPCC (2006; Appendix 10A.2).

Table 3.3 Summary of estimated global residual agricultural biomass flows by world region for year 2012.[a]

		Africa	Asia	Latin America	Europe	North America	Oceania	World
Overview[b]								
Population	(bn inhab.)	1.1	4.3	0.61	0.74	0.35	0.038	7.1
Land area	(Gha)	3.0	3.1	2.0	2.2	1.9	0.85	13
Crop residues[c] **(Pg DM)**		**0.252**	**1.938**	**0.413**	**0.551**	**0.524**	**0.072**	**3.758**
Cereals	(Tg DM)	187	1544	205	468	400	56	2860
Pulses	(Tg DM)	14	28	0.28	1.9	5.7	6.2	63
Oil crops[d]	(Tg DM)	27	289	128	60	113	10.5	123
Sugar crops	(Tg DM)	7.7	57	72	10.5	3.9	2.1	153
Tuber and potato	(Tg DM)	16	21	3.3	7.1	1.5	0.24	49
Animal manure (Pg DM)		**0.338**	**1.168**	**0.527**	**0.251**	**0.159**	**0.072**	**2.515**
Cattle	(Tg DM)	169	441	422	173	115	50	1371
Pigs	(Tg DM)	4.6	79	12	27	11	0.73	134
Poultry and birds	(Tg DM)	17	122	30	14	21	0.64	205
Sheep and goat	(Tg DM)	102	172	17	26	1.5	20	338
Others	(Tg DM)	45	354	47	11	9.9	0.38	468

[a] Inconsistencies in the numbers are due to rounding. The number of significant digits (maximum 4) does not reflect precision and was selected only in order to avoid ambiguity.

[b] World population retrieved from FAO (2014f), where bn is billion, that is, 10^9. Land area was retrieved from FAO (2014a), but data are for 2011.

[c] An estimation on other eventual crop residues (e.g., fruit and citrus production, vegetable, tree nuts) could not be included here; as a result, the total figures presented in this row may be underestimated. On the other hand, it must be remembered that an overestimation may have occurred with the oil crops (see note d).

[d] These values may be overestimated. The aggregated total provided by FAO for the world is indeed about three times lower. The authors suspect this may be due to a mistake in the FAO dataset for soybeans.

that half of the non-dairy cattle manure is excreted on pastures (Table 3.2), thus not collectable and excluded from any alternative use, the manure from non-dairy cattle still remains the largest source of DM from animal excreta. Another highlight from the figures presented in Tables 3.1–3.3 is the colossal importance of Asia as a supplier of agricultural residues.

Figure 3.1 presents the trends in amount of residues supplied by cereals and non-cattle dairy manure, per world region. It highlights Asia as a dominant supplier of agricultural waste biomass, providing about 50% of the residues from cereals since the mid-1990s, and a significant share of the DM from non-dairy cattle manure. In the latter case, however, the greatest potentials are found in Latin America since the early 1980s. Figure 3.1b also shows a continuous decrease of the amount of non-dairy cattle manure excreted in Europe since the early 1990s, at the benefit of both the Asian and Latin American regions.

Table 3.3 presents an overall potential of 6.3 Pg DM year^{-1} from global agricultural residues (the crop residues representing 60% of this, and animal manure 40%), or 116 EJ year^{-1}, which corresponds to about 22% of the current primary energy consumed worldwide in 2013 (533 EJ year^{-1}; BP, 2014). However, this 116 EJ year^{-1} figure, which is close to the earlier mentioned 100 EJ potential forecast from "residues" in 2050, represents the maximal potential.

There are a number of limitations that must be considered in relation to the estimates presented in Tables 3.1 and 3.2 and Figure 3.1; these estimates were obtained from an accounting exercise derived from best available data and therefore should be seen more as indicators of magnitude rather than as absolute values. In an attempt to minimize the uncertainty involved by such an approach, most data were retrieved from the same database, in this case the FAO database. The reliability of the data reported to FAO can be questioned, as shown by Grainger (2008) in the case of tropical forest areas. Remote sensing using satellite imagery, could possibly improve the accuracy of the estimates presented herein. Also, although average harvest index figures were found for each crop type, the crop residue estimates could be further refined by using specific harvest indexes for each world region. Manure estimates could also be improved through the use (and availability) of excretion rates in terms of DM per head, rather than the indirect approach used in this study where the volatile solids (VS) excretion rates per head reported by the IPCC (2006) are used, together with data on the proportion of VS in each manure type. As a consequence of these unavoidable uncertainties involved when estimating global potentials, there are discrepancies between the values presented in Tables 3.1–3.3 and the potentials estimated from approaches based on national inventories, as reflected in this chapter for Denmark, China, and the USA. Finally, no detailed estimates on the global production of agro-industrial residues could be found, and thus reported herein.

3.2.2   China

Over the past 30 years the amount of organic wastes including crop residue and livestock manures, have increased significantly with agricultural intensification, urbanization, and industrial development in China (Sun *et al.*, 2005). The conversion of

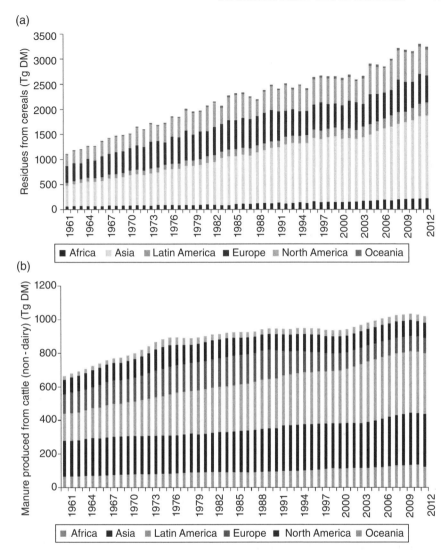

Figure 3.1   Development (1961–2012) of the amount of estimated DM produced worldwide from (a) crop residues resulting from the cereals production and (b) manure resulting from the non-dairy cattle production. Data source: as in Tables 3.1 and 3.2, with data from FAO (2014b,d) retrieved for the full 1961–2012 period.

backyard farming to large-scale farming in the last decade has resulted in an increase in discharge of organic waste from livestock farming to surface waters (Qiu *et al.*, 2013). The discharge is due to low efficiency in the recycling and use of organic waste in industrial livestock farming with inadequate land application, especially in the eastern regions or the suburbs of big cities in China.

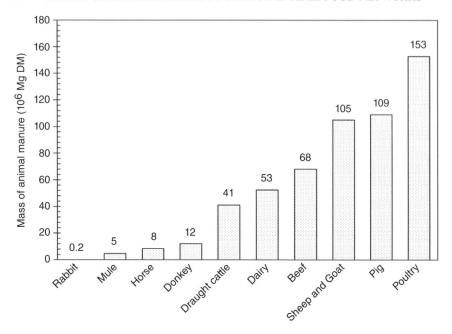

Figure 3.2   Mass of animal manure production in China in 2012. Draught cattle are stock required for agricultural operations, and include yellow cattle and buffalo.

The annual production of organic wastes has been estimated using the coefficients of crop harvested and livestock excretion based on the data given in national statistics for 2012 (Figures 3.2 and 3.3). In 2012, a total of 554 Tg DM of livestock manure was collected in China. The largest amount of manure was collected from cattle (including draught cattle, beef, and dairy) representing 44.5% of the total amount of manure. The amount of manure from pig, sheep+goat, and poultry accounted for 19.7, 19.0, and 27.6% of the total manure, respectively. It is estimated that 277 Tg DM of the collected manure is produced on intensive large-scale systems, and 278 Tg DM from traditional backyard systems.

The estimated crop residue produced in China for the year 2012 is 767 Tg DM, with 29.5% of the residues originating from maize, 29.0% from rice, and 21.6% from wheat. Residues from oil, tubers, and beans crops accounted for 8.7, 3.3, and 3.3% of the total amount of residues, produced during 2012.

Qiu *et al.* (2013) analyzed the fate of organic waste collected on 718 livestock farms in the rural regions of the Jilin, Sichuan, Zhejiang, Anhui, and Hebei provinces (Table 3.4). The survey showed that main livestock manure was applied to fields with or without composting, and the proportions of livestock manure application to fields in the backyard livestock production system (extensive production) were relatively higher than those in the concentrated animal feeding operations (CAFOs) system (intensive production). However, the study showed that the amount of manure applied to fields declined from 2010 in both the backyard and CAFOs production systems.

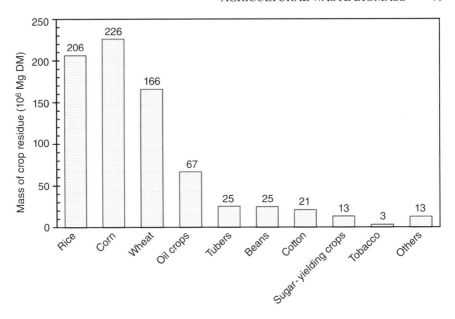

Figure 3.3   Mass of crop residue production in China in 2012. "Others" comprise crude fiber crops and other grain crops.

Table 3.4   Fate of pig and poultry manure in the Chinese provinces of Jilin, Sichuan, Zhejiang, Anhui, and Hebei in 2005 and 2010 (%).

Fate of feces and urine	Pig backyard system		Pig CAFOs production		Poultry backyard system		Poultry CAFOs production	
	2005	2010	2005	2010	2005	2010	2005	2010
Samples (*n*)	356	305	49	76	392	477	25	45
Land application	93.7	87.1	57.4	49.5	71.4	66.0	51.1	48.8
Biogas processing	3.5	8.5	12.0	15.0	1.4	2.8	2.2	1.4
Sale to market	0	0	5.5	9.1	0	0	31.5	29.8
Feeds	0.8	1	0	1.2	0.4	0.2	0	5.8
Abandoned[a]	1.9	3.5	25.1	25.3	26.9	31.0	15.2	14.2

The table is a modified version of data presented by Qiu *et al.* (2013).

[a] Disposed of through discharge to surface waters or informal land fill type community "dumping piles." This may, for example, be straw collected and stacked and then never used or manure collected from storage pits and left in "dumping piles."

Table 3.5   Manure handling practices in China: field survey 2006 and 2007 (Wang *et al.* 2010).

Types	Samples	Collection (%)			Fate of the collected manure (%)		
		Pig	Cattle	Poultry	Manure application	Anaerobic digestion	Making organic fertilizers
CAFOs	60	32	20	66	76	14	10
Backyard system	2000	90	50	20	—	—	—

Table 3.6   The fate of crop residues in China in 2009.

Fate of organic waste	End use of crop residues in 2009 (%)	Dry matter biomass (10^6 Mg DM)
Animal feed	30.7	235
Energy	18.7	144
Land application	14.8	113
Production of edible mushroom	2.1	16
Industry feedstock	2.4	18
Abandoned	31.3	240

Data from the Ministry of Agriculture of China (2010).

Table 3.5 shows the management of livestock manure in 2006 and 2007 on 60 CAFOs and 2000 backyard production systems situated in 20 counties in 10 provinces, retrieved from a study by Wang *et al.* (2010). A very little amount of manure was collected in CAFOs pig farms, the amount not collected was discharged to surface waters. In 2006 and 2007 most of the manure was applied on land, composted or used for biogas production. In 2008, 72.4% of the total manure produced on intensive livestock farms in China (CAFOs production system) was treated and recycled to farmland using traditional composting (stockpiling, 49.9%), industrial composting (8.7%), and biogas fermentation (13.7%) (Yang *et al.*, 2010).

In another survey carried out by the Chinese Ministry of Agriculture, it was estimated that a total of 820 Tg of crop residue (15% DM) was produced in 2009, as shown in Table 3.6 (Ministry of Agriculture of China, 2010). This was an increase from the 760 Tg DM crop residue produced in 2006 (Gao *et al.*, 2009). In seven western provinces, Gansu, Shaanxi, Inner Mongolia, Ningxia, Tibet, Qinghai, and Xinjiang, 88.2 Tg DM crop residues were produced in 2009 (Bao *et al.*, 2014). In these provinces the straw was mainly used as fuel (33.8%), feed (29.3%), industrial materials (5.2%), as substrate for soil culture (1.8%), or returned to directly to the field (13.5%). In addition, the amount of straw burnt was 11.1% (Bao *et al.*, 2014). About 5.3% of the straw collected was never used.

3.2.3   Denmark

In Denmark the largest source of agricultural waste biomass produced in 2010 was harvested straw from cereals followed by animal manure, grass, and oilseed rape production (Figure 3.4). In total, the amount of agricultural residues produced per year was 10.3 Tg of DM. In this calculation the amount of straw being used as strewing material was not included, because it contributed to the amount of manure produced. Most dairy and fattening pig houses are equipped with slurry removal systems, and in beef housing deep litter is frequently used. In old production systems liquid manure and farmyard manure (FYM) are produced. Manure slurry is produced in poultry houses with layers, and solid manure in poultry houses with broilers.

The beet tops from beet harvesting accounted, in 2010, for about 0.2 Tg of biomass (DM), bearing in mind that 36% of the beet harvested consists of plant leaves (Triolo *et al.*, 2014).

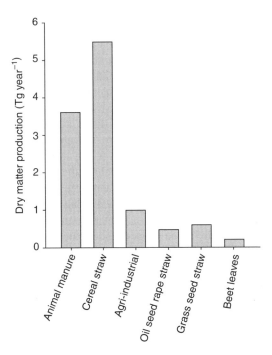

Figure 3.4   The most important agricultural waste products available for collection and harvesting in 2010 (Jørgensen et al., 2013) and agro-industrial waste from dairies, the sugar industry, potato processing, oil-mill, the margarine industry, and abattoirs in Denmark (Petersen, Kaysen, and Hansen, 2014).

Agro-industrial organic waste from food and feed production is estimated to about 6 Tg (wet) biomass (Petersen, Kaysen, and Hansen, 2014). The dairy and potato industry contributes about 4.5 Tg (wet), of which most of the dairy effluent is used for feed substitution in the livestock sector and the very dilute waste water from the potato industry is used as a fertilizer. Given as DM, the amount of Danish agro-industrial waste is about 1 Tg.

For comparison, the annual amount of household, sewage, and industrial organic waste produced is about 6.1 Tg fresh weight per year (Jørgensen *et al.*, 2013), which is nearly 1/10th of the amount of agricultural organic waste produced.

In Denmark, manure biomass is considered a valuable fertilizer and is mostly applied to the soil. This was not the case before 1980, when manure was considered a waste. The poor management of manure caused massive water and air pollution with nitrate (NO_3^{-1}) and NH_3. As a consequence, regulations focusing on reducing pollution of the environment are now forcing farmers to use manure efficiently through increased manure storage capacity and better field application. As a result, nitrogen fertilizer consumption has been reduced by about 50% (Sommer, 2013) from 1980 to 2010, without a reduction in crop DM production. In 2010, about 7% of the manure was used for biogas production (Figure 3.5) and the political goal is to increase this use to 50% of the manure DM by 2020, although reaching this target by 2020 seems unrealistic (Danish Energy Agency, 2014). The current use of manure for biogas is constrained by the low or negative profitability of such operations. There is a need to apply more energy-rich materials to enhance biogas production, yet, available easily degradable materials (i.e., slaughterhouse waste, food industry waste, etc.) are already used for biogas production and thus are no longer in surplus and cheaply available. There are, therefore, considerations of using other plant waste products for biogas, such as straw and beet leaves.

Of the plant residues produced, about 0.7 Tg of the straw biomass is used for strewing in animal houses and feed (Figure 3.5). Governmental directives have previously encouraged farmers and power plants to use the straw for power and heat production – the low water content of straw making the biomass an efficient energy producer – but due to the high alkaline salt content large corrosion problems in incinerators have emerged. Therefore, a shift toward using wood chips and pellets for bioenergy in power plants has evolved. The Danish energy company DONG with support from the government has developed a pilot bioethanol production unit using cereal straw as feedstock. The state-of-the-art technology has been demonstrated at pilot scale, but the company has not yet established a commercial production unit due to the high production costs.

The major part of the beet leaves are being incorporated into the soil after harvest of the roots. These leaves provide an excellent source of organic material in fermentation processes producing biogas or bioethanol.

It has been noted that there is unused biomass available for bio-refinery processing. In addition, new harvest technologies and new crop varieties can increase the amount of available plant residues. New harvesting machines which collect the leaves and chaff have the potential to increase DM collection by 12–20% and rib harvesting could provide 25% more straw DM than the existing harvesting

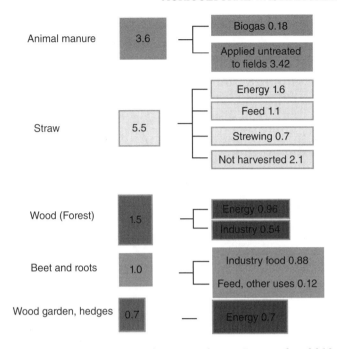

Figure 3.5   Use of animal manure and crop residues in Denmark in 2010–2012 (Tg DM year[1]). Manure is co-digested with organic waste from slaughter houses, and so on, and the digestate produced is applied to fields. The data have been calculated using data from Statistics Denmark (2012), Jørgensen et al. (2013), and Poulsen (2013).

technologies (Jørgensen *et al.*, 2013). Further, it is shown that new cereal varieties may increase straw biomass production by 57% without a reduction in grain production. Options for harvesting cover crops/catch crop for use in the waste streams may be available and therefore there are big opportunities for increasing biomass harvesting and providing bio-refinery manufacturers with biomass from the existing farming systems.

3.2.4   USA

The US produces one-quarter of the global economic output through a complex, market-driven economic system. In the US economy, plants and livestock in industrial agriculture form an effective system for using and recycling agricultural waste biomass. Although agriculture is just about 1% of the US GDP, the US is one of the largest agriculture producers in the world (FAO, 2013). In addition, the 9.1×10^6 km^2 that make up the US territory range from subtropical to temperate climates, which creates opportunities for highly specialized industrial agricultural production.

Figure 3.6 depicts the production and use of the most important agricultural waste streams in the US, based on data from the 2012 Census of Agriculture and the

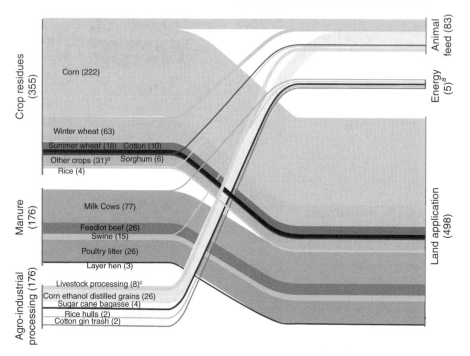

Figure 3.6   Agricultural waste generation and end use (in Tg DM). Adapted from the 2012 Census of Agriculture and the US Billion-Ton biomass report by the US Department of Energy (United States Department of Agriculture, 1992, 2014; MacDonald et al., 2009; ASAE, 2010; Nelson, 2011; United States Department of Energy, 2011; BSR, 2013; Kosseva and Webb, 2013; Swisher, 2013; Busby, Wells, and Hyman, 2014). [a]Energy includes anaerobic: digestion (0.4 Tg DM) and combustion (4.6 Tg DM). [b]Other crops include: oats (1.6 Tg DM), barley (6 Tg DM), soybeans (11 Tg DM), sugar beet (2 Tg DM), sugar cane (1 Tg DM), potatoes (5 Tg DM), orchard, and vineyard trimmings (5 Tg DM). [c]Livestock processing are rendered products from animal meat processing turned into components of animal feed rations.

Billion-Ton biomass report by the US Department of Energy (USDA, 2014; USDOE, 2011). The amount of agricultural waste produced in the US was estimated to be 584 Tg DM, with 91% (531 Tg) composed of crop residues and collected manure. Crop residues accumulated 355 Tg DM, of which 91% (323 Tg DM) were returned back into the soil for erosion prevention and nutrient recycling. As the largest crop in the US, the crop residues of maize (corn grain only, excluding silage maize) account for more than half of the crop residues (222 Tg DM, 63%), while winter and summer wheat make up the second largest fraction (82 Tg DM, 23%). Most of the crop residues were left in the field (some incorporated into the soil) for erosion protection and soil nutrient recycling. The amount of crop residues harvested depends on many factors including the type of crop produced, tillage practices (tillage, no tillage), climate, soil condition, and the cost-benefit of harvesting versus leaving the residue on the soil as nutrient supply for future crops (Andrews, 2006; USDOE, 2011). For example, sugar cane field residues which previously were burned before harvest are now 100% harvestable biomass, while soybean residues (stems, leafs, pods) are all left in the field to provide protection against the risk of soil erosion. The average amount of harvestable crop residue is estimated to be 10% of the total residue DM. The current emphasis on energy production derived from biomass has increased interest in sustainable harvest of crop residues and resulted in research initiatives directed toward increasing the proportion of crop residues harvested as biomass feed-stock (Andrews, 2006; USDOE, 2011).

Animal manure production was estimated at 176 Tg DM and almost all (175 Tg DM) was applied directly to cropland for nutrient recycling, with less than 1.0% directed to energy production through anaerobic digestion or combustion. The estimate for manure includes collectable manure, but not manure deposited during grazing on pastures or rangelands. Manure from dairy cattle excluding bedding material accounted for 44% (77 Tg DM) of manure residues – a conservative estimate given that replacement heifers, calves, and dry cows were excluded, and just a fraction of the manure generated by farms with less than 20 milking cows was included. Feedlot beef cattle manure accounted for 15% (26 Tg DM), and 31% (55 Tg DM) is contributed by poultry litter, which included the bedding material and excluded close to 2.8 Tg DM manure from laying hens.

The most abundant waste biomass resulting from industrial processing of agri-cultural products include corn ethanol distiller grains, sugar cane bagasse, cotton gin trash, potato waste, and animal processing residues, which together accounted for 8.3% (48 Tg DM) of the total agricultural waste. A large fraction of the industrial biomass waste was recycled as animal feed and bedding, or burned for energy pro-duction, as is common practice with sugar cane bagasse. The main by-products of processing milk into dairy products such as cheese, butter, and cream, are whey and waste water. Whey was not included as agricultural waste because it is generally used as a feedstock for other industrial products, and is only discarded in small amounts through wastewater (Danalewich et al., 1998; Milani, Nutter, and Thoma, 2011; Aguirre-Villegas et al., 2012). Although dairy processing wastewater has high biological oxygen demand, its total solids concentration is in general very low; there-fore, it is not considered a significant component of agricultural residues (Danalewich et al., 1998; Milani, Nutter, and Thoma, 2011).

At the other end of food production, an estimated 60 Tg (fresh weight) of bio-mass resulted from food waste at the manufacturing and consumer level, and close to 30% was recycled into the agricultural production system through animal feed com-ponents and land application to crops (BSR, 2013; Busby, Wells, and Hyman, 2014).

The industrialization of US agriculture – driven by optimization of production and economic efficiency – has resulted in a reduction in the number of agricultural production operations, with smaller family farms disappearing and large consolidated farms increasing in size. This in turn has increased the concentration of agricultural waste generation streams that are greater than the capacity of near available land to receive it, creating nutrient imbalances with a high risk of water and air pollution. Although the US production system as a whole is very efficient in recycling agricultural waste, regional imbalances in agricultural production, and waste stream generation are an important geographic component of the system (Kellogg *et al.*, 2000; MacDonald, Korb, and Hoppe, 2013). A good example of the interconnectedness and efficiency of the US agricultural production system in recycling biomass is illustrated by recent changes in the production system brought about by the emphasis on renewable energy. The US government established economic incentives for the production of ethanol to be used in liquid fuels, which resulted in an increased maize harvest – the intended goal of placing the incentives in the first place – but also in a reduced availability of maize for animal feed, which in turn was partially offset by the incorporation of distillers grains and other residues from ethanol production into animal feed rations (Mathews and McConell, 2009; Beckman, Borchers, and Jones, 2013).

3.3   Biorefinery Processing of Agricultural Waste Products

The amounts of biomass waste and crop residues produced worldwide are considerable (Table 3.3), and the demands for biomass to be used for energy production, biomaterials, and biochemicals are also increasing. Most of the chemical products and fuels are currently derived from finite fossil sources, the mining and use of which causes increasing pollution (shale gas, tar sand, etc.) and contribute to climate change. Therefore, there is a need to substitute fossil energy sources with renewable energy, and biomass is one of the few readily available alternatives. We refer to biorefinery as a process that converts biomass into bio-based chemicals and energy carriers as well as direct energy and feed production. Existing biorefineries producing building materials, pulp, paper, and so on, are being challenged by emerging biorefineries producing biofuels, bioenergy, biochemicals, bioplastics, and so on. Here we present the most common bioenergy and biochemical products, and relate these to biomass characteristics.

3.3.1   Physiochemical Properties and Organic Composition of Agricultural Waste and Residue

Knowing the composition of biomass is particularly important in the development of processes for production of energy and valuable chemicals, as well as choosing the best use of the biomass in a biorefinery concept. While water content is the most important property in choosing between a thermal or biochemical process (e.g., biogas, bioethanol), the organic composition is the most important parameter in the biochemical process and in the physical and chemical fractionation of the biomass. The efficiency of the microbial transformation of organic carbon in biochemical processes is related to the biodegradability of organic substances, and for chemical and physical processes, fiber characteristics, and concentration of specific organic

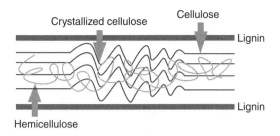

Figure 3.7 Simplified lignocellulose matrix composed of three biopolymers. (Copyright Jin Mi Triolo).

components are the most important factors determining the outcome of the production process.

The Main organic components in agricultural residues have a rigid structured plant cell wall (Figure 3.7) which is called "lignocellulose." Lignocellulose concentration in plant agricultural residues typically composes 80–90% of total organic matter, and represents the most abundant renewable carbon source (Figure 3.7). Straw residue is a representative of the category of lignocellulosic residual biomass (83.8% of organic matter; Triolo *et al.*, 2011). Lignocellulose comprises three main biopolymers, hemicellulose, cellulose, and lignin, which are strongly intermeshed in the lignocellulosic matrix. Within these three biopolymers, cellulose is most abundant and its hydrolytic degradability is limited by lignin and hemicellulose. While cellulose and hemicellulose (hollocelluloses) are C6 and C5 carbohydrates, lignin is an insoluble polymer containing an aromatic phenolic group. The degradability of biomass declines with the concentration of lignin or lignocellulose and the presence of lignin is apparently the most important parameter affecting energy conversion (Monlau *et al.*, 2013).

The high content of lignocellulose in animal manure (Figure 3.8) is due to the disintegrated cell wall being accumulated during animal digestion. Consequently, biodegradability of the manure varies from 0.38 to 0.93% depending on animal diet (Triolo *et al.*, 2011). Cattle manure has a lower energy potential and digestibility than pig slurry, with its higher lignin and lignocellulose content limiting bioenergy production.

3.3.2 Bioenergy Production

Currently biomass contributes approximately 7–10% (or 45±10 EJ) to the world's energy supply (Krausmann *et al.*, 2008). In industrialized countries, biomass contributes 9–14% to the total energy supply, while biomass contribution is much higher in developing countries, being one-fifth to one-third (Khan *et al.*, 2009). Notably, much of the biomass in developing countries is agricultural waste products used for cooking in wood-stoves.

The energy outcome of combustion is high from biomass with a high lignin content due to the low content of oxygen in lignin, thus, biomass with a high lignin

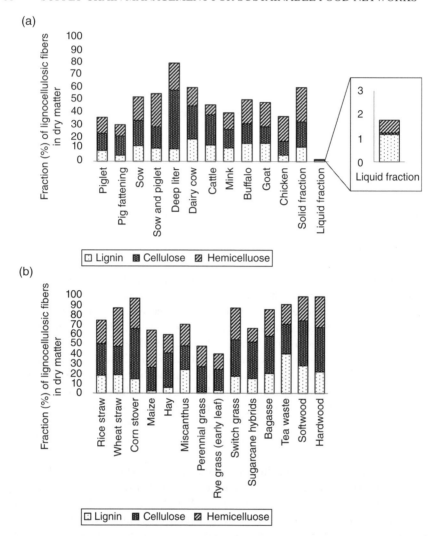

Figure 3.8   The amount of lignin, cellulose, and hemicellulose in animal manure (a) and crop residues (b) (Demirbas, 1996; Lemus et al., 2002; Saxena, Adhikari, and Goyal, 2009; Abbasi and Abbasi, 2010; Saidur et al., 2011; Triolo et al., 2011, 2013; Sommer et al., 2013; Cu et al., 2015).

content has a higher calorific value than a biomass with a low content of lignin (Jørgensen and Jensen, 2009; Thygesen and Johnsen, 2012). Combustion is only efficient if the DM content is higher than 45% (Al Seadi *et al.*, 2008; Sommer *et al.*, 2013). Consequently the most useful biomasses for energy production through combustion are wood chips or pellets, straw, bagasse, and corn stalks. Solid manure produced by separation of animal slurry or by collection of excreta may be used for combustion; however, because of the high water content of separated manure fiber (above 60% water of wet weight), the resulting net calorific value is below 4 MJ kg^{-1}, and the combustion process does not produce much energy (Westborg, Johansen, and Christensen, 2010).

Biomass with high water content can be used for the production of energy carriers such as biogas, bioethanol, and biodiesel through biochemical conversion, chemical treatment, or mechanical extraction. Biogas is produced through anaerobic digestion of agricultural residues with animal waste being the most commonly used feedstock (Triolo et al., 2013; Thygesen, Triolo, and Sommer, 2014). Manure provides macro- and micronutrients for microbial growth, has a strong buffer capacity which contribute to maintaining neutral pH, and is a good diluter of biomass toxicity (Fang et al., 2011). Biohydrogen can be produced by dark fermentation where biohydrogen is produced instead of biomethane by manipulating the operational conditions, that is, inhibiting methane (CH_4) formation, by manipulation of pH and retention time (Laureano-Perez et al., 2005).

Liquid fuel production is roughly divided into biodiesel production and bioethanol production. At present, the main raw materials for biodiesel production are lipid-rich substances, that is, vegetable oils and fat from animals, while for bioethanol production carbohydrate-rich raw materials such as sugar cane and maize are the main feedstock. To avoid the use of biomass that could have been used as food and feed, there is growing ambition to produce bioethanol using agricultural residues. The use of plant residues for bioethanol production is challenging as it requires pretreatment using, for example, acid, thermal steam, and enzymes to break down the lignocellulose complex, and enhance the availability of cellulose to the fungi.

Current energy production based on residual biomass is low and could be much higher, considering the large amounts of waste biomass being produced. One barrier is that bioenergy production is too costly compared with fossil energy sources. Bioenergy is therefore often the last produced in the chain of use of residual biomasses, where high value products are typically given the first priority.

3.3.3   Bio-based Chemical Production

Biorefinery platforms are developed for the use of biomass for manufacturing protein products for animal feed, lactic acid for plastic production, and fiber for a variety of end uses (O'Keeffe et al., 2011; Ecker et al., 2012). Wood, grass, corn, or organic waste are possible feedstocks for conversion to value added products. The manufacturing process is related to the raw materials and their composition, as indicated by the categories of production, that is, lignocellulosic feedstock biorefin- ery, whole-crop biorefinery, sugar platform biorefinery, or green biorefinery (GBR) (Ecker et al., 2012).

GBR is an alternative option for using agricultural waste residues (O'Keeffe et al., 2011; Triolo et al., 2014). The GBR concept accounts for the biomass management chain from harvest, one or a few times during the growth period, to storage and continuous use of feedstock manufacturing throughout the year. During transport and storage, part of the green biomass will inevitably be transformed, new components be produced, and valuable components may be lost. Silage, which has been developed for feed storage of roughages for feeding cattle, is an option that is often chosen as an efficient method for storing and also treating the green biomass.

Lactic acid and glycerin accounted for two-thirds of the bio-based chemical market in 2011, reflecting that lactic acid is needed for various applications within the food, chemical, cosmetic, and pharmaceutical industries (Abdel-Rahman, Tashiro, and Sonomoto, 2013). The world market demand for lactic acid has been growing steadily and rapidly by 5–8% each year since 2008 and was estimated to reach 300 000 t in 2013 (Anonymous, 2014; Xu and Xu, 2014). Most lactic acid is produced through fermentation of biomass using sugar or corn stovers as raw material. Alternatives to these high value products are, for example, glycerine, which is a main by-product from biodiesel production, which traditionally has been used in the pharmaceutical, cosmetics, and food industries. As a residue from biodiesel production, glycerine has been used for animal feed, a carbon source for fermentation, polymers, surfactants, and so on (Vicente *et al.*, 2008). Worldwide biodiesel production is forecast to reach 140 billion l by 2016 and about 23 billion l of crude glycerol is estimated to be produced (Fan, Burton, and Zhou, 2010).

3.4   Environmental and Land Use Issues

The previous sections of this chapter have shown that current food supply chains generate considerable amounts of potentially useful agricultural residues, and have provided insights on how these could be used more efficiently as a feedstock for energy, nutrient, material, and/or chemical production. Yet, there is a need for management strategies for these resources to enhance the overall environmental performance of current food supply chains.

Answering this question involves the notion of comparison: any assessment of the environmental performance of a given system must, indeed, be comparative, that is, the implications of a given residue management chain lie in the changes it induces compared with a reference situation (Hamelin and Wenzel, 2011).

Manure, crop, and agro-industrial residues contain, among others, carbon, nitrogen, phosphorus, and even heavy metals, which end up in the environment in different forms, depending on the management performed. Table 3.7 summarizes the key environmental impacts involved with regards to agricultural residue management, where the above-mentioned flows are affected (as well as other important flows). Keeping track of changes in these flows throughout the whole residue supply chain is one way to estimate the related emission flows. For example, collecting CH_4 from manure through anaerobic digestion (biogas) and subsequently burning it (for energy) involves the manure's easily degradable carbon being mostly emitted as carbon dioxide (CO_2), besides contributing to the production of renewable energy (thereby preventing fossil fuels being used). In a more conventional system where manure is simply stored with minimal cover until land application, part of the carbon is emitted as CH_4 to the atmosphere, without any energy benefits. Yet, as CH_4 has a global warming potential 34 times higher than that of CO_2 (from a 100 years perspective;

Table 3.7   Agricultural waste management contributes to environmental pollution at local (farm, village), regional (province, country, part of continent), and global scale, depending on the substances involved.

Substances	Scale of environmental impact			References
	Local	Regional	Global	
Odor, H_2S	+	—	—	Schiffman and Williams (2005)
Nitrate	+	+	—	Xiong et al. (2008) and Galloway et al. (2008)
Phosphorus	+	+	—	Schröder et al. (2011)
Ammonia	+	+	—	Sutton et al. (2011)
Greenhouse gases			+	Davidson (2009) and Galloway et al. (2008)
Heavy metals	+	+	—	Nicholson et al. (1999)
Pathogens	+	+	—	Albihn and Vinnerås (2007)

IPCC, 2013), having manure's degradable carbon emitted as CO_2 (instead of CH_4) and avoiding fossil fuels being used appear to be a much more sustainable strategy from a global warming perspective. The following sections detail such considerations, with focus on crop residues and manure management.

3.4.1   Manure Management

The prevailing manure management continuum, in the most intensive production systems, may be summarized as four main system stages, at which an intervention can be applied: (i) feed and feeding systems; (ii) housing systems including in-house manure management and storage; (iii) outdoor manure management and storage; and (iv) field application of manure (Hamelin and Wenzel, 2011).

In the feed system, the different crop (e.g., wheat, barley, soy, rape) and non-crop ingredients (e.g., enzymes, mineral supplements) are produced (stage i) and fed to the animals (stage ii). The portion of the consumed feed not respired, evaporated, or retained by the animals will end up as excreta (urine or feces). Any change in the feed system (e.g., change of any ingredient) would thus affect the composition of the excreted manure, and thus the manure potential for subsequent emission. Except for grazing or specific production systems (e.g., free range, organic production), urine and feces are generally excreted inside animal houses, and are eventually mixed with other inputs (e.g., water from spillage and/or washing, straw, wood chips, etc.). Depending on the specific management practices, the manure can remain in-house for varying periods of time: that is, from a very short period (e.g., less than a day) to being stored during the whole animal production duration (e.g., more than a month). In cases where manure is not stored in-house for the whole production duration, it is transferred from the animal house to an outdoor storage facility, where it can be stored (stage iii) until its use as a fertilizer (stage iv). All system stages are characterized by emissions to the environment, and these, in turn, induce changes in the manure composition. As manure composition is the basis for any emissions flows from the manure system, a changed manure composition at one stage of the continuum will in turn trigger changes in the emission flows occurring at other stages of the continuum.

It is commonly accepted that the reference manure management system, that is, what would happen "by default," is a conventional management system where manure is stored (in-house and outdoor; stages ii and iii) in a way that respect the minimal legislative requirements and then applied on land as a fertilizer when appropriate, without any prior treatment (Hamelin and Wenzel, 2011; Tonini *et al.*, 2012).

The key environmental issues with regards to conventional manure management relate to air emissions [NH_3, nitrous oxide (N_2O) and CH_4], nutrient losses (phosphorus and NO_3^{-1}) as well as heavy metal losses to soil and water, and these are detailed below. Other important issues include odor and pathogens.

3.4.1.1   Ammonia, Nitrous Oxide, and Methane

Livestock production contributes 70–80% of the anthropogenic NH_3 emission in Europe and 55% of global NH_3 emission (Bouwman, Boumans, and Batjes, 2002). In Europe the emission represents a loss of about 30% of the total nitrogen in animal manure. In the atmosphere, NH_3 readily combines with sulfate (SO_4^{2-}) and may combine with NO_3^- to form particulates containing ammonium (NH_4^+). Particulate NH_4^+, and to a lesser extent NH_3, may be transported over long distances. Close to the source, NH_3 gas is deposited rapidly on vegetation or soil, and up to 50% of the NH_3 emitted will be deposited within a radius of 2 km from the source (Sommer *et al.*, 2009a). Deposition of NH_3 or particulate NH_4^+ on land or in water may cause acidification and eutrophication of ecosystems. NH_3 emission plays a role in the formation of particulate matter <2.5 μm (PM 2.5) and of particulate matter <10 μm (PM 10), which are airborne micro-particulates that can be a health hazard (Erisman and Schaap, 2004). Ammonia in animal houses is a hazard to the health of farm workers and animals being housed, as it causes lung and respiratory diseases.

Livestock farming systems are major sources of N_2O and CH_4, which contribute considerably to the so-called greenhouse effect. These two gases have a large capacity to absorb the radiation that is reflected from the soil surface; notably, the global warming potential (from a 100 years perspective) of N_2O is 298 times that of CO_2 (IPCC, 2013).

Ammonia emission is favored by high slurry pH and temperature, and CH_4 emission by anaerobic conditions and a high temperature. The formation of N_2O is particularly favored by partial or transient anaerobic conditions. Thus, any management involving a change in manure pH, temperature, and oxygen status is likely to influence the emission flows of NH_3, CH_4 and N_2O. Similarly, measures such as using tightly covered storage would prevent NH_3 emission during the storage stage, low temperatures will reduce both NH_3 and CH_4 emissions, and N_2O emission may be reduced by avoiding low oxygen conditions in the field.

3.4.1.2 Pollution of Aquifers with Phosphorus and Nitrate

Surface- and groundwater may be polluted by leaching and runoff of manure nutrients applied to fields. These diffuse sources of plant nutrients give rise to an increased production of algae and water plants and when these decay the oxygen is used and eventually depleted, leading to fish mortality. In countries where groundwater is used without purifying treatment, the increase in NO_3 concentration requires water purification treatment for the provision of safe drinking water supplies. The direct discharges of liquid manure to receiving waters cause pollution or eutrophication of lakes, rivers, and oceans. In most European countries, direct discharge of animal manure to recipient waters is not allowed.

In the Baltic Sea region, about 90% of the 57 000 t of nitrogen annually lost to rivers originates from agriculture. Most nitrogen is leached through drains and aquifers and from there to the recipient waters. Surface runoff is the mechanism of transport for much of the phosphorus from agriculture. For example, Denmark discharges 1800 t of phosphorus to rivers and oceans, and 45% originates from its high density agriculture (Christensen, Hansen, and Ærtebjerg, 2004). Losses of plant nutrients from manure through leaching and erosion can be reduced through better management of the manure in the field and through removing the excess nitrogen and phosphorus from the manure with separation technologies (Hjorth *et al.*, 2010).

3.4.1.3 Heavy Metals

Contaminants such as heavy metals may be present in livestock feeds at much higher concentration than plants, because these heavy metal ions may be added to certain feeds as supplements for health and welfare reasons, or as growth promoters. A proportion of the metals in livestock manures, as well as those excreted directly onto grazing land, are recycled through the agricultural system in animal feeds grown and fed on-farm (Nicholson *et al.*, 1999). Copper (Cu) is added to the diet for pig growth in many intensive production systems as a cost-effective method of enhancing performance and is thought to act as an antibacterial agent in the gut. Zinc (Zn) is also used in diets for weaner pigs for the control of post-weaning scours. In poultry production both Zn and Cu are required in trace amounts. Other heavy metals may be present in livestock diets as a result of contamination of mineral supplements (e.g., limestone added to the feed for laying hens may contain relatively high levels of cadmium). For all livestock, the majority of heavy metals consumed in feed and drinking water are excreted in the feces or urine. In addition to heavy metals being ingested, sources of heavy metals (e.g. Zn) in animal manure may also be from bedding materials (e.g., straw), corrosion of the galvanized metal used to construct livestock housing, and the licking and biting of metal housing components . Footbaths containing Cu or Zn may be used as hoof disinfectants for sheep and cattle, and these may be disposed of into manure stores thus contributing to the heavy metal content of manures spread on land. A survey of manures collected from commercial farms in England and Wales in the mid-1990s (Nicholson *et al.*, 1999) found the highest concentrations of Zn and Cu in pig slurry and laying hen manure, reflecting higher levels of dietary supplementation for these types of livestock.

3.4.1.4 Pathogens

In the control of health risks from manure handling several pathways of pathogen transfer have to been identified. Transfer of zoonotic pathogens to farm staff and neighbors is a well-known problem, which includes direct transfer and transfer due to contamination of water and food crops that affects both humans and animals. Pathogens may likewise spread among and between livestock productions and therefore the risks of diseases spreading, such as foot and mouth disease, salmonella, and so on, should be carefully considered when establishing new manure handling methods (Albihn and Vinnerås, 2007).

Contamination of water is a major concern, especially if surface water is used untreated within the local community. The water can be contaminated through direct discharge of manure to waterways, for example, from pretreatment ponds, a common procedure in Asia (today). Pollution of shallow drinking water wells due to percolation of pathogens to groundwater, usually via cracks and fissures (macro-pores), is a considerable risk to human and animal health as livestock manure contains numerous pathogens (bacteria, viruses, parasites). Some of these are zoonotic, that is, they may be transmitted to humans where they may cause systemic or local infections. These pathogens cause an estimated 17 million cases of typhoid fever per year with approximately 600 000 deaths (World Health Organization, 2000). Findings on parasitic diseases including food-borne trematode zoonoses and cysticercosis have highlighted the risks of disease transmission through animal manure and human excreta (De et al., 2003).

Discharge of animal manure to rivers may pose a much greater risk of pathogen transfer than careful recycling on agricultural land with appropriate precautions during spreading. Highly contagious and pathogenic diseases, such as foot and mouth disease, swine fever, and Aujeszky's disease may spread with animal effluent through waterways. The health risk associated with the ingestion of field crops that have been treated with manure will depend on the extent and method of manure application to fields. The risk is particularly high if effluents are applied to the foliage of a growing crop where the leaves are consumed; such practices are not recommended. An example of the risk of diseases spreading with contaminated water was the VTEC (verotoxin-producing Escherichia coli) outbreak in Sweden in 2005. This was caused by surface runoff of manure to a canal from which, several kilometers downstream, water was used for irrigation of lettuce (Vännerås et al. in Petersen et al., 2007). In countries with direct discharge of animal manure to rivers there are indications that outbreaks of foot and mouth disease spread to farms downstream and it may be assumed that the spread is waterborne.

From a whole-farm perspective, the recycle of waste back into food production should be as short as possible to minimize environmental impacts and ensure high nutrient use efficiency. If manure from livestock is used on the same farm, the epidemiological risk for the animals does not increase, which is why animal manure is exempted from the strict EU rules of pretreating animal by-products used as fertilizers. A special situation arises if central storage units for liquid manure are used by several different farmers. Under such conditions the closed cycle of the farm no longer exists and the epidemiological risks are increased by the number of farms and animals involved. Therefore, special hygienic precautions must be taken in communal or co-digestion biogas plants where several livestock farms transport manure to be treated at the same site.

3.4.2 Crop Residues Management

Crop residues contain, among others, carbon, nitrogen, and phosphorus that can be lost to the environment in various ways. This section discusses these flows, throughout the residues management continuum, but focuses on the harvestable fraction of the residues only (i.e., which is referred to as "agricultural residue" that can be managed in various ways). Emissions related to, for example, crop fertilization, crop foliage (e.g., NH_3 emission), or decay of non-harvestable residues are not within the scope of this chapter and are thus not discussed here.

The management chain involved for crop residues is as follows:

- *Residues harvest:* The harvestable portion of residues may or may not be harvested. If not harvested, it may be left on the field to decay, burnt on-site (where legal), or incorporated into the soil.

- *Storage (when harvested):* The harvested residues can be stored as a stack, baled, or as silage.

- *Use (when harvested):* use of the residues as feed, bedding, energy (direct combustion, fermentation, etc.), biomaterials, biochemicals, and so on.

From an environmental perspective, burning the crop residues on the field is without doubt the least desirable method. First, it returns all the biomass carbon to the atmosphere in the form of CO_2, but some CH_4 and monoxide emissions may also be emitted (Hamelin, 2013) without producing any useful service (e.g., usable energy). The nitrogen is instead mostly lost to the atmosphere (nitrogen oxides, nitrous oxide), and so is the sulfur (sulfur dioxide). The latter notably contributes to the acidification of ecosystems, N_2O to global warming, and nitrogen oxides to photochemical ozone formation (smog).

The incorporation of the residues contributes to return the carbon and nutrients to the soil. Based on a complex dynamic, part of the carbon will end up in the soil organic pool (so called humified carbon, or slowly degradable carbon), the rest being essentially oxidized as CO_2 to the atmosphere (Hamelin, 2013). Part of the nutrients may end up becoming mineralized and available for plant uptake, and the rest may be lost to water compartments.

Losses will also occur during the storage phase, depending on the storage method and the management factors (e.g., exposure to rain, etc.). For example, when ensiling various biomass sources 10% of VS may be lost and 5% of the VS may be transformed to CH_4 (Weissbach, 2009). It has been estimated that 28.4% of VS were lost when ensiling sugar beet for nine months (Triolo *et al.*, 2014). Hafner *et al.* (2013) estimated significant emissions of alcohol (10% of VS) during ensiling of maize, which can contribute to ozone formation in the troposphere.

As for the use phase, most of the environmental impacts are through interactions with adjoining systems. For example, using residues with high nutritional value for feed avoids cropping of protein (e.g., soy) and carbohydrate (e.g., wheat, maize) crops, and thereby reduces the pressure on land. Such land use changes can lead to important improvement of the GHG balance of a system (Hamelin, 2013) due to the avoided CO_2 releases associated with land conversion. Anaerobic digestion of the residual biomass produces a renewable and storable gas, which avoids the use of so-called marginal energy (e.g., from fossil fuels). Similarly, the biomass from anaerobic digestion, that is, digestate, can be returned to soil, thereby recycling the slowly degradable carbon (i.e., carbon that has not been degraded to CH_4 during the anaerobic digestion process) as well as the plant nutrients (nitrogen, phosphorus, potassium). This means that the corresponding mineral fertilizers do not need to be produced and used (Hamelin, 2013). This latter advantage, that is, nutrient recycling, is typically the strength of anaerobic digestion compared with other energy conversion technologies. Emissions also occur during the processing of the biomass, as shown by GHG emission in the handling chain of vinasses from the bioethanol industry (Souza Moraes *et al.*, 2014).

3.4.3   Land Use Aspects

Just like reducing food waste along the food production chain, using agricultural waste as efficiently as possible is probably a key in addressing the tremendous twenty-first century challenge of feeding and supplying energy to a growing and richer population without destroying the planet.

Particularly, converting these residues to a valuable energy, feed, or material/chemical service reduces the demand for land. As a result of the increased population, diet shifts involving more meat consumption, and bioenergy demands, biomass production would need to roughly double in comparison with today's level (Running, 2012; Foley *et al.*, 2011). For example, recent studies (e.g., FAO, 2008; Nonhebel, 2012) forecast that the demand of cereals, only for food and feed needs (i.e., excluding bioenergy), will increase by about 50% within the next 20 years. Yet, the amount of available land for cultivation is physically limited to the planet boundary [besides being limited by other constraints, e.g. climatic (in particular water), defining the suitability for cultivation], and so is the potential for yield increases (Haberl *et al.*, 2013). To supply the increased biomass demands, an increased conversion of natural land to cultivated land is thus inevitable. According to FAO (2010), the conversion of forest ecosystems, today, is about 13 Mha year^{-1} (this includes the losses from natural causes).

These land use changes are a concern because of the considerable environmental cost they could involve, particularly if they lead to the expansion of crop production in carbon-rich ecosystems, causing the release of carbon that was stored in these ecosystems over long periods of time. In fact, as highlighted in recent studies (Gibbs *et al.*, 2008), biomes such as tropical or temperate forests are those that are likely to be the first converted following an increase in demand for crops, these being the biomass where the frontier between agriculture and nature is already moving.

According to the IPCC Assessment Report 5 (IPCC, 2006), net deforestation in tropical ecosystems today accounts for about 10% of the annual worldwide CO_2 emissions. This consequence is often termed the "C debt," reflecting the number of years of bioenergy (from land-dependent biomass) offsetting fossil fuels that are needed before balancing the amount of carbon released from the conversion of natural ecosystems to agriculture.

Therefore, any innovation allowing decoupling, fully or partly, the (green) energy, food, feed, chemical, and materials demand from the land demand is likely to be a key in addressing the sustainability challenge we are increasingly facing.

3.4.4 Whole Chain Management

The agricultural residues system is, as demonstrated throughout this chapter, a rather complex system to optimize. In fact, any intervention in the management of agricultural residues at any stage of the system may influence environmental impacts from the whole chain, typically from the point of application (of the given management practice) and downstream the chain. In some cases, an intervention in the system may even influence environmental aspects upstream from the point of application, as shown in Hamelin and Wenzel (2011). Therefore, holistic approaches (e.g., life cycle assessments, system analysis with cost assessment, energy, and mass balances) including the whole management chain, the whole spectrum of substance flows affected, the interaction with adjoining systems (e.g., avoided fossil energy) are necessary in order to capture most consequences and accurately quantify the environmental performance of a given management strategy as opposed to a reference situation.

To illustrate this, acidifying manure from in-house storage pits can be used as an example. This technique (i.e., lowering manure pH) is typically applied in order to reduce the NH_3 emission, from animal houses as in this example but also from outdoor storage and/or field. However, while a low pH reduces NH_3 emissions by pushing the equilibrium toward NH_4^+, it also inhibits the CH_4-producing methanogenic bacteria, which results in a reduction of CH_4 emissions as well. Acidification, thus, results in an overall reduction of these emissions from the animal houses when compared to the case without acidification. The same benefits are also seen for the subsequent outdoor storage (Wesnæs, Wenzel, and Petersen, 2009). Moreover, the benefits of acidified slurry comprise reduced NH_3 emission even further downstream, that is, from field application of the manure. However, when emissions from the manure are reduced, the remaining nitrogen content of the manure is increased. The higher nitrogen content of acidified manure combined with conditions favoring N_2O emission, may lead to slightly higher N_2O emissions than they would have been without the manure acidification. Moreover, application of acidified manure to fields is likely to lead to an increased need for application of lime (adjoining system). On the other hand, the increased ammonium nitrogen content of the manure leads to an increased crop availability of nitrogen potentially resulting in higher crop yield and/ or higher mineral fertilizer replacement (adjoining systems), all other parameters remaining equal.

As this example illustrates, only considering the consequences on the emission flows at the point of application (i.e., housing unit) does not capture the benefits of the management technique during outdoor storage and on the field and thereby does not provide a fair assessment of the technique. Similarly, the potential drawbacks occurring at the field stage (increased lime application and potential N_2O increases) would not be reflected by considering only the point of application. Further, accounting for NH_3 only (the targeted substance) would not capture the additional benefit of reducing CH_4 (and potentially N_2O) and would not provide a fair assessment of the full potential of the management technique.

In this example, the interdependence revealed additional benefits. Such interdependences may also reveal trade-offs. For instance, an air cleaning technique using a biofilter/bioscrubber to treat emissions from housing units could be very efficient at reducing NH_3 and odors, but could also contribute to the unintended emission of N_2O generated from the microbial processes in the filter.

Such system thinking in agreement with Chadwick *et al.* (2011) pinpoint that for an efficient and environmentally friendly use of the biomass, the whole management chain from collection to end use should be considered, and that through optimizing the technologies in a chain management approach, the outcome of the whole system can become economically sustainable by taking advantage of all valuable resources in the waste stream.

3.5   Conclusion

The key points of this chapter can be summarized as follows:

- Globally agriculture produces over 6.3 Pg DM year^{-1} organic waste products; this represents about 22% of today's primary energy consumed worldwide. The greatest share of agricultural residues worldwide is found in Asia.

- Today, much of livestock excreta is used as fertilizer and for soil amendment, but in countries such as China where agricultural production is being intensified and industrialized a lot of manure is discharged to rivers, lakes, or oceans. A significant proportion of animal manure is used for biogas production or composting in China. In Denmark about 7% is used in biogas production, whereas a very small proportion of biogas is produced from agricultural wastes in the US.

- Crop residues in the US and Denmark, and to some extent in China, are being incorporated into the soil to avoid erosion and for soil amelioration. In Denmark straw is incinerated for power and heat production and in China plant biomass is used for cooking and heating.

- Other uses of agricultural residual biomass include the production of liquid fuels (bioethanol, biodiesel), biohydrogen, or biorefinery pathways where a variety of protein supplements (e.g., for animal feed), lactic acid (e.g., for plastic production0, solid fuel (e.g., for combined heat and power), and so on are produced.

- The physiochemical properties and chemical composition of the biomass are key determinants for selecting the most appropriate use of the biomass.

- There is a demand for developing new recycling technologies to support the use of agricultural waste products on the specialized farms emerging worldwide. Although technologies have been developed, they are not economically viable and are only being used when regulation demands it or incentives support them. These regulations and incentives which are aimed at reducing pollution will also provide valuable products such as energy carriers.

- A whole-system approach is essential to evaluate the most sustainable strategies for the management of agricultural residues in future food supply chains.

References

Abbasi, T. and Abbasi, S.A., 2010. Biomass energy and the environmental impacts associated with its production and utilization. *Renewable and Sustainable Energy Reviews*, **14**, 919–937.

Abdel-Rahman, M.A., Tashiro, Y., and Sonomoto, K., 2013. Recent advances in lactic acid production by microbial fermentation processes. *Biotechnology Advances*, **31**, 877–902.

Aguirre-Villegas, H.A., Milani, F.X., Kraatz, S., and Reinemann, D.J., 2012. Life cycle impact assessment and allocation methods development for cheese and whey processing. *Transactions of the ASABE*, **55**, 613–627.

Albihn, A. and Vinnerås, B., 2007. Biosecurity and arable use of manure and biowaste – Treatment alternatives. *Livestock Science*, **112**, 232–239.

Al Seadi, T., Rutz, D., Prassl, H., Köttner, M., Finsterwalder, T., Volk, S., and Janssen, R., 2008. *Biogas Handbook*. Esbjerg: University of Southern Denmark.

American Society of Agricultural Engineers (ASAE), 2003. *Manure Production and Characteristics*, Publication No. D384.1 FEB03. Michigan: ASAE.

American Society of Agricultural Engineers (ASAE), 2010. *Manure Production and Characteristics*, Publication No. D384.2 MAR2005 (R2010). Michigan: ASAE.

Andrews, S., 2006. *Crop Residue Removal for Biomass Energy Production: Effects on Soils and Recommendations*, Agronomy Technical Note 19. Greensboro, NC: Soil Quality National Technology Development Team-Natural Resources Conservation Service, United States Department of Agriculture.

Anonymous, 2014. Global and China Lactic Acid and Derivative Industry Report, 2014–2016, Research in China.

Bao, J.C., Yu, J.H., Feng, Z., Chen, B.H., Lei, C., and Yang, J., 2014. Situation of distribution and utilization of crop straw resources in seven western provinces, China. *Chinese Journal of Applied Ecology*, **25**, 181–187.

Beckman, J., Borchers, A., and Jones, C.A., 2013. *Agriculture's Supply and Demand for Energy and Energy Products*, Economic Information Bulletin No. 112. Washington, DC: USDA, Economic Research Service.

Birkmose, T.F., Hjort-Gregersen, K., and Stefanek, K., 2013. *Biomasse til biogasanlæg i Danmark – på kort og langt sigt (Biomass to Biogas Plants in Denmark: on a Long- and Short-term Perspective)*. Aarhus: Agrotech A/S.

Bouwman, A.F., Boumans, L.J.M., and Batjes, N.H., 2002. Estimation of global NH_3 volatilization loss from synthetic fertilizers and animal manure applied to arable lands and grasslands. *Global Biogeochemical Cycles*, **16**, 8-1–8-14.

BP, 2014. BP Statistical Review of World Energy, 63rd edn, bp.com/en/global/corporate/about-bp/energy-economics/statistical-review-of-world-energy.html (accessed June 23, 2015).

Bruun, S., Jensen, L.S., Vu, T.K.V., and Sommer, S.G., 2014. Small-scale household biogas digesters: an option for global warming mitigation or a potential climate bomb? *Renewable and Sustainable Energy Reviews*, **33C**, 736–741.

BSR, 2013. Analysis of U.S. Food Waste Among Food Manufacturers, Retailers, and Wholesalers,http://www.foodwastealliance.org/full-width/fwra-publishes-toolkit-reducing-food-waste/ (accessed June 23, 2015).

Busby, J., Wells, H., and Hyman, J., 2014. *The Estimated Amount, Value, and Calories of Postharvest Food Losses at the Retail and Consumer Levels in the United States*, Economic Information Bulletin Number 121. Washington, DC: USDA Economic Research Service.

Chadwick, D., Sommer, S.G., Thorman, R., Fangueiro, D., Cardenas, L., Amon, B., and Misselbrook, T., 2011. Manure management: implications for greenhouse gas emissions. *Animal Feed Science and Technology*, **166–167**, 514–531.

Christensen, P.B., Hansen, O.S., and Ærtebjerg, G., 2004. Iltsvind (Oxygen Depletion). Hovedlandet, Aarhus: Miljøbiblioteket, p. 132.

Cu, T.T.T., Nguyen, T.X., Triolo, J.M., Pedersen, L., Le, V.D., Le, P.D., and Sommer S.G., 2015. Biogas production from Vietnamese animal manure, plant residues and organic wastes: Influence of biomass composition on the methane yield. *Asian-Australasian Journal of Animal Sciences*, **28**, 280–289.

Danalewich, J.R., Papagiannis, T.G., Belyea, R.L., Tumbleson, M.E., and Raskin, L., 1998. Characterization of dairy waste streams, current treatment practices, and potential for biological nutrient removal. *Water Research*, **32**, 3555–3568.

Danish Energy Agency, 2014. Biogas i Danmark – Status, barrierer og perspektiver (Biogas in Denmark – Status, Barriers and Perspectives). Copenhagen: Danish Energy Agency, http://www.ens.dk/info/publikationer/biogas-danmark-status-barrierer-perspektiver (accessed June 23, 2015).

Davidson, E.A., 2009. The contribution of manure and fertilizer nitrogen to atmospheric nitrous oxide since 1860. *Nature Geoscience*, **2**, 659–662.

De, N.V., Murrell, K.D., Cong, L.D., Cam, P.D., Chau, L.V., Toan, N.D., and Dalsgaard, A., 2003. The foodborne trematode zoonoses of Vietnam. *The Southeast Asian Journal of Tropical Medicine and Public Health*, **34**(Suppl. 1), 12–35.

Demirbas, A., 1996. Calculation of higher heating values of biomass fuels. *Fuel*, **76**, 431–434.

Dornburg, V., van Vuuren, D., van de Ven, G., Langeveld, H., Meeusen, M., Banse, M., van Oorschot, M., Ros, J., van den Born, G.J., Aiking, H., Londo, M., Mozaffarian, H., Verweij, P., Lysen, E., and Faaij, A., 2010. Bioenergy revisited: key factors in global potentials of bioenergy. *Energy & Environmental Science*, **3**, 258–267.

Ecker, J., Schaffenberger, M., Koschuh, W., Mandl, M., Böchzelt, H.G., Schnitzer, H., Harasek, M., and Steinmüller, H., 2012. Green biorefinery upper Austria – Pilot plant operation. *Separation and Purification Technology*, **96**, 237–247.

Erisman, J.W. and Schaap, M., 2004. The need for ammonia abatement with respect to secondary PM reductions in Europe. *Environmental Pollution*, **129**, 159–163.

Fan, X., Burton, R., and Zhou, Y., 2010. Glycerol (byproduct of biodiesel production) as a source for fuels and chemicals e mini review. *The Open Fuels & Energy Science Journal*, **3**, 17–22.

Fang, C., Boe, K., and Angelidaki, I. 2011. Anaerobic co-digestion of by-products from sugar production with cow manure. *Water Research*, **45**, 3473–3480.

FAO, 2008. *The State of Food and Agriculture: Biofuels: Prospects, Risks and Opportunities*. Rome: Food and Agriculture Organization of the United Nations.

FAO, 2010. *Global Forest Resources Assessment 2010* Main Report. FAO Forestry Paper 163. Rome: Food and Agriculture Organization of the United Nations.

FAO, 2013. *FAO Statistical Yearbook. World Food and Agriculture*. Rome: Food and Agriculture Organization of the United Nations.

FAO, 2014a. FAOSTAT. Inputs, Land, http://faostat3.fao.org (accessed July 29, 2014).

FAO, 2014b. FAOSTAT. Production, Crops, http://faostat3.fao.org (accessed May 31, 2014).

FAO, 2014c. FAOSTAT. Production, Live Animals, http://faostat3.fao.org (accessed July 29, 2014).

FAO, 2014d. FAOSTAT. Emissions – Agriculture, Enteric Fermentation, http://faostat3.fao.org (accessed July 29, 2014).

FAO, 2014e. FAOSTAT. Emissions – Agriculture, Manure Left on Pasture, http://faostat3.fao.org (accessed July 29, 2014).

FAO, 2014f. FAOSTAT. Population, Annual Population, http://faostat3.fao.org (accessed July 29, 2014).

Feedipedia, 2014. Animal Feed Resources Information System, http://www.feedipedia.org (accessed June 23, 2015).

Foley, J.A., Ramankutty, N., Brauman, K.A., Cassidy, E.S., Gerber, J.S., Johnston, M., Mueller, N.D., O'Connell, C., Ray, D.K., West, P.C., Balzer, C., Bennett, E.M., Carpenter, S.R., Hill, J., Monfreda, C., Polasky, S., Rockström, J., Sheehan, J., Siebert, S., Tilman, D., and Zaks, D.P.M., 2011. Solutions for a cultivated planet. *Nature*, **478**, 337–342.

Galloway, J.N., Townsend, A.R., Erisman, J.W., Bekunda, M., Cai, Z., Freney, J.R., Martinelli, L.A, Seitzinger, S.P., and Sutton, M.A., 2008. Transformation of the nitrogen cycle: recent trends, questions, and potential solutions. *Science*, **320**, 889–892.

Gao, L., Ma, L., Zhang, W., Wang, F., Ma, W., and Zhang, F., 2009. Estimation of nutrient resource quantity of crop residue and its utilization situation in China. *Transactions of the Chinese Society of Agricultural Engineering*, **25**, 173–179.

Gibbs, H.K., Johnston, M., Foley, J.A., Holloway, T., Monfreda, C., Ramankutty, N., and Zaks, D., 2008. Carbon payback times for crop-based biofuel expansion in the tropics: the effects of changing yield and technology. *Environmental Research Letters*, **3**, 034001.

Grainger, A., 2008. Difficulties in tracking the long-term global trend in tropical torest area. *Proceedings of the National Academy of Sciences*, **105**, 818–823.

Haberl, H., Erb, K.-H., Krausmann, F., Running, S., Searchinger, T.D., and Smith, W.K., 2013. Bioenergy: how much can we expect for 2050? *Environmental Research Letters*, **8**, 031004.

Hafner, S.D., Howard, C., Muck, R.E., Franco, R.B., Montes, F., Green, P.G., Mitloehner, F., Trabue, S.L., and Rotz, C.A., 2013. Emission of volatile organic compounds from silage: compounds, sources, and implications. *Atmospheric Environment*, **77**, 827–839.

Hamelin, L., 2013. *Carbon management and environmental consequences of agricultural biomass in a Danish renewable energy strategy.* PhD thesis. University of Southern Denmark, http://www.ceesa.plan.aau.dk/Publications/PhD+dissertations/ (accessed July 21, 2015).

Hamelin, L., Naroznova, I., and Wenzel, H., 2014. Environmental consequences of different carbon alternatives for increased manure-based biogas. *Applied Energy*, **114**, 774–782.

Hamelin, L. and Wenzel, H., 2011. Methodological aspects of environmental assessment of livestock production by LCA (Life Cycle Assessment). In: H. Döhler, B. Erichh-Menden, E. Grimm, M. Hofman, U. Schultheiss, and S. Wulf, eds. *Emissionen der Tierhaltung. Treibhausgase, Umweltbewertung, Stand der Technik.* Darmstadt: KTBL, pp. 34–53.

Hjorth, M., Christensen, K.V., Christensen, M.L., and Sommer, S.G., 2010. Solid–liquid separation of animal slurry in theory and practice: a review. *Agronomy for Sustainable Environment*, **30**, 153–180.

Intergovernmental Panel on Climate Change (IPCC), 2006. Emissions from livestock and manure management. In: H.S. Eggleston, L. Buendia, K. Miwa, T. Ngara, and K. Tabane, eds. *2006 IPCC Guidelines for National Greenhouse Gas Inventories.* Tokyo: Institute for Global Environmental Strategies, Chapter 10.

Intergovernmental Panel on Climate Change (IPCC), 2013. In: T.F. Stocker, D. Qin, G.-K. Plattner, M. Tignor, S.K. Allen, J. Boschung, A. Nauels, Y. Xia, and P.M. Midgley, eds. *Climate change 2013: the physical science basis. Contribution of Working Group I to the IPCC. Fifth Assessment Report of the Intergovernmental Panel on Climate Change.* Cambridge: Cambridge University Press.

Johnston, A.E., Poulton, P.R., and Coleman, K., 2009. Soil organic matter: its importance in sustainable agriculture and carbon dioxide fluxes. *Advances in Agronomy*, **101**, 1–57.

Jørgensen, U., Elsgaard, L., Sørensen, P., Olsen, P., Vinther, F.P., Kristensen, E.F., Ejernæs, R., Hygaard, B., Krogh, P.H., Bruhn, A., Rasmussen, M.B., Johansen, A., Jensen, S.K., Gylling, M., and Bojesen, M., 2013. Biomasseudnyttelse i Danmark – Potentielle ressourcer og bæredygtighed (Use of Biomass in Denmark – Potential Resources and Sustainability). DCA Rapport 033, Aarhus University, Aarhus.

Jørgensen, K. and Jensen, L.S., 2009. Chemical and biochemical variation in animal manure solids separated using different commercial technologies. *Bioresource Technology*, **100**, 3088–3096.

Kellogg, R.L., Lander, C.H., Moffitt, D.C., and Gollehon, N. (2000) *Manure Nutrients Relative to the Capacity of Cropland and Pastureland to Assimilate Nutrients: Spatial and Temporal Trends for the United States*, Economic Research Service nps00-0579. Washington, DC: United States Department of Agriculture, Natural Resources Conservation Service.

Khan, A.A., Jonga, W.D., Jansens, P.J., and Spliethoff, H., 2009. Biomass combustion in fluidized bed boilers: potential problems and remedies. *Fuel Processing Technology*, **90**, 21–50.

Kosseva, M. and Webb, C., 2013. *Food Industry Wastes: Assessment and Recuperation of Commodities.* San Diego, CA: Academic Press.

Krausmann, F., Erb, K.-H., Gingrich, S., Lauk, C., and Haberl, H., 2008. Global patterns of socioeconomic biomass flows in the year 2000: a comprehensive assessment of supply, consumption and constraints. *Ecological Economics*, **65**, 471–487.

Lal, R., 2005. World crop residues production and implications of its use as a biofuel. *Environment International*, **31**, 575–584.

Laureano-Perez L., Teymouri, F., Alizadeh, H., and Dale, B.E., 2005. Understanding factors that limit enzymatic hydrolysis of biomass: characterization of pretreated corn stover. *Applied Biochemistry and Biotechnology*, **121**, 1081–1099.

Lemus, R., Brummer, E.C., Moore, K.J., Molstad, N.E., Burras, C.L., and Barker, M.F., 2002. Biomass yield and quality of 20 switchgrass populations in southern Iowa, USA. *Biomass and Bioenergy*, **23**, 433–442.

MacDonald, J., Korb, P., and Hoppe, R., 2013. Cropland Consolidation and the Future of Family Farms. Economic Research Report Number 152. Economic Research Service, U.S. Department of Agriculture, Washington, DC.

MacDonald, J., Ribaudo, M., Livingston, M., Beckman, J., and Huang, W., 2009. *Manure Use for Fertilizer and for Energy: Report to Congress*. Washington, DC: Economic Research Service, U.S. Department of Agriculture.

Mathews, K. H. and McConell, M., 2009. *Ethanol Co-Product Use in U.S. Cattle Feeding Lessons Learned and Considerations*, FDS-09D-01. Washington, DC: Economic Research Service, U.S. Department of Agriculture.

Milani, F.X., Nutter, D., and Thoma, G., 2011. Invited review: environmental impacts of dairy processing and products: a review. *Journal of Dairy Science*, **94**, 4243–4254.

Ministry of Agriculture of China, 2010. *The National Crops Residue Resource Survey and Evaluation Report*. Beijing: MOA.

Møller, J., Thøgersen, R., Kjeldsen, A.M., Weisbjerg, M.R., Søegaard, K., Hvelplund, T., and Børsting, C.F., 2000. Fodermiddeltabel. Sammensætning og foderværdi af fodermidler til kvæg (Feed Table. Composition and Nutritional Value of Feedstuff for Cattle). Rapport 91, Landbrugets Rådgivningscenter, Landskontoret for kvæg, Aarhus.

Monlau, F., Barakat, A., Trably, E., Dumas, C., Steyer, J.-P., and Carrere, H. 2013. Lignocellulosic materials into biohydrogen and biomethane: impact of structural features and pretreatment. *Critical Reviews in Environmental Science and Technology*, **43**, 260–322.

Nelson, M.L., 2010. Utilization and application of wet potato processing coproducts for finishing cattle. *Journal of Animal Science*, **88**(Suppl. E), E133–E142.

Nguyen, Q.S., Le, T.T.H., and Dinh, V.B. (2000) Manure from rabbits, goats, cattle and buffaloes as substrate for earthworms. Making Better Use of Local Feed Resources" SAREC-UAF Workshop-Seminar, Ho Chi Minh City, Vietnam, January 18–20, 2000, SIDA-SAREC (Swedish International Development Cooperation Agency, Department for Research Cooperation-SAREC) and UAF (University of Agriculture and Forestry, National University of Ho Chi Minh City), http://mekarn.org/sarpro/frontp.htm (accessed June 23, 2015).

Nicholson, F.A., Chambers, B.J., Williams, J.R., and Unwin, R.J., 1999. Heavy metal contents of livestock feeds and animal manures in England and Wales. *Bioresource Technology*, **70**, 23–31.

Nonhebel, S. 2012. Global food supply and the impacts of increased use of biofuels. *Energy*, **37**, 115–121.

O'Keeffe, S., Schulte, R.P.O., Sanders, J.P.M., and Struik, P.C., 2011. I. Technical assessment for first generation green biorefinery (GBR) using mass and energy balances: scenarios for an Irish GBR blueprint. *Biomass and Bioenergy*, **35**, 4712–4723.

Pain, B. and Menzi, H., 2011. *RAMIRAN (Recycling Agricultural, Municipal and Industrial Residues in Agriculture Network) Glossary of Terms on Livestock and Manure Management 2011*, 2nd edn. Darmstadt: Kuratorium für Technik und Bauwesen in der Landwirtschaft e.V. (KTBL).

Petersen, C., Kaysen, O. and Hansen, J.P., 2014. *Organiske restprodukter – vurdering af potentiale og behandlet mængde (Organic Residues – Assessment of Potential and*

Throughput. Environmental), Miljøprojekt nr. 1529 (Project no. 1529). Copenhagen: Miljøstyrelsen (Environmental Protection Agency), p. 82.

Petersen, S.O., Sommer, S.G., Béline, F., Burton, C., Dach, J., Dourmad, J.Y., Leip, A., Misselbrook, T., Nicholson, F., Poulsen, H.D., Provolo, G., Sørensen, P., Vinnerås, B., Weiske, A., Bernal, M.-P., Böhm, R., Juhász, C., and Mihelic, R., 2007. Recycling of livestock manure in a whole-farm perspective – Preface. *Livestock Science*, **112**, 180–191.

Poulsen, H.D., 2013. *Normtal for husdyrgødning – 2013 (Normative Manure Composition Data – 2013)*. Aarhus: Aarhus University.

Qiu, H.G., Jing, Y., Liao, P.S., and Cai, Q.Y., 2013. Environmental pollution of livestock and the effectiveness of different management policies in China. *China Environmental Science*, **33**, 2268–2273.

Running, S., 2012. A measurable planetary boundary for the biosphere. *Science*, **337**, 1458–1459.

Saidur, R., Abdelaziz, E.A., Demirbas, A., Hossain, M.S., and Mekhilef, S., 2011. A review on biomass as a fuel for boilers. *Renewable and Sustainable Energy Reviews*, **15**, 2262–2289.

Saxena, R.C., Adhikari, D.K., and Goyal, H.B., 2009. Biomass-based energy fuel through biochemical routes: a review. *Renewable and Sustainable Energy Reviews*, **13**, 167–178.

Schiffman, S.S. and Williams, C.M. 2005. Science of odor as a potential health issue. *Journal of Environmental Quality*, **34**, 129–138.

Schröder, J.J., Smit, A.L., Cordell, D., and Rosemarin, A., 2011. Improved phosphorus use efficiency in agriculture: a key requirement for its sustainable use. *Chemosphere*, **84**, 822–831.

Sommer, S.G., 2013. *Ammonia volatilization from livestock slurries and mineral fertilizers*. PhD thesis. University of Southern Denmark, p. 128.

Sommer, S.G., Christensen, M.L., Schmidt, T., and Jensen, L.S. 2013. *Animal Manure – Treatment and Management*, 1st edn. Chichester: John Wiley & Sons, Ltd.

Sommer, S.G., Olesen, J.E., Petersen, S.O., Weisbjerg, M.R., Valli, L., Rohde, L., and Béline, F., 2009a. Region-specific assessment of greenhouse gas mitigation with different manure management strategies in four agroecological zones. *Global Change Biology*, **15**, 2825–2837.

Sommer, S.G., Østergård, H.S., Løfstrøm, P., Andersen, H.V. and Jensen, L.S., 2009b. Validation of model calculation of ammonia deposition in the neighbourhood of a poultry farm using measured NH_3 concentrations and N deposition. *Atmospheric Environment*, **43**, 915–920.

de Souza Moraes, B., Petersen, S.O., Zaiat, M., Sommer, S.G., and Triolo, J., 2014. Reduction in greenhouse gas emission from Brazilian vinasse through anaerobic digestion. *Bioresource Technology*, **190**, 227–234.

Statistics Denmark, 2012. Husdyr (Livestock) 1982–2011. Danmarks Statistik, www.dst.dk (accessed June 23, 2015).

Steinfeld, H., Gerber, P., Wassenaar, T., Castel, V., Rosales, M., and de Haan, C., 2006. *Livestock Long Shadow – Environmental Issues and Options*. Rome: Food and Agriculture Organization of the United Nations.

Sulaiman, F., Abdullah, N., Gerhauser, H., and Shariff, A., 2011. An outlook of Malaysian energy, oil palm industry and its utilization of wastes as useful resources. *Biomass and Bioenergy*, **35**, 3775–3786.

Sun, Y.M., Li, G.X., Zhang, F.D., Shi, C.L., and Sun, Z.J., 2005. Status quo and developmental strategy of agricultural residues resources in China. *Transactions of the Chinese Society of Agricultural Engineering*, **21**, 169–173.

Sutton, M.A., Oenema, O., Erisman, J.W., Leip, A., van Grinsven, H., and Winiwarter, W., 2011. Too much of a good thing. *Nature – Comment*. **472**, 159–161.

Swisher, K., 2013. Market Report. Render Magazine (April 2013), pp. 10–17.

Syakti, A.D., Asia, L., Kanzari, F., Umasangadji, H., Lebarillier, S., Oursel, B., Garnier, C., Malleret, L., Ternois, Y., Mille, G., and Doumenq, P., 2013. Indicators of terrestrial biogenic hydrocarbon contamination and linear alkyl benzenes as land-base pollution tracers in marine sediments. *International Journal of Environmental Science and Technology*, **12**, 581–594.

Thygesen, O. and Johnsen, T., 2012. Manure-based energy generation and fertilizer production: determination of calorific value and ash characteristics. *Biosystems Engineering*, **113**, 166–172.

Thygesen, O., Triolo, J.M., and Sommer, S.G., 2014. Anaerobic digestion of pig manure fibres from full-scale commercial manure separation units. *Biosystems Engineering*, **123**, 91–96.

Tonini, D., Hamelin L., Wenzel, H., and Astrup, T., 2012. Bioenergy production from perennial energy crops: a consequential LCA of 12 bioenergy chains including land use changes. *Environmental Science & Technology*, **46**, 13521–13530.

Triolo, J.M., Amon, T., Moraes, B.S., Abildgaard, L., Hafner, S., and Sommer, S.G., 2014. An analysis of the benefits of and loss of energy in the sugar beet supply chain for biogas production. *Journal Agricultural Systems*.

Triolo J.M., Sommer S.G., Møller H.B., Weisbjerg M.R., and Jiang X.Y. 2011. A new algorithm to characterize biodegradability of biomass during anaerobic digestion: influence of lignin concentration on methane production potential. *Bioresource Technology* **102**, 9395–9402.

Triolo, J.M., Ward, A.J., Pedersen, L., and Sommer, S.G., 2013. Characteristics of animal slurry as a key biomass for biogas production in Denmark. In: M.D. Matovic, ed. *Biomass Now – Sustainable Growth and Use*. Rijeka: InTech, pp. 307–326.

United States Department of Agriculture, 1992. *Weights, Measures, and Conversion Factors for Agricultural Commodities and Their Products* Agricultural Handbook No. 697. Washington, DC: Economic Research Service in cooperation with the Agricultural Marketing Service, the Agricultural Research Service, and the National Agricultural Statistics Service, US Department of Agriculture.

United States Department of Agriculture, 2014. *2012 Census of Agriculture, United States, Summary and State Data*, Geographic Area Series, vol. **1**, Part 51, AC-12-A-51. Washington, DC: US Department of Agriculture.

United States Department of Agriculture and Natural Resources Conservation Service, 2008. Agricultural waste characteristics. *Agricultural Waste Management Field Handbook*, Part 651 210-VI-AWMFH. Washington, DC: US Department of Agriculture.

United States Department of Energy, 2011. *U.S. Billion-Ton Update: Biomass Supply for a Bioenergy and Bioproducts Industry*. Oak Ridge, TN: US Department of Energy, Energy Efficiency and Renewable Energy, Office of the Biomass Program, Oak Ridge National Laboratory.

Vänneras, B., 2013. Sanitation and hygiene in manure management. In: S.G. Sommer, M.L. Christensen, T. Schmidt, and L.S. Jensen, eds. *Animal Manure Recycling – Treatment and Management*. Chichester: John Wiley & Sons, Ltd, pp. 91–104.

Vicente, G., Martínez, M., and Aracil, J., 2008. Optimisation of integrated biodiesel production. Part I. A study of the biodiesel purity and yield. *Bioresource Technology*, **98**,
1724–1733.

Wang, F., Dou, Z., Ma, L., Ma, W., Sims, J.T., and Zhang, F., 2010. Nitrogen mass flow in
China's animal production system and environmental implications. *Journal of Environmental
Quality*, **39**, 1537–1544.

Weissbach, F., 2009. Losses of methane forming potential of pulpified sugar beets stored in
open ground basins. *Landtechnik*, **68**, 50–57.

Wesnæs, M., Wenzel, H., and Petersen, B.M. 2009. *Life Cycle Assessment of Slurry
Management Technologies*, Environmental Project no. 1298. Copenhagen: Danish Ministry
of the Environment, Environmental Protection Agency.

Westborg, S., Johansen, L.P., and Christensen, B.H., 2010. *Fuel Characteristics of Manure*.
Environmental Project no. 1339. Copenhagen: Danish Environmental Protection Agency.

World Health Organization, 2000. *The World Health Report 2000 – Health Systems: Improving
Performance*. Geneva: WHO.

Xiong, Z.Q., Freney, J.R., Mosier, A.R., Zhu, Z.L., Lee, Y., and Yagi, K., 2008. Impacts of
population growth, changing food preferences and agricultural practices on the nitrogen
cycle in East Asia. *Nutrient Cycling in Agroecosystems*, **80**, 189–198.

Xu, K. and Xu, P., 2014. Efficient production of L-lactic acid using co-feeding strategy based
on cane molasses/glucose carbon sources. *Bioresource Technology*, **153**, 23–29.

Yang, F., Li, R., Cui, Y., and Duan, Y.H., 2010. Utilization and develop strategy of organic
fertilizer resources in China. *Soil and Fertilizer Sciences in China*, **4**, 77–82.

4

Maintaining Momentum: Drivers of Environmental and Economic Performance, and Impediments to Sustainability

Marcus Wagner

Chair of Management, Innovation and International Business, Augsburg University, Universitätsstraße 16, Augsburg, Germany

4.1 Introduction

Corporate sustainability management has increasingly moved into the focus of organizational analysis in recent years (Lindgreen, Swaen, and Johnston, 2009; Scherer and Palazzo, 2011). With regard to environmental management the "pays-to-be-green" debate has received much attention (Pava and Krausz, 1996; Orlitzky, Schmidt, and Rynes, 2003; Ambec and Lanoie, 2008), and corporate social responsibility has similarly become a major issue for firms (Graafland, van de Ven, and Stoffele, 2003; Smith, 2003; Jamali, 2008). Still, no unequivocal picture emerged, which has been attributed to moderating factors such as an industry's pollution intensity (Clarke, 2001; Dixon-Fowler *et al.*, 2013). For example, the textile industry in Germany recently refused to join the relevant ministry in self-regulation efforts concerning negative social and environmental effects in the value chain, with some firms even being

Supply Chain Management for Sustainable Food Networks, First Edition. Edited by Eleftherios Iakovou, Dionysis Bochtis, Dimitrios Vlachos and Dimitrios Aidonis.
© 2016 John Wiley & Sons, Ltd. Published 2016 by John Wiley & Sons, Ltd.

accused of benefiting from problematic labor practices whilst at the same time claiming to have a sustainability strategy.

This chapter addresses this debate and specifically the role of moderators using structural equation modeling (SEM) by focusing on the question of whether firms empirically sustain efforts to simultaneously improve environmental and economic performance and in doing so will hopefully also provide generalizable insights for sustainability at large that help academics to focus future research and managers to design well-informed strategies.

The relevance of the manufacturing sector and its products has often been emphasized (Jackson, 1996). As its negative impacts have increased due to a continuing process of globalization, especially multinational firms increasingly face demands from stakeholders to reduce their environmental impacts (Scherer and Palazzo, 2011).

In accordance with stakeholder theory, stakeholder demands are seen as an important motivating factor for the environmental and societal activities of firms (Henriques and Sadorsky, 1999; Johnstone, 2007). Various studies have explored this relationship (Kassinis and Vafeas, 2006; Etzion, 2007; Delmas and Toffel, 2008; Rueda-Manzanares, Aragon-Correa, and Sharma, 2008), and stakeholder theory can help to classify demands more systematically, for example, as originating from either within the firm or beyond its boundaries in the value chain or the public domain at large (Clarkson, 1995; Donaldson and Preston, 1995; Frooman, 1999; Doh and Guay, 2006).

Given stakeholder demands affect firm conduct, they should also relate to their economic performance, at least according to the structure–conduct–performance paradigm (Berman and Wicks, 1999; McWilliams, Siegel, and Wright, 2006). At the same time, given that organizational actions range across a wide spectrum from lobbying activities to the implementation of environmental management systems and environmental technologies, a positive relationship between activities aimed at corporate sustainability and environmental performance (i.e., reduced environmental impacts, and by analogy also social performance) seems a less certain result of stakeholder demands addressed to firms, which again raises issues about how firms can sustain their efforts to protect public goods over the long term.

Specific gaps in the literature that emerge from these considerations and which the chapter addresses are whether integration of sustainability with other dimensions of the firm benefits economic performance and environmental performance. Especially for the latter, empirical evidence is scarce (Florida and Davidson, 2001; Thornton, Kagan, and Gunningham, 2003; Potoski and Prakash, 2005; Hertin *et al.*, 2008). This lack of evidence is also a major impediment for maintaining current and developing further corporate sustainability efforts in private firms.

Two theories are frequently involved in framing the response of firms to stakeholder demands for reducing their environmental impact, namely institutional theory and the natural-resource-based view. They can inform the link of stakeholder demands, environmental activities, and their integration and environmental and economic performance, respectively, and help to explain how stakeholder demands lead to the integration of environmental activities, and to improved economic and environmental performance.

The next two sections first introduce relevant theories that motivate the structural model and then develop hypotheses. This is followed by a section on data and methodology and the results section. The final section provides conclusions and a discussion.

4.2   Literature Review

Stakeholder demands, organizational activity, and performance outcomes can be linked through different theoretical mechanisms (Davis, 2006). A first prominent theory in this respect is institutional theory. It predicts the adoption of specific activities by firms as a consequence of demands by stakeholders that represent the institutional context of a firm (DiMaggio and Powell, 1983; Oliver, 1991). Increasingly, such firm-external demands relate to the way firms deal with the natural environment and social issues (Hoffman, 1999; Hoffman and Ventresca, 1999; Bansal and Clelland, 2004), and consequently firms pursue activities that address such demands (Meyer and Rowan, 1977). Due to this, environmental activities and corporate sustainability management at large are often seen as ceremonial activities in the context of institutional theory which benefit from asymmetric information and are aimed at addressing stakeholder concerns with or without changes in the actual performance of firms (Aldrich and Fiol, 1994; McWilliams, Siegel, and Wright, 2006).

A second important theory that informs scholars about the link of stakeholders, activities, and performance and which has gained increasing prominence in recent years is the natural-resource-based view. The resource-based view (Barney, 1991; Wernerfelt, 1984) transfers the idea that "resources are firm-specific assets that are difficult if not impossible to imitate. [...] Such assets are difficult to transfer among firms because of transaction costs, and because the assets may contain tacit knowledge" (Teece, Pisano, and Schuen, 1997, p. 516) to the context of environmental and social sustainability. More specifically, Hart (1995) and Aragon-Correa and Sharma (2003) developed a natural-resource-based view of the firm ultimately proposing three inter-related strategies for improving the environmental performance of firms that simultaneously lead to the development of capabilities that ultimately lead to resources with the characteristics captured in the quote above by Teece, Pisano, and Schuen (1997).[1] These characteristics turn them into strategic resources which, according to the resource-based view enable a sustained competitive advantage.[1] In the natural-resource-based view therefore the demands of stakeholders are a precursor for the development of proactive environmental activities, an interpretation empirically supported in Henriques and Sadorsky (1999) and in analysis on environmental marketing (Leonidou *et al.*, 2013).

The two theories just presented provide an overarching structure–conduct–performance framework and in doing so relate to a longstanding debate on the social issues in management and organizations and the natural environment literature,

---

[1] The three strategies (pollution prevention, product stewardship, and sustainable development) strongly relate to the three fundamental categories of innovation, that is, process innovation, product innovation, and organizational innovation.

namely the empirical "pays-to-be-green" literature, which also connects with the strategic management literature in general. Pava and Krausz (1996), Margolis and Walsh (2001, 2003) and Walsh, Weber, and Margolis (2003), as well as Orlitzky, Schmidt, and Rynes (2003) and Ambec and Lanoie (2008), provide recent reviews and meta-studies summarizing empirical work on the relationship of environmental and social performance with economic performance. Of the 95 studies reviewed by Margolis and Walsh (2001) and discussed in more detail in Margolis and Walsh (2003), a considerable share still finds a non-significant link. Orlitzky, Schmidt, and Rynes (2003) analyzing the relationship of environmental to economic performance found that there is significant variation across individual studies, ranging from negative to non-significant to moderately or even strongly positive relationships of environmental and economic performance, and similar findings apply to social performance.

What these studies suggest is that combining both of the aforementioned theories can lead to a comprehensive structural model linking stakeholder demands (i.e., firm-exogenous structures), conduct (e.g., in terms of environmental or social management activities), and performance (environmental and economic) that can be a sound basis for empirical analysis. Specifically, Judge and Douglas (1998) show that environmental issue integration positively relates to both economic and environmental performance, in turn suggesting integration as a capability. Integration is at the same time determined by demands outside the boundaries of the firm as reflected by different stakeholder domains and this suggests integration as an indispensable mediator variable between stakeholder demands and performance dimensions, in light of the firm heterogeneity observed empirically. Given that integration across corporate functions (such as quality assurance, corporate development, personnel/H&S) and the integration of sustainability with administrative (e.g., H&S), engineering (e.g., quality), and entre-preneurial (i.e., corporate strategy) dimensions of the firm have been identified as crucial elements of a proactive environmental strategy (Aragon-Correa and Sharma, 2003), it is useful to link these to the notion of integration as a capability.

4.3   Hypothesis Development

Hypothesis development initially addresses the relationship between stakeholder demands and integration of environmental activities with other corporate functions and then the link of integration with economic and environmental performance.

4.3.1   The Link between Stakeholder Demands and Integration of Environmental Activities

With respect to the link between stakeholder demands and integration, the literature has proposed different stakeholder typologies. For example, Henriques and Sadorsky (1999) developed 4 categories based on 12 individual stakeholders, namely regulatory stakeholders (such as governments), organizational stakeholders (such as employees), community stakeholders (such as local residents), and mass media. Similarly, Buysse and Verbeke (2003) developed 4 categories out of 14 individual stakeholders, namely regulatory stakeholders, external primary stakeholders (such as suppliers), internal primary stakeholders (such as owners), and secondary stakeholders (such as non-governmental organizations). This is also in line with the constitution of external stakeholders in legitimacy theory (Suchman, 1995) and internal stakeholders in social norm theory (Akerlof, 1980).

Finally, Murillo-Luna, Garces-Ayerbe, and Rivera-Torres (2008) identify 5 categories out of 14 individual stakeholders (which overlap with, but are not completely identical with, those of the earlier two studies): corporate government stakeholders, internal economic stakeholders, external economic stakeholders, regulatory stakeholders, and external social stakeholders. Overlaying the three classification schemes results in four generally accepted stakeholder categories, namely internal stakeholders, regulatory stakeholders, value chain stakeholders, and public stakeholders.

Institutional theory proposes an association of demands from typical stakeholder categories relate to integration because it maintains that mainly stakeholder demands trigger the development of environmental activities in the firm (Howard-Grenville and Hoffman, 2003; Garcia de Madariaga and Valor, 2007; Jamali, 2008). This is also evidenced by the aforementioned example of the German textile industry, where government (stakeholder) pressure led several firms to join an initiative on self-regulation in the textile supply chain. Based on these considerations, the following hypotheses are examined:

H1: Internal stakeholder demands positively correlate with the level of integration of environmental considerations within the firm.

H2: Regulatory stakeholder demands positively correlate with the level of integration of environmental considerations within the firm.

H3: Value chain stakeholder demands positively correlate with the level of integration of environmental considerations within the firm.

H4: Public stakeholder demands positively correlate with the level of integration of environmental considerations within the firm.

4.3.2 The Link between Integration and Economic and Environmental Performance

Implied in the above arguments is that integration is mediating firm-external and firm-internal dimensions, and in this context specifically stakeholder and performance categories. Without mediation, the assumption would be that stakeholder demands uniformly relate to economic and environmental performance of firms. This would be inconsistent with the heterogeneous resource endowments and capabilities of firms that can be observed (Fryxell and Vryza, 1998; Aragon-Correa, Matias-Reche, and Senise-Barrio, 2003; Aragon-Correa and Sharma, 2003).

Another reason that integration is an indispensable mediating variable is, as highlighted earlier, that the empirical link of environmental and economic performance is unequivocal. This suggests the possibility of a third variable jointly influencing environmental and economic performance (Garcia-Castro, Arino, and

Canela, 2009). Integration (i.e., the coupling of environmental management aspects with other managerial tasks and activities to enable cross-functional coordination) has been proposed as such a variable (Klassen and McLaughlin, 1996; Christmann, 2000; Jansson, Nilsson, and Rapp, 2000).

The natural-resource-based view holds that integration is achieved through a process based on tacit capabilities which are difficult to imitate (e.g., quality management activities or corporate strategy development) turning integration into a strategic resource (Hart, 1995; Aragon-Correa and Sharma, 2003; Branco and Rodriguez, 2006). In line with this, Judge and Douglas (1998) show that environmental issues integration relates positively to both economic and environmental performance and therefore supports the notion of firm idiosyncrasies in the case of integration. Therefore, firms that voluntarily engage in increasing integration beyond the minimum level legally implied can realize competitive advantages and an improvement of environmental performance (in terms of lower resource inputs or emissions) and economic performance (in terms of the dimensions identified in earlier research such as in Dyllick (1999) and Sharma (2001), especially market competitiveness, efficiency, risk management, and corporate image). This means that integration generates resources and capabilities that are thought in the resource-based view to create various competitive advantages (Leonidou *et al.*, 2013). Based on these considerations, the following hypotheses can be proposed:

H5: The level of integration of environmental considerations within the firm positively correlates with market-related economic performance.

H6: The level of integration of environmental considerations within the firm positively correlates with efficiency-related economic performance.

H7: The level of integration of environmental considerations within the firm positively correlates with risk-related economic performance.

H8: The level of integration of environmental considerations within the firm positively correlates with image-related economic performance.

H9: The level of integration of environmental considerations within the firm positively correlates with input-related environmental performance.

H10: The level of integration of environmental considerations within the firm positively correlates with output-related environmental performance.

4.3.3 The Role of Moderating Factors

Moderating factors have been suggested as one main explanation for the heterogeneous evidence of whether corporate sustainability pays off economically (Dixon-Fowler *et al.*, 2013). For example, these concern firm-related factors such as corporate governance (e.g., in terms of whether the head of the environmental or sustainability department of the firm is a member of the corporate board or not) or the type of firm (e.g., sole proprietorship or being part of another firm), but also structural ones, such as the type of industry (e.g., high- versus low-polluting industries).

More specifically, it can be expected that with respect to corporate governance, having a board member being tasked with sustainability responsibilities increases the effects of integration on environmental performance (Rothenberg, 2007). This is because a board member with such responsibilities can ensure better that environmental and social aspects are considered at the board level. On the other hand, a board member with such responsibilities has more difficulty in accepting a priority of economic performance aspects, since he or she can be held accountable for not sufficiently carrying out his or her sustainability-related duties.

Secondly, as concerns firm type, institutional theory, and the natural-resource-based view jointly suggest that firms held in sole proprietorship exhibit more heterogeneity in the relationships as concerns both, the link of stakeholder pressure with integration and the link of integration with environmental performance (Sharma and Henriques, 2005; Parmigiani *et al.*, 2011; Scherer and Palazzo, 2011). This is because sole proprietorship allows the owner of a firm to act regardless of headquarter guidelines. Hence the individual proprietor's sustainability orientation becomes pivotal. If, for example, the environmental concerns are high, then the relationship between environmental performance and integration should be more positive, as should be the link between certain stakeholder pressures and integration. Overall, given that individual proprietor preferences are heterogeneous, the effect can go in both directions or even cancel out, so that actual differences are an empirical question.

Actual behavior of firms with regard to corporate sustainability management in an interrelated world of multinational firms, international institutions, and non-governmental organizations is multi-causal (de Lange, 2010). Still, as concerns the effect of industry differences, the general notion, and evidence is that for firms in higher polluting industries it is more difficult to bring about a positive relationship between environmental and social performance on the one hand and economic performance on the other hand than for firms in low-polluting industries (Wagner, 2005). This also applies in the case that integration is a joint third variable influencing both performance aspects. These considerations lead to the following moderation hypotheses:

H11: The level of integration of environmental considerations within the firm correlates more negatively with market-related, efficiency-related, risk-related, and image-related economic performance if a board member is assigned sustainability responsibilities.

H12: The level of integration of environmental considerations within the firm correlates more positively with input- and output-related environmental performance if a board member is assigned sustainability responsibilities.

H13: The level of integration of environmental considerations within the firm correlates differently with market-related, efficiency-related, risk-related, and image-related economic performance if a firm is held in sole proprietorship.

H14: The level of integration of environmental considerations within the firm correlates differently with input- and output-related environmental performance if a firm is held in sole proprietorship.

H15: The level of integration of environmental considerations within the firm correlates differently with internal, regulatory, value chain, and public stakeholder demands if a firm is held in sole proprietorship.

H16: The level of integration of environmental considerations within the firm correlates more negatively with market-related, efficiency-related, risk-related, and image-related economic performance if a firm is in a high-polluting industry.

H17: The level of integration of environmental considerations within the firm correlates more negatively with input- and output-related environmental performance if a firm is in a high-polluting industry.

4.4 Data and Method

4.4.1 Sample Description

The empirical data used for hypothesis testing was collected during the 2001 European Business Environment Barometer (EBEB) survey. This survey was carried out in several European countries based on a standard mail questionnaire in order to assess the state of environmental management. The questionnaire asked firms to self-assess their main environmental effects in detailed categories as well as stakeholder demands for a large range of stakeholder groups. Also the effects on different competitive benefit aspects from environmental management and the level of integration between social, quality, H&S and corporate management and strategy was evaluated (Baumast and Dyllick, 2001). Extant research has shown that managerial perceptions on these areas produce valid assessments (Murillo-Luna, Garces-Ayerbe, and Rivera-Torres, 2008).

The sample for the survey was based on random sampling of the firm population (the total number of firms in the manufacturing sectors of the eight countries surveyed). The survey questionnaire was targeted to the general or environmental manager of a company but it asked that the person most knowledgeable should answer it. In some cases, the questionnaire was thus completed by quality managers, whereas especially in small firms, often the general manager or managing director him- or herself completed the questionnaire. The full questionnaires for the German and British survey are available from the author on request. After collecting the survey responses, in line with extant work, missing values were treated with the expected maximization method (Schafer and Graham, 2002). For this purpose, missing values in the data collected were estimated for all firms using the Missing Value Analysis tool in SPSS to complete the data set. Expected maximization is considered (under the assumption that data are missing at random) the most suitable method and state-of-art to substitute missing values with estimated values (Schafer and Graham, 2002).

The data used here are from two European countries, namely Germany and the UK, and represents 503 firm responses in the manufacturing industries. Comparing the breakdown of the respondents by country and industry sector with data from Eurostat and the German Federal Agency for Employment shows that the breakdown is largely representative for the sectoral distribution of manufacturing industries in both countries. The response rates varied across the two countries with the one for

Germany being 16.7%, and that for the UK being 10.7%. The lower response rate in the UK is partly also explained by the fact that no second mailing was sent to those firms who did not respond to the first invitation to participate in the survey. Since EU regulation in the period considered in the survey was converging and unifying, similar regimes can be assumed for both countries which is why their data are pooled for the analysis (however, a moderator analysis is also analyzed to confirm this).

Concerning response bias beyond country differences, and given the topic, it could be that the replies represent over-proportionally many firms that are significantly more active in terms of environmental and sustainability management since such firms are more interested in the subject and thus more likely to respond. Such bias is cited as a frequent problem for surveys based on written questionnaires (Armstrong and Overton, 1977). However, no strong bias was found in that for the German responses, the characteristics and response behavior of early respondents was not significantly different from the late replies, based on comparison of means for all variables between the first and last 10% of respondents. Also, broad variability is found in the responses, indicating that the data also include environmentally less active firms. One bias evident in the data is that smaller firms are under-represented in both countries, which is why the finding cannot be generalized prima facie to small firms. Unfortunately, this is a frequent issue in empirical research on manufacturing firms and thus hard to rectify since smaller firms inherently have more limited resources to participate in surveys.

Next to response bias, common method bias (whilst generally being below average in the field studied here) might exist, even though self-assessment does neither necessitate its existence nor its homogeneity since method-related variance can deflate or inflate the true relationship (Cote and Buckley, 1987; Podsakoff *et al.*, 2003). For the EBEB survey data used here a number of procedural and statistical steps were taken to ensure that common method bias is minimized or reduced. For example, in terms of procedures, the anonymity of respondents was guaranteed, different response formats were used, question order was counter-balanced and scale items were improved in the survey pre-test. All these actions aimed at reducing item ambiguity as well as socially desirable responses. Since the survey ensured the anonymity of respondents, it could not directly implement separating measurements and obtaining assessments from different respondents as two other procedural steps. However, the way the survey was implemented and its instructions in principle enabled these two latter remedies.

4.4.2 Variable Descriptions

As a first step of the empirical analysis, several indices measuring the relevant constructs for testing the hypotheses in the empirical analysis were calculated. First, the level of integration was calculated based on three items concerning the integration of environmental with quality, H&S, and strategy aspects to identify different levels of integration. The item questions asked respondents to rate on a five-point Likert scale ranging from "not at all" to "fully integrated" the level of integration of environmental with quality, H&S, and strategy aspects, respectively. Prior to calculating the index, a factor analysis yielded only one factor, hence confirming unidimensionality.

Dimensions of environmental and economic performance are the dependent variables in the regression analysis. Economic performance is defined in this chapter based on multiple items including, for example, corporate image, market share, employee satisfaction, or the ability to recruit excellent staff. Survey participants were asked to rate the effect of the totality of their environmental management activities on these different items relating to different economic performance aspects. Since independently verified data on economic performance are not available for many of the firms analyzed here (e.g., because they are private limited companies in either Germany or the UK), the most suitable approach for validly measuring it is self-assessment. This approach is also pursued by others who also provide further theoretical arguments for the validity of this approach (Sharma, 2001; Sharma and Henriques, 2005).

A factor analysis was carried out on the individual items used in the survey, identifying three different factors of (environmentally related) economic performance. The first factor refers to competitive advantage, product image, sales, market share, and new market opportunities. Therefore, it was labeled "market-related dimensions of economic performance" since it predominantly relates to the market- and product-related benefits of a company's activities. The relevant items for the second factor are corporate image, owner/shareholder satisfaction, management satisfaction, worker satisfaction, and recruitment and staff retention. Therefore, this factor was labeled "image-related economic performance" since it mainly refers to internally oriented satisfaction and image benefits from a company's environmental activities. For the third factor identified, the items short-term and long-term profits, cost savings, and productivity are particularly relevant. These predominantly refer to the profitability of a company and the factor was therefore labeled "efficiency-based economic performance." The two remaining items, namely "improved insurance conditions" and "better access to bank loans" could not be assigned to one of the above factors, but looking at them, it becomes clear that they potentially represent a fourth factor, since both are linked to the financial exposure of a company due to its (low) level of environmental management activities and it was therefore decided to interpret these two items as a fourth factor labeled "(financial) risk-related economic performance."

Environmental performance is measured through assessing the environmental impacts of the firms in a number of detailed categories (such as energy or water use or harmful emissions), each measured by a separate item variable that was identified in extant literature (Belz and Strannegard, 1997). For each of the items, the survey asked about the degree of the company's environmental impact relative to the industry average. Respondents provided answers on a five-point Likert scale indicating if their impacts were "much lower," "lower," "the same," "higher," and "much higher" compared with the industry average. Given that independently verified data on environmental performance to date cannot be obtained reliably for firms from different European countries, the most suitable approach for validly measuring environmental performance seemed to be the use of self-assessment by firms. Also, rating environmental aspects in this manner on an ordinal scale has been shown to be valid in the past (Sharfman, 1996).

Stakeholder demands are classified based on an evaluation of 23 stakeholder groups used in extant literature (Henriques and Sadorsky, 1999; Buysse and

Verbeke, 2003; Murillo-Luna, Garces-Ayerbe, and Rivera-Torres, 2008) of which 13 are usable as indicator variables based on the results of the confirmatory factor analysis.[2] These were the owning company, employees, trade unions, distributors, corporate buyers, consumers, consumer associations, insurance companies, national legislators, European legislators, the press/media, scientific institutes, and local communities. A joint high rating with regard to stakeholder demands from some of these groups (measured on a five-point Likert scale ranging from "none" via "average" to "very strong" demands) was interpreted as a latent variable representing a specific class of shareholders and was labeled accordingly. This approach has been used before (Murillo-Luna, Garces-Ayerbe, and Rivera-Torres, 2008) and thus was deemed feasible.

4.4.3   Statistical Estimation

To test the hypotheses derived, SEM is employed. SEM combines factor analyses and linear regression models and in doing so is more powerful and efficient than regression models or other approaches that separate the operationalization of concepts and analysis of relationships between the latter (Williams, Vandenberg, and Edwards, 2009). It is also unique in that it allows inclusion of so-called latent variables, which is of particular relevance in the context of empirical corporate sustainability research due to low levels of reporting standardization (Rao, 2004). All analyses were carried out with AMOS (Arbuckle, 1999) with raw data as inputs and full information maximum likelihood estimation of the parameters. The usual marker variable strategy of fixing the loading of one of the items for each latent variable to 1 was used for purposes of model identification (Ullman, 2001).

4.5   Results

The evaluation of measurement model fit was based on the overall model as summarized in Figure 4.1 and chi-square values as well as common fit indices (Bentler and Bonnett, 1980; Fornell and Larcker, 1981; Bagozzi and Yi, 1988; Bentler, 1990; Hu and Bentler, 1999; Hult et al., 2006;) suggest good fit.

Table 4.1 gives details for the moderation effects in the SEM used for hypothesis testing.

As a sensitivity analysis, moderator analyses for industry type and firm type were also carried out on a sample of Dutch and German firms. This is reported in Table 4.2.

With respect to testing of the hypotheses derived, Table 4.3 provides a summary overview, based on the results in both Tables 4.1 and 4.2.

As can be seen, support is strongest for the hypotheses on the link between integration and economic performance (H5–H8). More specifically, significant positive associations of integration with the latent market-, risk-, image-, and efficiency-related sub-dimensions of economic performance, respectively.

---

[2] Confirmatory factor analysis is considered superior to assess measurement validity compared with other measures such as the Cronbach Alpha (Bagozzi, Yi, and Phillips, 1991).

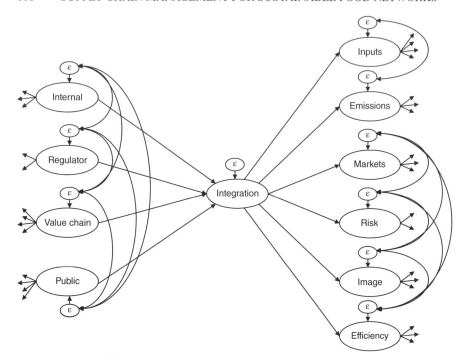

Figure 4.1   Structural equation model analyzed.

The hypotheses on the link between stakeholder demands and integration (H1–H4) draw lesser support. Strong support is only found for H1 stating a positive relationship between the level of internal stakeholder demands and integration. Opposed to this it becomes clear that H2 and H3 which state a positive relationship between levels of regulatory and value chain stakeholder demands with integration are not supported. However, H4 stating a positive relationship between public stakeholder pressure and integration receives at least some support.

With respect to the link between integration and environmental performance, the hypotheses H9 proposing a positive association of integration with input-oriented environmental performance and H10 stating a positive relationship of integration with output-oriented environmental performance (i.e., emissions) are only partly supported, so that again the evidence is weaker.

Finally, with respect to the moderation hypotheses (H11–H17), there is overall limited evidence of moderation effects. More specifically, the evidence is strongest for H14 on the moderating effect of sole proprietorship on the link of integration of environmental considerations with input- and output-related environmental performance. Also a weak moderating effect of sole proprietorship on the link of integration of environmental considerations within the firm and internal, regulatory, value chain, and public stakeholder demands and the one of integration with economic performance is found. Therefore, whilst the hypothesis system

Table 4.1 Group comparison (values <0.001 in the *P* columns indicate significance).

Moderator	Firm type	Beta	Critical ratio sol.	P sol.	Beta	Critical ratio n. sol.	P n. sol.	Chi squared	P
Integration	Regulator	0.032	0.453	0.651	−0.044	−0.626	0.531	0.305	0.581
Integration	Value chain	−0.250	−0.913	0.362	−0.112	−0.672	0.501	0.308	0.579
Integration	Public	0.074	0.535	0.593	0.302	2.340	0.019	1.800	0.180
Integration	Internal	0.618	2.506	0.012	0.489	3.922	<0.001	0.788	0.375
Efficiency	Integration	0.101	2.216	0.027	0.199	3.738	<0.001	2.151	0.142
Image	Integration	0.615	5.913	<0.001	0.584	6.187	<0.001	0.131	0.717
Risk	Integration	0.107	2.710	0.007	0.255	4.569	<0.001	3.937	0.047
Markets	Integration	0.145	3.342	<0.001	0.238	4.338	<0.001	1.241	0.265
Emissions	Integration	−0.338	−3.100	0.002	−0.160	−1.802	0.072	1.864	0.172
Inputs	Integration	−0.333	−2.992	0.003	−0.122	−1.236	0.217	1.827	0.176

Moderator	Industry	Beta	Critical ratio dirt.	P dirt.	Beta	Critical ratio cln	P cln	Chi squared	P
Integration	Regulator	−0.047	−0.715	0.475	0.009	0.110	0.913	0.153	0.696
Integration	Value chain	−0.287	−1.455	0.146	0.008	0.033	0.974	1.073	0.300
Integration	Public	0.299	2.487	0.013	0.003	0.019	0.985	2.376	0.123
Integration	Internal	0.613	3.709	<0.001	0.523	2.594	0.009	0.405	0.525
Efficiency	Integration	0.140	3.183	0.001	0.168	2.899	0.004	1.421	0.233
Image	Integration	0.589	6.734	<0.001	0.654	5.237	<0.001	1.414	0.234
Risk	Integration	0.201	4.367	<0.001	0.164	2.938	0.003	0.009	0.924
Markets	Integration	0.155	4.018	<0.001	0.251	3.360	<0.001	1.962	0.161
Emissions	Integration	−0.301	−3.554	<0.001	−0.042	−0.307	0.759	2.619	0.106
Inputs	Integration	−0.287	−3.065	0.002	−0.093	−0.783	0.433	0.901	0.342

(continued overleaf)

Table 4.1 (*Continued*)

Moderator	Governance	Beta	Critical ratio brd	P brd	Beta	Critical ratio n. brd	P n. brd	Chi squared	P
Integration	Regulator	0.022	0.428	0.669	−0.068	−0.465	0.642	0.193	0.660
Integration	Value chain	−0.051	−0.367	0.714	−0.550	−1.103	0.270	1.523	0.217
Integration	Public	0.149	1.417	0.156	0.130	0.730	0.465	0.025	0.875
Integration	Internal	0.279	2.488	0.013	1.381	1.652	0.099	4.284	0.038
Efficiency	Integration	0.300	3.198	0.001	0.080	0.977	0.329	1.897	0.168
Image	Integration	0.911	4.609	<0.001	0.469	3.286	0.001	1.892	0.169
Risk	Integration	0.393	3.712	<0.001	0.152	1.956	0.050	1.131	0.288
Markets	Integration	0.295	3.205	0.001	0.166	2.060	0.039	0.407	0.524
Emissions	Integration	−0.473	−2.918	0.004	−0.253	−1.258	0.208	0.569	0.451
Inputs	Integration	−0.526	−2.918	0.004	−0.341	−1.691	0.091	0.131	0.718

brd, board member with sustainability responsibility; cln, clean industry; dirt., dirty/polluting industry; n. brd, no board member with sustainability responsibility; n. sol., no sole proprietorship; sol., sole proprietorship.

Table 4.2 Sensitivity group comparison (values <0.001 in the *P* columns indicate significance).

Moderator	Firm type	Beta	Critical ratio sol.	P sol.	Beta	Critical ratio n. sol.	P n. sol.	Chi squared	P
Integration	Regulator	−0.057	−0.732	0.464	0.120	1.897	0.058	0.005	0.942
Integration	Value chain	−0.482	−2.230	0.026	−0.297	−1.423	0.155	1.648	0.199
Integration	Public	0.468	2.858	0.004	−0.066	−0.642	0.521	9.665	0.002
Integration	Internal	0.599	4.506	<0.001	0.753	4.350	<0.001	0.160	0.689
Efficiency	Integration	0.049	1.265	0.206	0.060	2.170	0.030	0.059	0.808
Image	Integration	0.526	6.001	<0.001	0.466	6.831	<0.001	1.954	0.162
Risk	Integration	0.161	3.716	<0.001	0.055	2.628	0.009	0.128	0.720
Markets	Integration	0.115	2.774	0.006	0.119	3.714	<0.001	0.173	0.678
Emissions	Integration	−0.040	−0.500	0.617	0.090	1.350	0.177	3.085	0.079
Inputs	Integration	0.064	0.730	0.465	0.045	0.673	0.501	0.462	0.497

Moderator	Industry	Beta	Critical ratio dirt.	P dirt.	Beta	Critical ratio cln	P cln	Chi squared	P
Integration	Regulator	0.024	0.460	0.645	0.129	1.246	0.213	1.017	0.313
Integration	Value chain	−0.307	−2.016	0.044	−0.527	−1.794	0.073	2.210	0.137
Integration	Public	0.213	2.171	0.030	−0.022	−0.159	0.874	1.904	0.168
Integration	Internal	0.551	5.107	<0.001	0.969	3.536	<0.001	1.232	0.267
Efficiency	Integration	0.061	1.966	0.049	0.083	2.439	0.015	2.112	0.146
Image	Integration	0.552	7.105	<0.001	0.322	5.308	<0.001	0.199	0.656
Risk	Integration	0.116	3.630	<0.001	0.107	3.666	<0.001	1.706	0.192
Markets	Integration	0.086	3.078	0.002	0.150	3.393	<0.001	0.784	0.376
Emissions	Integration	−0.021	−0.298	0.765	0.137	1.790	0.073	0.724	0.395
Inputs	Integration	0.003	0.043	0.966	0.104	1.325	0.185	0.208	0.649

For abbreviations see Table 4.1.

Table 4.3   Summary of hypothesis testing.

Hypothesis	Results of Table 4.1, columns 3–6	Results of Table 4.1, columns 7–10	Results of Table 4.2, columns 3–6	Results of Table 4.2, columns 7–10
H1	+/+/+[a]	+/+/+	+/+	+/+
H2	O/O/O	O/O/O	O/O	+/O
H3	O/O/O	O/O/O	–/–	O /–
H4	O/+/O	+/O/O	+/+	O/O
H5	+/+/+	+/+/+	+/+	+/+
H6	+/+/+	+/+/O	+/+	+/+
H7	+/+/+	+/+/+	+/+	+/+
H8	+/+/+	+/+/+	+/+	+/+
H9	+/+/+	O/O/+	O/O	O/O
H10	+/+/+	+/O/O	+/O	O/–
H11	No significantly different paths		n.a.	
H12	No significantly different paths		n.a.	
H13	1 of 4 significantly different (public stronger for firms in sole proprietorship)		No significantly different paths	
H14	No significantly different paths		1 of 2 significantly different (emissions lower for firms in sole proprietorship)	
H15	No significantly different paths		1 of 4 significantly different (public stronger for firms in sole proprietorship)	
H16	No significantly different paths		No significantly different paths	
H17	No significantly different paths		No significantly different paths	

[a] First sign refers to firm type, second sign to industry, and third sign to governance (Table 4.1 only).
+, Hypothesis confirmed; O, coefficient insignificant; –, hypothesis rejected; n.a., not available.

is in majority supported, specific deviations stand out, especially with regard to environmental performance and the dominant role of firm type as a moderator.

4.6   Discussion and Conclusions

This chapter addresses the debate about whether firms can simultaneously sustain efforts to improve environmental and economic performance. It does so using empirical data and applying SEM analyses and specifically addresses the role of possible moderator variables. It therefore moves beyond macro-level "black-box"-type studies on the link between environmental and economic performance.

Overall it finds that integration of environmental management with other areas of the firm is more consistently related to economic than to environmental performance. This suggests to some degree that firms see sustainability management as a ceremonial function and the conventional "core" of the firm and the environmental impact linked to it is consequently "buffered" and therefore remains largely unchanged (Meyer and Rowan, 1977).

Also, with respect to moderation effects, the one strong factor standing out in the analysis is firm type (operationalized as a firm being solely owned or not). Beyond this, as expected no significantly different effects of stakeholder pressure between high- and low-polluting industries are found and only one significantly different path is found for governance effects. This latter effect concerns the relationship between internal stakeholder pressure and integration, where the path is relatively less strong in firms where a board member is assigned responsibility for sustainability. Beyond this and more generally, stakeholder pressure does not seem to be a major driver of corporate sustainability management (especially if internal stakeholders are not considered exogenous).

Based on these findings, this chapter confirms earlier research by Marcus and Anderson (2006) in the US food industry that suggests, based on the notion of integration as a capability, that it is not one general capability that jointly matters for economic and environmental/social performance, but that different capabilities matter for business and corporate social responsibility objectives. This means that the capabilities assumed to be developed (as proposed by the natural-resource-based view) as a correlate to integration are still largely focused on business objectives.

An alternative interpretation is that a more comprehensive theoretical framework encompassing elements from both institutional theory and the natural-resource-based view, as well as possibly additional theories such as the ones raised in Scherer and Palazzo (2011), is needed to explain the actual behavior of firms with regard to sustainable management in an interrelated world of multinational firms, international institutions, and non-governmental organizations (de Lange, 2010). This could be a focus of future research which probably also answers part of the question when extending the natural-resource-based view is feasible.

One promising approach in this respect could be to extend the natural-resource-based view using recent conceptualizations of dynamic capabilities (Helfat *et al.*, 2007) in order to arrive at a comprehensive framework that integrates institutional and evolutionary aspects with the natural-resource-based view, as has partly been attempted by Rueda-Manzanares, Aragon-Correa, and Sharma (2008). Such an approach would also help to reconcile the internal and external locus of theories and hence would resound with recent arguments suggesting that a cautious stance by firms may not necessarily imply a reactive approach (Busch and Hoffmann, 2009).

References

Akerlof, G., 1980. A theory of social customs, of which unemployment may be one consequence. *Quarterly Journal of Economics* **94**, 749–775.

Aldrich, H.E. and Fiol, C.M., 1994. Fools rush in? The institutional context of industry creation. *Academy of Management Review*, **19**, 645–670.

Ambec, S. and Lanoie P., 2008. Does it pay to be green? A systematic overview. *Academy of Management Perspectives*, **22**, 45–62.

Aragon-Correa, J.A., Matias-Reche, F., and Senise-Barrio, M.A., 2003. Managerial discretion and corporate commitment to the natural environment. *Journal of Business Research*, **57**, 964–975.

Aragon-Correa, J.A. and Sharma S., 2003 A contingent resource-based view of proactive corporate environmental strategy. *Academy of Management Review*, **28**, 71–88.

Arbuckle, J.L., 1999. *Amos 4*. Chicago, IL: SmallWaters Corporation.

Armstrong, J.S. and Overton, T.S., 1977. Estimating nonresponse bias in mail surveys. *Journal of Marketing Research*, **14**, 396–402.

Bagozzi, R.P. and Yi, Y., 1988. On the evaluation of structural equation models. *Journal of the Academy of Marketing Science*, **16**, 74–94.

Bagozzi, R.P., Yi, Y., and Phillips, K.W., 1991. Assessing construct validity in organizational research. *Administrative Science Quarterly*, **36**, 421–458.

Bansal, P. and Clelland, I., 2004. Talking trash: legitimacy, impression management, and unsystematic risk in the context of the natural environment. *Academy of Management Journal*, **47**, 93–103.

Barney, J., 1991. Firm resources and sustained competitive advantage. *Journal of Management*, **17**, 99–120.

Baumast, A. and Dyllick, T., 2001. *Umweltmanagement-Barometer 2001*. St. Gallen: Institute for Economy and the Environment, University of St. Gallen.

Belz, F. and Strannegard, L., 1997. *International Business Environmental Barometer 1997*. Oslo: Cappelen Akademisk Forlag.

Bentler, P.M., 1990. Comparative fit indexed in structural models. *Psychological Bulletin*, **107**, 238–246.

Bentler, P.M. and Bonnett, D.G., 1980. Significance test and goodness of fit in the analysis of covariance structures. *Psychological Bulletin*, **88**, 588–606.

Berman, S. and Wicks, A., 1999. Does stakeholder orientation matter? The relationship between stakeholder management models and firm financial performance. *Academy of Management Journal*, **42**, 488–506.

Branco, M.C. and Rodriguez, L.L., 2006. Corporate social responsibility and resource-based perspectives. *Journal of Business Ethics*, **69**, 111–132.

Busch, T. and Hoffmann, V.H., 2009. Ecology-driven real options: an investment framework for incorporating uncertainties in the context of the natural environment. *Journal of Business Ethics*, **90**, 295–310.

Buysse, K. and Verbeke, A., 2003. Proactive environmental strategies: a stakeholder management perspective. *Strategic Management Journal*, **24**, 453–470.

Christmann, P., 2000. Effects of 'best practices' on environmental management. *Academy of Management Journal*, **3**, 663–680.

Clarke, T., 2001. Balancing the triple bottom line: financial, social and environmental performance. *Journal of General Management*, **26**, 16–27.

Clarkson, M.B.E., 1995. A stakeholder framework for analyzing and evaluating corporate social performance. *Academy of Management Review*, **20**, 92–117.

Cote, J.A. and Buckley, M.R., 1987. Estimating trait, method and error variance: generalizing across 70 construct validation studies. *Journal of Marketing Research*, **24**, 315–318.

Davis, G., 2006. Mechanisms and the theory of organizations. *Journal of Management Inquiry*, **15**, 114–118.

Delmas, M. and Toffel, A., 2008. Organizational responses to environmental demands: opening the black box. *Strategic Management Journal*, **29**, 1027–1055.

DiMaggio, P.J. and Powell, W.W., 1983. The iron cage revisited: institutional isomorphism and collective rationality in organizational fields. *American Sociological Review*, **48**, 147–160.

Dixon-Fowler, H.R., Slater, D.J., Johnson, J.L., Elstrand, A.E., and Romi, A.M., 2013. Beyond 'Does it pay to be green?' A meta-analysis of moderators of the CEP–CFP relationship. *Journal of Business Ethics*, **112**, 353–366.

Doh, J.P. and Guay, T.R., 2006. Corporate social responsibility, public policy, and NGO activism in Europe and the United States: an institutional-stakeholder perspective. *Journal of Management Studies*, **43**, 47–73.

Donaldson, T. and Preston, L.E., 1995. The stakeholder theory of the corporation: concepts, evidence, and implications. *Academy of Management Review*, **20**, 65–91.

Dyllick, T., 1999. Environment and competitiveness of companies. In: D. Hitchens, J. Clausen, and K. Fichter (eds). *International Environmental Management Benchmarks. Best Practice Experiences from America, Japan and Europe*. Berlin: Springer-Verlag, pp. 55–69.

Etzion, D., 2007. Research on organizations and the natural environment, 1992-present: a review. *Journal of Management*, **33**, 637–664.

Florida, R. and Davidson, D., 2001. Why do firms adopt environmental practices (and do they make a difference?). In: G. Coglianese and J. Nash (eds). *Regulating From the Inside: Can Environmental Management Systems Achieve Policy Goals?* Washington, DC: RFF Press, pp. 82–104.

Fornell, C. and Larcker, D.F., 1981. Evaluating structural equation models with unobservable variables and measurement error. *Journal of Marketing Research*, **18**, 39–50.

Frooman, J., 1999. Stakeholder influence strategies. *Academy of Management Review*, **24**, 191–205.

Fryxell, G.E. and Vryza, M., 1998. Managing environmental issues across multiple functions: an empirical study of corporate environmental departments and functional coordination. *Journal of Environmental Management*, **55**, 39–56.

Garcia-Castro, R., Arino, M.A., and Canela, M.A., 2009. Does social performance really lead to financial performance? Accounting for endogeneity. *Journal of Business Ethics*, **92**, 107–126.

Garcia de Madariaga, J. and Valor, C., 2007. Stakeholders management systems: empirical insights from relationship marketing and market orientation perspectives. *Journal of Business Ethics*, **71**, 425–439.

Graafland, J., van de Ven, B., and Stoffele, N., 2003. Strategies and instruments for organising CSR by small and large businesses in the Netherlands. *Journal of Business Ethics*, **47**, 45–60.

Hart, S.L., 1995. A natural-resource-based view of the firm. *Academy of Management Review*, **20**, 874–907.

Helfat, C., Finkelstein, S., Mitchell, W., Peteraf, M., Singh, H., Teece, D., and Winter, S., 2007. *Dynamic Capabilities*. New York, NY: Wiley-Blackwell Publishing.

Henriques, I. and Sadorsky, P., 1999. The determinants of an environmentally responsive firm: an empirical approach. *Journal of Environmental Economics and Management*, **30**, 381–395.

Hertin, J., Berkhout, F., Wagner, M., and Tyteca, D., 2008. Are EMS environmentally effective? The link between environmental management systems and environmental performance in European companies. *Journal of Environmental Planning and Management*, **51**, 255–280.

Hoffman, A., 1999. Institutional evolution and change: environmentalism and the U.S. chemical industry. *Academy of Management Journal*, **42**, 351–371.

Hoffman, A. and Ventresca, M., 1999. The institutional framing of policy debates. *American Behavioral Scientist*, **42**, 1368–1392.

Howard-Grenville, J. and Hoffman, A., 2003. The importance of cultural framing to the success of social initiatives in business. *Academy of Management Learning and Executive*, **17**, 70–84.

Hu, L. and Bentler, P.M., 1999. Cutoff criteria for fit indexes in covariance structure analysis: conventional criterial versus new alternatives. *Structural Equation Modeling*, **6**, 1–55.

Hult, G.T.M., Ketchen, D.J., Shaojie Cui, A., Prud'homme, A.M., Seggie, S.H., Stanko, M., Shichun Xu, A. and Cavusgil, S.T., 2006. An assessment of the use of structural equation modeling in international business research. In: D.J. Ketchen and D. Bergh (eds). *Research Methodology in Strategy and Management*. New York, NY: Emerald Publishing, pp. 385–410.

Jackson, T., 1996. *Material Concerns. Pollution, Profit and Quality of Life*. London: Routledge.

Jamali, D., 2008. A stakeholder approach to corporate social responsibility: a fresh perspective into theory and practice. *Journal of Business Ethics*, **82**, 213–231.

Jansson, A., Nilsson, F., and Rapp. B., 2000. Environmentally driven mode of business development: a management control perspective. *Scandinavian Journal of Management*, **16**, 303–333.

Johnstone, N., 2007. *Environmental Policy and Corporate Behaviour*. Cheltenham: Edward Elgar.

Judge, W.Q. and Douglas, T.J., 1998. Performance implications of incorporating natural environmental issues into the strategic planning process: an empirical assessment. *Journal of Management Studies*, **35**, 241–262.

Kassinis, G. and Vafeas, N., 2006. Stakeholder pressures and environmental performance. *Academy of Management Journal*, **49**, 145–159.

Klassen, R.D. and McLaughlin, C.P., 1996. The impact of environmental management on firm performance. *Management Science*, **42**, 1199–1214.

de Lange, D., 2010. *Power and Influence: The Embeddedness of Nations*. London: Palgrave Macmillan.

Leonidou, L.C., Leonidou, C.N., Fotiadis, T.A., and Zertiti, A., 2013. Resources and capabilities as drivers of hotel environmental marketing strategy: implications for competitive advantage and performance. *Tourism Management*, **35**, 94–110.

Lindgreen, A., Swaen, V., and Johnston, W.J., 2009. Corporate social responsibility: an empirical investigation of U.S. organizations. *Journal of Business Ethics*, **85**, 303–323.

Marcus, A. and Anderson, M., 2006. A general dynamic capability: does it propagate business and social competencies in the retail food industry? *Journal of Management Studies*, **43**, 19–46.

Margolis, J. and Walsh, J., 2001. *People and Profits: The Search for a Link between a Company's Social and Financial Performance*. Mahwah, NJ: Lawrence Erlbaum Associates.

Margolis, J. and Walsh, J., 2003. Misery loves companies: rethinking social initiatives by business. *Administrative Science Quarterly*, **48**, 268–305.

McWilliams, A., Siegel, D., and Wright, P., 2006. Corporate social responsibility: strategic implications. *Journal of Management Studies*, **43**, 1–18.

Meyer, J.W. and Rowan, B., 1977. Institutionalized organizations: formal structure as myth and ceremony. *American Journal of Sociology*, **83**, 340–363.

Murillo-Luna, J.L., Garces-Ayerbe, C., and Rivera-Torres, P., 2008. Why do patterns of environmental response differ? A stakeholders pressure approach. *Strategic Management Journal*, **29**, 1225–1240.

Oliver, C., 1991. Strategic responses to institutional processes. *Academy of Management Review* **16**, 145–179.

Orlitzky, M., Schmidt, F., and Rynes, S., 2003. Corporate social and financial performance: a meta-analysis. *Organization Studies*, **24**, 403–441.

Parmigiani, A., Klassen, R.D., and Russo, M.V., 2011. Efficiency meets accountability: Performance implications of supply chain configuration, control, and capabilities. *Journal of Operations Management*, **29**, 212–223.

Pava, M.L. and Krausz, J., 1996. The association between corporate social responsibility and financial performance: the paradox of social cost. *Journal of Business Ethics*, **15**, 321–357.

Podsakoff, P.M., MacKenzie, S.B., Lee, J.Y., and Podsakoff, N.P., 2003. Common method biases in behavioral research: a critical review of the literature and recommended remedies. *Journal of Applied Psychology*, **88**, 879–903.

Potoski, M. and Prakash, A., 2005. Covenants with weak swords: ISO14001 and facilities environmental performance. *Journal of Policy Analysis and Management*, **24**, 745–769.

Rao, P., 2004. Greening production: a South-East Asian experience. *International Journal of Operations & Production Management*, **24**, 289–320.

Rothenberg, S., 2007. Environmental managers as institutional entrepreneurs: the influence of institutional and technical pressures on waste management. *Journal of Business Research*, **60**, 749–757.

Rueda-Manzanares, A., Aragon-Correa, J., and Sharma, S., 2008. The influence of stakeholders on the environmental strategy of service firms: the moderating effects of complexity, uncertainty and munificence. *British Journal of Management*, **19**, 185–203.

Schafer, J.L. and Graham, J.W., 2002. Missing data: Our view of the state of the art. *Psychological Methods*, **7**, 147–177.

Scherer, A. and Palazzo, G., 2011. The new political role of business in a globalized world – a review of a new perspective on CSR and its implications for the firm, governance, and democracy. *Journal of Management Studies*, **48**, 899–931.

Sharfman, M., 1996. A concurrent validity study of the KLD social performance ratings data. *Journal of Business Ethics*, **15**, 287–296.

Sharma, S., 2001. Different strokes: regulatory styles and environmental strategy in the North-American oil and gas industry. *Business Strategy and the Environment*, **10**, 344–364.

Sharma, S. and Henriques, I., 2005. Stakeholder influences on sustainability practices in the Canadian forest products industry. *Strategic Management Journal*, **26**, 159–180.

Smith, N.C., 2003. Corporate social responsibility: whether or how? *California Management Review*, **45**, 52–76.

Suchman, M.C., 1995. Managing legitimacy: strategic and institutional approaches. *Academy of Management Review*, **20**, 571–610.

Teece, D., Pisano, G., and Schuen, A., 1997. Dynamic capabilities and strategic management. *Strategic Management Journal*, **18**, 509–533.

Thornton, D., Kagan, R., and Gunningham, N., 2003. Sources of corporate environmental performance. *California Management Review*, **46**, 127–141.

Ullman, J.B., 2001. Structural equation modeling. In: B.G. Tabachnick and L.S. Fidell (eds). *Using Multivariate Statistics*. Needham Heights, MA: Allyn & Bacon, pp.653–771.

Wagner, M., 2005. How to reconcile environmental and economic performance to improve corporate sustainability: corporate environmental strategies in the European paper industry. *Journal of Environmental Management*, **76**, 105–118.

Walsh, J., Weber, K., and Margolis, J., 2003. Social issues and management: our lost cause found. *Journal of Management*, **29**, 859–881.

Wernerfelt, B., 1984. A resource-based view of the firm. *Strategic Management Journal*, **5**, 171–180.

Williams, L J., Vandenberg, R.J., and Edwards, J.R., 2009. Structural equation modeling in management research: a guide for improved analysis. *Academy of Management Annals*, **3**, 543–604.

5

A Hierarchical Decision-Making Framework for Quantitative Green Supply Chain Management: A Critical Synthesis of Academic Research Efforts

Rommert Dekker,[1] Jacqueline Bloemhof-Ruwaard,[2] and Ioannis Mallidis[3]

[1] *Department of Econometrics, Erasmus School of Economics, Erasmus University Rotterdam, Rotterdam, The Netherlands*
[2] *Department of Operations Research and Logistics, Wageningen University, Hollandseweg 1, Wageningen, The Netherlands*
[3] *Laboratory of Quantitative Analysis, Logistics and Supply Chain Management, Department of Mechanical Engineering, Aristotle University of Thessaloniki, Thessaloniki, Greece*

Supply Chain Management for Sustainable Food Networks, First Edition. Edited by Eleftherios Iakovou, Dionysis Bochtis, Dimitrios Vlachos and Dimitrios Aidonis.
© 2016 John Wiley & Sons, Ltd. Published 2016 by John Wiley & Sons, Ltd.

5.1   Introduction

The environmental and social impacts of global warming have changed consumers' perception regarding the environmental performance of globalized corporate supply chain networks. In this respect, companies are not only concerned about the profitability of their supply chains, but also about their environmental impacts as these may affect their consumers' preferences (Cyprus Ministry of Labour, 2006). Quantitative methodologies are usually employed for reducing the costs of implementing supply chain management decisions. However, recently they have also proven effective for reducing their environmental impacts. In this regard, supply chain stakeholders must be able to have access to up-to-date modeling methodologies employed for quantifying the cost, as well as the environmental impacts, of implementing supply chain management decisions.

This chapter contributes to the literature by updating and extending the work of Dekker, Bloemhof, and Mallidis (2012). More specifically, this work first updates the work of Dekker, Bloemhof, and Mallidis (2012) through the development of a hierarchical green supply chain management (GSCM) framework that aims, through an extensive literature review of recent and representative academic research efforts, at assisting supply chain stakeholders in: (i) providing an outline of the main decisions that may affect the environmental performance of all three main physical drivers of a supply chain, namely products, facilities, transportation, while identifying the quantitative models employed for quantifying this performance; (ii) identifying how these decisions could transcend the strategic, tactical, and operational decision phases; and (iii) providing a critical synthesis of the reviewed academic research efforts. Finally, this chapter extends the work of Dekker, Bloemhof, and Mallidis (2012) through the consideration of additional GSCM decisions, namely: (i) product design; (ii) manufacturing design and planning; (iii) transportation planning; (iv) product quality control; (v) equipment maintenance; and (vi) green work methods.

A recent review of green quantitative models is also provided by Brandenburg *et al.* (2014). The authors present an extensive sustainable supply chain management literature review, focusing on quantitative policies classified by objective function, research methodology employed, and the type of the sustainability element of the supply chain. Our chapter provides a completely different approach to quantitative models for GSCM, as it identifies all GSCM decisions associated with products, facilities, and transportation, while identifying how these decisions could transcend the strategic, tactical, and operational decision phases. To this end, a critical synthesis of academic research efforts is conducted in order to identify gaps and overlaps of key issues tackled by the existing literature.

Figure 5.1 depicts the proposed quantitative GSCM framework under study.

The rest of the chapter is organized as follows. In Section 5.2, we classify the literature on quantitative GSCM associated with each driver, in strategic, tactical, and operational phases. In Section 5.3, we further provide a critical synthesis of the reviewed literature. Finally, we summarize the findings of our research in Section 5.4.

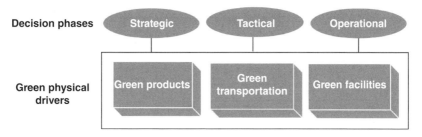

Figure 5.1 Framework for quantitative green supply chain management. Adapted from Dekker, Bloemhof, and Mallidis, 2012. Reproduced with permission of Elsevier.

5.2   Hierarchy of the Decision-Making Process

The design, planning, and control of green supply chains involve a complex hierarchy of the decision-making process. In the following sections the hierarchy of the decision-making process for the management of the three main physical drivers of a supply chain is analyzed. Moreover, the GSCM quantitative research efforts will be thoroughly discussed, while providing their taxonomy on the strategic, tactical, and operational levels of the hierarchy.

5.2.1   Strategic GSCM Decisions

Strategic design decisions are long-term decisions, usually with duration of more than a year, and involve decisions such as green product, packaging, manufacturing and facility design, transportation mode selection, facility location, and product recovery. In the following sections, these decisions are further discussed and properly classified to each one of the examined supply chain drivers of Figure 5.1.

5.2.1.1   Strategic Green Product Decisions

Products represent the main components of a supply chain. As such, their environmental impact is critical for their supply chain's overall performance. This impact depends mainly on the way they are designed, packaged, manufactured, and finally, whether their value can be recovered or not. These aspects will be thoroughly discussed in the following sections.

5.2.1.1.1   Product Design   Green product design involves the design of a product in a way that its toxic substances will be minimized during the manufacturing process, and its disassembly sequence will satisfy the regulated recycling and recovery rates in a cost efficient way (Chu *et al.*, 2009). Bovea and Belis (2012) provide an updated review of the appropriate methods employed for evaluating the environmental aspects of product design, along with the eco-design tools employed for

incorporating environmental aspects into product design. These eco-design tools are further classified according to the following six criteria: (i) the method applied for their environmental assessment; (ii) the product requirements that need to be integrated besides the environmental one (multi-criteria approach); (iii) whether the tool has a life cycle perspective (i.e., it considers all the stages of the life cycle of a product); (iv) the nature of the results (qualitative or quantitative); (v) the stages of the conceptual design process where the tool can be applied; and (vi) the methodology implemented as a basis for integrating environmental aspects into product design. Chen *et al.* (2012) propose a novel methodology for evaluating how well multiple product specifications and attributes can be combined into the design process. This methodology is called "design efficiency" and is based on the application of data envelope analysis. The authors illustrate the applicability of their proposed methodology through a data analysis of the key engineering specifications, product attributes, and emissions performances of the vehicle emissions, testing a database published by the US Environmental Protection Agency. Yangjian *et al.* (2013) deal with green modular design as a practice that enhances life-cycle material efficiency through component material reuse. The authors propose a constrained genetic algorithm for deciding jointly on technical system modularity and material reuse modularity. The applicability of the proposed methodology is illustrated through the case study of a refrigerator's green modular design, indicating that the proposed model is indeed an effective tool for supporting green modular design with material reuse.

Moon *et al.* (2013) propose guidelines for the design and production of sustainable energy-saving fashion products (ESFPs). The authors initially identify the main energy saving factors for apparel products along with consumers' preferences for ESFPs. Finally, they develop product design scenarios for sustainable fashion. The purpose of their research is to assist companies in identifying their customers' preferences for greener products, thus providing guidelines to the designers for producing products that meet their green consumers' requirements. Chen and Liu (2014) propose a two-stage modeling approach for evaluating pricing and design decisions for products with virgin and recycled material components in a duopoly market. In the first stage, each firm determines the mix of recycled and virgin material contents of its own product, while in the second stage, and after observing the other firm's product design choice, the two firms set their prices under a specific type of price leadership. Finally, Russo and Rizzi (2014) propose a computer-aid methodology, called Eco-OptiCAD, which integrates the Structural Optimization and Life Cycle Assessment methodologies. Their methodology develops a set of optimization strategies, namely Life Cycle Mapping (LCM), which enables product designers to choose the best triad shape–material–production, identifying the minimum environmental impact, and meeting both the structural and functional requirements of the product.

5.2.1.1.2 Packaging Design According to Palanivelu and Dhawan (2011), packaging represents approximately 23% of all waste weight from global supply chain operations. Specifically in the US, this demand is projected to increase by

3.9% annually, reaching US$41.7 billion in 2014, and resulting in the consumption of 58 billion pounds of material (Green Packaging, 2011). In addition, a small number of academic articles focus on green packaging. Indicatively, Zhang and Zhao (2012) define green packaging as packaging where the components can be reused and recycled, degraded, and corrupted and do not affect humans or the environment during its whole life cycle. In their research effort, the authors initially identify the negative impact of packaging on the environment, which is through waste, liquid, and gaseous pollution, and then move further into analyzing the concept of green packaging and green packaging management strategies at a governmental and corporate level. Barlo and Morgan (2013) evaluate the environmental impact of different packaging strategies applied in the meat and cheese industry in terms of waste and energy consumption. They conclude that even though the current packaging is somewhat wasteful, the environmental impact of alternative packaging strategies is even less desirable. They illustrate the applicability of their methodology in the case of polymer film packaging in the meat and cheese industry. Finally, Silva *et al.* (2013) conduct a technical and environmental analysis for comparing two packaging models, namely disposable and returnable packaging, used for the transportation of machine engine heads. Their purpose is to document that the practice of reverse logistics with returnable packaging may significantly improve a company's environmental performance. The authors also examine the usefulness of their employed methodology in the case of a company that produces machine heads.

5.2.1.1.3   Manufacturing Design   Based on Deif (2011), Green Manufacturing (GM) design incorporates: (i) the development of product processes that consume less material and energy; (ii) the replacement of toxic with non-toxic and non-renewable with renewable input materials; (iii) the minimization of pollution in the manufacturing process outputs; and (iv) the conversion of outputs to inputs (recycling). The authors investigate whether GM could constitute a competitive manufacturing advantage and evaluate possible ways to combine GM with sustainability. Then, they provide an extensive literature review on GM. Zhou *et al.* (2012) provide another definition of GM. The authors identify GM as the integration of environmental protection and energy conservation practices into production and service activities in order to reduce industrial waste, save energy and scarce resource, and minimize pollution of the natural environment and manufacturing costs. The authors propose a simulation-based methodology for capturing the stochastic behavior of production and distribution flows, while they combine this methodology with a robust search algorithm for evaluating and selecting optimal green production strategies. Duflou *et al.* (2012) provide a comprehensive literature review of the methodologies and technologies in the field of discrete part manufacturing that reduce the environmental impact of manufacturing operations. As climate change, energy independence, and energy costs have gained increasing attention in recent years, their paper mainly focuses on energy related issues. Plehn *et al.* (2012) propose a novel machine design and production planning methodology that could assist managers in identifying the environmental impact of manufacturing operations. Finally, Govindan, Diabat, and Shankar (2015) propose a fuzzy Multi-Criteria Decision-Making model, which is based on the Analytic Hierarchy Process (AHP) that could be used as an effective driver for the development of a quick and efficient GM process.

5.2.1.1.4 Product Recovery   As many products have some value at the end of their life cycle, their recovery may result in money and resource savings. On this basis, the practice of product recovery has attracted the research interest of numerous supply chain specialists and academics. Ting, Feng, and Bin (2014) propose a product recovery methodological framework that incorporates a method for evaluating the product's condition at the end of its life cycle. Their proposed framework can be employed by equipment manufacturers in order to quickly adopt an appropriate product recovery policy. Pal, Sana, and Chaudhuri (2013) propose a stochastic inventory model for determining optimal production quantities and their recovery rates after their end-of-life use. The main characteristic of the model is that at the beginning of one production cycle, the new production lot size is a combination of new and recovered components, and the defective products are remanufactured through the recovery of partly damaged products from previous cycles. The authors examine the applicability of their proposed methodology in the case of a crankshaft, which is a component in a refrigerator compressor that converts reciprocating linear piston motion into rotation. Johnson and McCarthy (2013) propose an integer-programming model for determining the optimal recovery plan of a product, which minimizes the costs of remanufacturing under Extended Producer Responsibility legislation. Their model's applicability is examined through the case of a Telecom manufacturer located in Eastern Canada and a subsidiary firm that carries out its asset recovery and remanufacturing operations. Ziout, Azab, and Atwan (2014) develop a holistic decision-making methodology, based on the AHP, for determining optimum product recovery options, while considering the PESTEL approach for identifying the factors with most influence for product recovery decision-making. The proposed method's usefulness is illustrated through the case of General Motors, and its commercial fuel cell powered car that will be available by 2022. Finally, Ondemir and Gupta (2014) propose a multi-objective advanced remanufacturing-to-order and disassembly-to-order (ARTODTO) model. The proposed model is based on mixed integer linear programming (MILP) methodology and determines the product recovery option that optimizes product quality levels and profits. The authors illustrate the practical usefulness of the model through a device-embedded (sensors and radio frequency identification tags) dryer ARTODTO system.

5.2.1.2 Strategic Green Transportation Decisions

Transportation represents the most polluting sector of global supply chain networks, resulting in the production of approximately 8% of global CO_2 emissions (Regmi and Hanaoka, 2009). There are three main decisions that may significantly affect these emissions. The first one involves the selection of environmentally friendly transportation modes, while the second one is associated with the use of environmentally friendly fuels. Finally, the third one involves the determination of the transportation mode's speed. These decisions will be thoroughly analyzed in the following three subsections.

5.2.1.2.1 Transportation Mode Selection Transportation mode selection decisions mainly depend on the type of the transported product. Therefore, for time sensitive products fast truck and air transportation are usually employed, while for slow moving items, rail, inland transportation, and pipelines (for gas and oils) are used. However, with respect to the environment, these modes exhibit different performance characteristics. Table 5.1 summarizes the environmental performance characteristics of different transportation modes.

The recent literature on green transportation mode selection decisions includes the paper of Jin, Marulanda, and Down (2013). The authors propose a MILP model for the transportation mode selection problem under a cap and trade and carbon tax policy. They evaluate the applicability of their modeling methodology through Walmart's supply chain. Soysal, Bloemhof-Ruwaard, and Van der Vorst (2014) develop a multi-objective linear programming (MOLP) model for transportation mode selection decisions, under cost and greenhouse gas (GHG) emissions minimization objectives. The applicability of their proposed methodology is illustrated through a real-world fresh-chilled beef logistics network operating in Brazil and exporting beef to the European Union (EU).

5.2.1.2.2 Green Transportation Fuel Production The choice of fuel may significantly affect the environmental impact of transportation operations. Today, modern gasoline types are carefully refined to reduce their lead additives. Moreover, multi-nationals now progressively reorganize their transport fleet engines to adapt to biofuels and natural gas. A typical example is that of UPS, which has expanded its fleet of heavy-duty trucks that use liquefied natural gas, or LNG, from 112 to 800 in 2014 (Cardwell and Krauss, 2013). On this basis, Janic (2014) examines the potential

Table 5.1   Energy use and emissions of alternative transportation modes.

Energy use/ emissions (g/t/km)	Container vessel (11 000 TEU)	Container vessel (6600 TEU)	Rail electric	Rail-diesel	Heavy truck	Boeing 747–400
kWh/t/km	0.014	0.018	0.043	0.067	0.18	2.00
CO_2	7.48	8.36	18	17	50	552
SOx	0.19	0.21	0.44	0.35	0.31	5.69
NOx	0.12	0.162	0.1	0.00005	0.00006	0.17
PM	0.008	0.009	n.a.	0.008	0.005	n/a

n.a., not available; SOx, sulfur oxides.
Adapted from Dekker, Bloemhof, and Mallidis, 2012. Reproduced with permission of Elsevier.

of using liquid hydrogen for greening air transportation. They provide a detailed analysis of the types of GHG emissions generated from air transportation, along with a review of measures for reducing them. Finally, they document the use of liquid hydrogen fuel as a breakthrough for addressing the environmental impacts of air transportation. Yang *et al.* (2014) provide an extensive literature review of biofuel types that could be used as transportation fuel along with the analysis of the existing technologies for transforming these biofuels to transportation fuel. Finally, Raslavičius *et al.* (2014) examine the potential of transforming algae biofuels to transportation fuel. The authors provide a detailed review of papers that analyze the biological attributes of algae, along with a review of research efforts on technologies for transforming biofuels from algae into transportation fuel.

5.2.1.2.3 Transportation Speed Design   The economic crisis of 2008 resulted in a surplus capacity of container ships, which has continued to increase as preceding ship orders are being implemented and thus, new ships enter the market. An aftermath of this surplus capacity is the significant reduction of container freight rates (Haralambides and Thanopoulou, 2014). This further dictates the need of reducing ship-operating costs as a prerequisite for the conservation of the ship owner's profitability and thus the owner's market share. As fuel costs represent a critical cost factor of ship operations today, their reduction could result in significant cost savings for the carrier. An effective way to do so is to reduce the ship's speed. This practice is called "slow steaming" and has been applied by numerous carriers resulting in significant fuel cost savings and thus emissions. There are numerous studies that analyze the cost and CO_2 emissions impacts of slow steaming through the development of speed models in maritime transportation. To this end, Wang and Meng (2012) develop a mixed integer nonlinear programming model for determining the optimum ship's speed, which minimizes its operating costs. Maloni, Paul, and Gligor (2013) evaluate the cost impact of optimum speed reduction on carriers and suppliers, through the simulation of a high volume Asia–North America container trade line. Psaraftis and Kontovas (2013) provide an extensive literature review of speed optimization models, while Psaraftis and Kontovas (2014) determine close form solutions for determining optimum ship speed and routing decisions that minimize ship costs and emissions, using various cargo value scenarios.

5.2.1.3 Strategic Green Facility Decisions

Facilities are the third driver in supply chains. The strategic facility related decisions involve decisions on the selection of their location and their design.

5.2.1.3.1 Facility Location   Decisions on the number of operating distribution centers may have a large impact on a network's transport efficiency as their increase can significantly reduce outbound transportation distances (Dekker, Bloemhof, and Mallidis, 2012). Consequently, the environmental impact of strategic facility location decisions has received significant attention by researchers. On this basis, Harris *et al.* (2011)

consider both logistics costs and CO_2 emissions in supply chain optimization. They simultaneously evaluate the impact of strategic decisions on depot location, and operational decisions on the utilization rates of trucks, focusing on inventory and transportation costs versus transportation and inventory CO_2 emissions. The calculation of CO_2 emissions from transportation involves different vehicle types and utilization levels (of 90, 75, and 60%) and their approach is examined through a simulation model for the European automotive industry.

Elhedhli and Merrick (2012) examine a three-echelon multi-objective (cost and CO_2 emissions) optimization model, where transportation CO_2 emissions are affected by vehicle weight. The model's objective is to jointly minimize logistics costs and CO_2 emissions by strategically locating facilities within the network. The model is developed as a MILP model and is decomposed by echelon and warehouse location using a Lagrangian Relaxation (LR) approach.

Mallidis, Dekker, and Vlachos (2012) propose a multi-objective MILP methodology in order to decide on: (i) the number and location of distribution centers (DCs) and entry Ports (EPs); (ii) the capacity of the operating distributions centers; (iii) the type of transportation modes employed between the nodes of the network under study; and (iv) the associated flows between the nodes of the network under study. The authors apply their model in a specific supply chain for the distribution of white goods produced in the Far East, and in the emerging region of Southeastern Europe. The results indicated that in most cases, using shared warehouses with Third Party Logistics operators improves both the cost and the environmental performance of a company. In all cases, shared use of transportation operations minimizes the amount of CO_2 and particulate matter emissions generated, while dedicated use minimizes costs. Pishvaee, Razmi, and Torabi (2014) determine optimal material flows between facilities, their processing technology, and their capacities as also locations, through the development of a Benders' decomposition algorithm. Their model optimizes: (i) the network's total logistics costs, namely fixed facility, transportation, and order processing costs; (ii) its environmental impact, namely damage to human health, the ecosystem, and the resources; and (iii) its social impacts, that is, the creation of job opportunities, the reduction of consumers' risk, the damage to workers' health, and the value of local development to communities. Finally, Harris, Mumford, and Naim (2014) determine the optimum number of operating facilities and the optimum allocation of customers to these facilities that minimize transportation and facility costs as also CO_2 emissions. The authors determine these decisions through the employment of a novel methodological framework, based on the multi-objective evolutionary algorithm (MOEA) with an LR approach.

5.2.1.3.2 Sourcing Decisions on the sourcing location of products may significantly affect transportation requirements and thus emissions. Companies that aim to achieve lower production costs have offshored their production operations to Asian countries, Turkey and Mexico, where wages are low (Jensen, Larsen, and Pedersen, 2013; Schmeisser, 2013). However today, and as fuel prices have increased, the combination of off-shoring and near-shoring or even re-shoring solely have emerged as new sourcing alternatives that result in shorter transportation times, increased flexibility, and the maintenance of lower inventories (Zhang *et al.*, 2013; Fratocchi *et al.*, 2014; Tate, 2014).

5.2.1.3.3 Facility Design The environmental efficiency of facilities can be mainly achieved through the use of energy efficient lighting systems and cargo-handling equipment along with the employment of temperature maintenance measures (McKinnon *et al.*, 2010). For example, several zero emissions warehouses have been built by installing energy saving installations and using sophisticated lighting systems and solar cells (Dekker, Bloemhof, and Mallidis, 2012). An extensive literature review of green facility design aspects is provided in Zuo and Zhao (2014). The authors provide a review of green building assessment tools and sustainable green building design aspects. They identify three critical components for an effective green facility design, namely: (i) utilization of renewable energy technological innovations, such as solar heat water, solar photovoltaic (PV) systems, small-scale wind turbines, and geothermal heat pumps; (ii) the control of waste during demolition or construction; and (iii) the use of recycled and reused demolition waste. Moreover, and with respect to the first component, Abdallah, Diabat, and Rigter (2013) develop a modeling methodology based on Net Present Value Analysis in order to estimate the cost of PV systems installed on facility rooftops. To this end, the authors evaluate the applicability of their methodology through a sensitivity analysis of different carbon credit prices, PV generation costs, and feed-in tariff prices.

5.2.2   Tactical GSCM Decisions

On the tactical side, with time horizons of a quarter to a year, the main GSCM decisions involve green procurement, manufacturing, inventory, facility, and transportation planning, as well as revenue management.

5.2.2.1 Tactical Green Product Decisions

Tactical green product planning decisions involve green procurement, manufacturing and inventory planning, and finally revenue management.

5.2.2.1.1 Procurement Green procurement decisions mainly involve supplier selection decisions based on their environmental performance (see for example Kannana *et al.* 2014). Walker *et al.* (2012), McMurray *et al.* (2014), and Correia *et al.* (2013) provide an extensive literature review of sustainable procurement practices. Walker *et al.* (2012) document the growing interest of researchers on sustainable procurement, along with the challenges that this research field has to face. Finally, they propose a sustainable procurement framework that will assist researchers in identifying all aspects of sustainable procurement. McMurray *et al.* (2014) identify the most important sustainable procurement practices along with the opportunities and barriers of their implementation. They document their findings by examining the state of sustainable procurement in Malaysian organizations. Finally, Correia *et al.* (2013) provide a literature review of low carbon procurement research efforts, identify the challenges for low carbon procurement along with the future research perspectives. Kannana *et al.* (2013) develop a comprehensive green supplier selection and order-allocation modeling methodology, according to economic and environmental criteria and based on the fuzzy multi-attribute utility theory and multi-objective programming methodology. The applicability of the proposed methodology is illustrated through a realistic case study of an automobile manufacturing company, thus assisting companies to develop supplier selection and order allocation problems in realistic cases. Finally, Su and Lin (2014) propose a fuzzy multi-objective linear programming (FMOLP) model to solve integrated, procurement/production planning problems for recoverable manufacturing systems. Their model jointly minimizes production costs and lead times to multiple component sources, vendors, and machines. The authors examine the applicability of their proposed methodology on the product case of a laser regeneration printer cartridge.

5.2.2.1.2 Manufacturing Planning   Manufacturing planning decisions indicate how well production is organized. Additionally, they determine the optimum quantities produced, the average levels of inventories reserved, and the optimal planning of manufacturing equipment operations. All the above operations result in considerable amounts of emissions and waste and thus, their optimization could result in significant environmental savings. In this respect, Mirzapour Al-e-hashem, Baboli, and Sazvar (2013) propose an aggregate production planning model that determines the amount of products produced, the number of workers hired or fired, the inventory levels of the product, the production assigned to each plant, the number of products shipped from node to node, and the number of vehicles type "g" employed for transportation between node to node under an allowed amount of CO_2 emissions and waste produced. Ohara *et al.* (2014) propose an optimization approach for hot rolling planning that optimizes the energy consumption of the sheet manufacturing process, while preserving the sheet's mechanical properties. Liu *et al.* (2014) propose a production planning bi-objective modeling approach of the so-called seru s production systems. The model determines the start time for producing a product type, and the production quantity of the product in seru s, which minimizes its production energy consumption and CO_2 emissions. Finally, Newman *et al.* (2012) provide a review on computer aided process planning, and multi-criteria process planning methodologies, which are further adjusted to additionally identify environmentally conscious GM planning. The applicability of their proposed methodology is illustrated through the process of finish cutting of aluminum.

5.2.2.1.3 Inventory Planning   As in the case of manufacturing planning, inventory-planning decisions determine the amount of stocks reserved at distribution facilities and demand points, along with the number of transportation consignments delivered between the nodes of a supply chain network. Thus, the optimization of these decisions leaves room for significant energy savings and thus environmental improvements. Several authors consider inventory planning decisions under a carbon cap and trade regulatory framework, where emissions due to single inventory decisions are limited. On this basis, Hua *et al.* (2011) propose a green economic order quantity (EOQ) inventory planning methodology for determining a company's optimal order quantity, under the carbon emissions trading scheme. The model's objective is to minimize on the one hand the CO_2 emissions, and on the other hand, the total costs of logistics operations. Bouchery *et al.* (2012) extend the model of Hua *et al.* (2011) by additionally incorporating social impact optimization objectives in their EOQ model. They refer to their model as the sustainable order quantity (SOQ) model, and further extend it for application to multi-echelon logistics networks. Sazvar *et al.* (2014) determine the inventory and shortage levels, the order quantity, the total sales, and the number of different types of vehicles for transporting a specific deteriorating product, which minimizes on the one hand the ordering, holding, recycling, and transportation costs, and on the other hand, the transportation, production, and recycling GHG emissions. Their employed methodology is a multi-echelon bi-objective MILP model. Konur (2014) determines the optimum order quantity, truckload, and truck type, which minimizes purchase, setup costs, and inventory holding and transportation costs. Their employed methodology is the single echelon EOQ model further solved under a constraint of inventory holding and transportation CO_2 emissions. Konur and Schaefer (2014) determine the optimum order quantity considering Less than Truckload (LTL) and Full Truckload (FTL) transportation, which minimizes the holding, ordering, transportation costs, and CO_2 emissions, under alternative green regulatory policies. Finally, Arıkan and Jammernegg (2014) determine the optimum order quantity of a newsvendor from a supplier, which maximizes the newsvendor's profits under a constraint on the transportation, warehousing emissions, and on the additional emissions of the products not sold. The authors employ a single period newsboy inventory planning model and further extend it by considering a second emergency supply channel.

5.2.2.1.4 Revenue Management Pricing   Revenue management was initially developed for airlines and its goal is to maximize the passenger's yield by varying prices based on the demand. Higher utilization of transportation modes results in less air deliveries and thus reduced emissions. The same applies for cargo. On this basis, Lovrić *et al.* (2013) propose a multi-agent, simulation-based modeling methodology that deals with a sustainable revenue management approach, which incorporates economic, social, and environmental aspects. The economic dimension includes the number of traveling passengers, the travel distance, the total revenue, and the vehicle's average load factor. The social dimension involves the passenger's convenience and comfort during transportation. Finally, the environmental dimension incorporates the evaluation of the impact of changing transit ridership in CO_2 emissions savings.

However, today revenue management is also employed for promoting sustainable corporate practices. Panda (2014) proposes a Manufacturer–Stackelberg game setting considering a revenue sharing contract in a manufacturer–retailer supply chain. The author evaluates the impact of implementing corporate social responsibility (CSR) practices on the price of the revenue sharing contract. On the same basis, Hsueh (2014) examines the supply chain option of one manufacturer and a retailer. The manufacturer signs a revenue sharing contract with the retailer and charges the retailer a wholesaler's price. The author determines the order quantity and unit cost of CSR that maximizes the profit and CSR performance of a centralized and a decentralized supply chain. Finally, Govindan and Popiuc (2014) manage a three-echelon closed loop supply chain under revenue sharing. They develop an analytical modeling methodology that quantifies the impact of a discount offered by a retailer on its customer's willingness to return its obsolete products. They illustrate the applicability of their proposed methodology through the case study of Apple Inc., which collects used devices from customers in return for a coupon that can be used to purchase another Apple product directly from the Apple web site or from Apple retail stores.

5.2.2.2 Transportation Planning

Green transportation planning decisions aim at minimizing the environmental impact of transportation operations by: (i) determining the optimum number and capacity of transportation equipment; and (ii) by coordinating alternative transportation consignments in a route. On this basis, Rosskopf, Lehner, and Gollnick (2014) propose a multi-objective fleet optimization problem, which determines: (i) the number of aircraft owned/leased; (ii) the number of aircraft bought/sold; (iii) the cash surplus/ deficit; and (iv) the aircraft type introduced in the fleet, which maximizes on the one hand the asset value of the fleet and minimizes the nitrogen oxides (NOx) emissions produced from their flight operations on the other. They illustrate the applicability of their proposed methodology through a real-world case of a European network carrier, which owns 270 aircraft and conducts 400 000 flights per year. Díaz-Madroñero, Peidro, and Mula (2014) deal with a three-level supply cahin in the automobile industry sector, which consists of a second-tier supplier, a first-tier supplier, and an automobile assembler. Supply between tiers occurs by truck transportation. The authors propose a FMOLP model that minimizes the total number of trucks utilized, and the total amount of inventory reserved. Khoo and Teoh (2014) propose a bi-objective fleet-planning model for determining the quantity, type of aircraft to be purchased or leased, and the quantity ordered in order to optimize the airline's environmental performance and profitability. Their model's usefulness is illustrated through the case of five aircraft types, namely B737–400, B737–800, B777–200, A330–300, and A380, which are considered for 38 Origin–Destination pairs. These aircraft were selected based on the fleet composition of Malaysia Airlines, which are used for serving international routes. Tsai *et al.* (2012) propose a mixed activity-based costing decision (MABCD) model for green airline fleet planning under the Emissions Trading Scheme. The model maximizes the total profit from airline operations, which consists of the total revenue minus the total operating, leasing, and CO_2 emissions costs. The model's applicability is illustrated through a case of both B747-400 and A380 aircraft flights between Hong Kong and Paris airports. Finally, Jeon, Amekudzi, and Guensler (2013) provide a literature review of quantitative models for assessing the sustainability of transportation planning operations, while analyzing the environmental, social, and economic sustainability indicators of transportation operators.

5.2.2.3   Fleet Management

Fleet management decisions are closely related to transportation planning decisions as they mainly deal with decisions on transportation mode utilization levels (Harris *et al.*, 2011) and fleet number changes. In this respect, Biellia, Biellib, and Rossic (2011) stress the difficulty of addressing fleet management problems, as these require the joint implementation of vehicle routing, scheduling, planning, and network design decisions. They indicate the main problems in the fleet management of different transportation modes and conduct a review of the modeling methodologies and algorithms employed for dealing with these problems. Pan, Ballot, and Fontane (2011) describe a case where transportation equipment is pooled between several companies in order to increase load factors (which are stated to be only 70%). To this end, they quantify the effect of pooling on cost and CO_2 emissions. Bae, Sarkis, and Yoo (2011) develop a two-stage game theoretic model that assists supply chain stakeholders in evaluating the impacts of greening their transportation fleets. In the first stage, the company determines the percentage of its fleet that must go green, while in the second stage, the optimal price for green transportation services is defined. Numerous parameters affect these decisions such as cost of fuel, regulatory compliance requirements, adaptation costs, adjusting tax policies, and so on. Finally, Stasko and Gao (2012) propose a customized stochastic approximate dynamic programming (ADP) model for determining vehicle purchase, resale, and retrofit policies that minimize expected discounted net costs under environmental regulation constraints.

5.2.2.4   Facility Planning

Green facility planning decisions involve the planning of the facility's resources in a way that their energy requirements are minimized. Muellera, Cannataa, and Herrmann (2014) identify heat recovery potentials from equipment, decentralized disposal chips, filtration, and by using natural conditions for light intense operations. To this end, they propose a modular green-factory planning approach that will assist planners to improve the environmental performance of their facility. Finally, Somplák *et al.* (2013) propose a multi-objective, two-stage stochastic programming model, which determines the waste treatment capacity and the heat or electricity-oriented operation of a facility, which maximizes the return on investment and minimizes waste production.

5.2.3   Operational GSCM Decisions

Although it may seem that strategic choices determine most of the environmental impacts of logistics operations, in daily operations there is also lots of scope for environmental improvement. Aspects such as quality control, work methods, vehicle routing, and speed control are addressed in the following.

5.2.3.1 Product Quality Control

Shelf life discard of products is a process where the batch size of a product is rejected if one item is out of date or defective. On this basis, accurate shelf-life estimation is initially required and then, quality control-based methods should be employed. Giménez, Ares, and Ares (2012) propose a trained assessors' panel or a group of experts, in order to check the products' quality during storage time at the warehouse. Through this method, it will be eventually identified whether the products have reached their shelf life and are therefore not salable. Etaio *et al.* (2011) deal with commercialized products that have quality labels. They propose a sensory certification approach that employs generic score cards for testing the quality of these products. They evaluate the applicability of their proposed methodology through the product case of the Bizkaiko Txakolina wine. However, quality control is also employed for reducing defective production. On this basis, Bettayeb, Bassetto, and Sahnoun (2014) propose a quality control process that determines the optimal quality control plan of a machine and minimizes the total number of the plan's controls, under the constraint of the machine's inspection capacity limitation and risk exposure objectives. The purpose of their proposed quality control process is to prevent the production of an excessive amount of scrap. Finally, Zhang *et al.* (2012) provide a novel final product quality control method for batch operations. Their proposed model integrates the midcourse correction (MCC) and the batch-to-batch approaches. MCC is used during the batch's development to reduce variations in final quality, while the batch-to-batch control improves the performance of future batch productions using results from previous batches produced.

5.2.3.2 Transportation

Operational control decisions for transportation involve decisions on equipment maintenance, vehicle routing, and speed control.

5.2.3.2.1 Equipment Maintenance   On a daily basis, a trained driver should carefully observe the vehicle's control panel-operating characteristics, in order to identify potential functional problems that may reduce the vehicle's fuel efficiency and thus environmental performance (McKinnon *et al.*, 2010). On the same basis, preventive maintenance is also a practice that could significantly improve the vehicle's fuel efficiency. Vujanović *et al.* (2012) document the significance of proper vehicle maintenance on its energy efficiency. They identify three significant indicators of an effective maintenance process. The first indicator evaluates the manufacturing process, the second indicator evaluates the transportation process, and finally, the third indicator evaluates the impact of the maintenance process on the environmental performance of the vehicles. To this end, the authors determine the interdependence and weighted importance of each indicator by combining the Decision

Making Trial and Evaluation Laboratory (DEMATEL) and the Analytic Network Process (ANP) methodology. The proposed methodology has been implemented in the case of numerous transportation companies with their own vehicle fleets in Serbia. Finally, Go, Kim, and Lee (2013) develop a MILP model with containership constraints in order to address the preventive maintenance scheduling problem for containerships. Their model determines the due-date and the actual start time of each maintenance activity, under a constraint of the workforce availability, working time limitation, and inter-maintenance time.

5.2.3.2.2 Vehicle Routing   A vehicle routing policy may reduce the total kilometer distances traveled and thus the environmental impact. Consequently recent research efforts have been published aiming at evaluating the impact of these decisions on the environment. An extensive literature review of green vehicle routing is presented in Lin et al. (2014). The authors identify green vehicle routing as the practice that deals with the optimization of the energy consumption of vehicles. Their review identifies the current studies on green vehicle routing after 2007, while it provides further future research directions. Demir, Bektas, and Laporte (2014b) provide an extensive literature review of: (i) vehicle emissions calculation models; and (ii) green vehicle routing papers that incorporate these models in their optimization process. Erdogan and Miller-Hooks (2012) also deal with the green vehicle routing problem. They employ a MILP model, which determines the route that a vehicle will take, its minimum fuel level reserve, and its optimum time of arrival. The model's objective is to minimize the total transportation distances traveled. Jabali, Van Woensel, and de Kok (2012) propose an Emissions-based Time-Dependent Vehicle Routing Problem. Their model differs from the traditional vehicle routing problem as it also considers time-dependent travel times and emissions cost. Thus, this model decides on the optimal upper limit of the vehicle's travel speed, besides the optimal route selection, which minimizes the driver's costs and vehicles fuel and CO_2 emissions costs. Cirovic, Pamucar, and Bozanic (2014) deal with the vehicle routing of light-duty vehicles. They propose a neuro-fuzzy model and a modified Clarke–Wright (CW) algorithm for determining the optimum travel routes of light-duty vehicles, which minimize their costs, emissions, and noise. Bing et al. (2014) deal with the problem of collecting plastic waste. The problem is modeled as a vehicle routing problem and is solved through a tabu search algorithm. Their model determines the routes that minimize transportation and labor costs along with the vehicle's emissions costs when driving, and when idle. Finally, Demir, Bektas, and Laporte (2014a) propose a multi-objective optimization methodology for dealing with the pollution routing problem. The proposed model is based on an enhanced adaptive large neighborhood search (ALNS) and a specialized speed optimization algorithm and is employed to jointly minimize fuel consumption and driving time.

5.2.3.2.3 Speed Control   Speed control reduces fuel consumption and thus emissions and could be achieved by maintaining braking and speed uniformity. On this basis, Yun et al. (2011) examine the environmental impact of speed uniformity in rail transportation. The results indicate that energy savings up to 6.8% can be achieved by keeping speed uniform with the traveling time of the train exhibiting very small increases. Moreover, the continuous technological improvements have resulted in the development of sophisticated vehicle control systems such as adaptive cruise control systems that have been shown to effectively manage speed uniformity. To this end, Li *et al.* (2014) examine the cruise control scheduling of a high-speed train based on sampled data. The authors model the dynamics of a high-speed train and design the sampled-data control, which guarantees that the high-speed train tracks the desired speed, and that the relative spring displacement between the two neighboring cars is stabilized to the equilibrium state with respect to the wind gust disturbance. Finally, Liu, Han, and Lu (2013) propose a high-speed railway fuzzy control system, which optimizes the train's energy consumption, comfort during travel, high speed, and safety. The system's evaluation has been made through simulation experimentation and the results indicated that the fuzzy control system is effective and accurate in the high-speed railway control process.

5.2.3.3   Facilities

Finally, operation control decisions for facilities involve the development of order-picking processes that reduce distances traveled by the facilities' mechanical equipment, and the training of employees in order to ensure the product's quality during its life cycle.

5.2.3.3.1   Control of Order Picking   The efficient use of a facility's order-picking process is critical for its environmental efficiency as it reduces the travel times of its equipment and thus, the energy requirements. On this basis, Andriansyah *et al.* (2011) develop a novel simulation modeling methodology for a miniload-workstation order-picking system. The model evaluates the impact of different control heuristics and numbers of miniloads in the system's time performance and its applicability is evaluated through a realistic industrial scale distribution center. Moellera (2011) proposes a Line Sequence Optimization Approach, which involves an optimization routine that determines the line sequence for a given batch that minimizes travel times. The author illustrates the applicability of its proposed methodology to the case of an electrical devices' distributor, while the derived results indicate that order-picking process times could be improved by an average of 7.4%. Finally, Zheng, Mohr, and Yoon (2014) develop an order-picking scheduling methodology for a food product distribution center. Their proposed methodology is based on a MILP model and determines optimum order-picking routes and picking truck assignments.

5.2.3.3.2   Work Methods   Work methods involve training employees to focus on maintaining the product's quality and thus avoid product discards. Soon and Baines (2012) examine issues associated with hand hygiene practices adopted by fresh product farm workers, along with the development of farm food safety educational and training materials. Teixeira, Jabbour, and de Sousa Jabbour (2012) discuss the environmental training guidelines of employees in Brazil, based on the ISO 14001 and ISO 10015 standards. Some of these practices involve the training of internal agents to identify environmental causes, participating in international fairs to exchange effective social–environmental practices, and environmental training as certification provisions. Finally, Jabbour (2013) provides an extensive literature review on environmental training in organizations. He further identifies the research gaps in the literature and proposes future research perspectives. Through the proposed framework, the author reveals the characteristics and constraints of a successful environmental training process in the organizational sector.

5.3   Critical Synthesis of Academic Research Efforts

The above analysis has clearly demonstrated that GSCM is a rapidly evolving research area. Table 5.2 summarizes the academic research efforts associated with the strategic, tactical, and operational decision phases of each one of the three main supply chain drivers, namely products, transportation, and facilities. This taxonomy can be further used to identify gaps and overlaps of key issues tackled by the existing literature.

The main observations derived from our critical synthesis are summarized as follows:

- Strategic supply chain management decisions have attracted the research interest of most researchers. Such an outcome was anticipated, as these decisions have a long-term impact on a supply chain's environmental performance.

- From a strategic design perspective, there is a gap in the academic research efforts that deal with the design of green buildings. As the environmental impacts of facilities are mainly attributed to the energy consumption due to their lighting and heating requirements, their improved efficiency can be mainly achieved through technological interventions, rather than through the use of quantitative models.

- From a tactical planning perspective, there is a gap in the literature associated with green facility planning. As these decisions mainly involve the optimization of equipment capacities and processes, this further highlights potential research areas that could be evaluated through the use of quantitative optimization methodologies.

- From an operational control perspective, there is a gap in the literature associated with the environmental impact of proper vehicle and equipment maintenance. On this basis, an optimum preventive maintenance schedule that optimizes the vehicles' emissions performance could be a very interesting approach to the problem.

- The majority of academic research efforts address green product related problems, while not much attention has been focused on green facility related problems.

Table 5.2 Classification of the literature by driver and decision.

Drivers \ Decisions	Products (number of papers: 42)	Facilities (number of papers: 19)	Transportation (number of papers: 31)
Strategic (number of papers: 38)	**Product design:** 1. Bovea and Belis (2012) 2. Chen *et al.* (2012) 3. Yangjian *et al.* (2013) 4. Chen and Liu (2014) 5. Russo and Rizzi (2014) **Packaging design:** 6. Palanivelu and Dhawan (2011) 7. Zhang and Zhao (2012) 8. Barlo and Morgan (2013) 9. Silva *et al.* (2013) **Manufacturing design:** 10. Deif (2011) 11. Zhou *et al.* (2012) 12. Duflou *et al.* (2012) 13. Plehn *et al.* (2012) 14. Govindan, Diabat, and Shankar (2015)	**Facility design:** 1. Abdallah, Diabat, and Rigter (2013) 2. Zuo and Zhao (2014) **Facility location:** 3. Harris *et al.* (2011) 4. Elhedhli and Merrick (2012) 5. Mallidis, Dekker, and Vlachos (2012) 6. Pishvaee, Razmi, and Torabi (2014) 7. Harris, Mumford, and Naim (2014) **Sourcing:** 8. Jensen, Larsen, and Pedersen (2013) 9. Schmeisser (2013) 10. Tate (2014) 11. Fratocchi *et al.* (2014)	**Transportation mode selection:** 1. Jin, Marulanda, and Down (2013) 2. Soysal, Bloemhof-Ruwaard, and Van der Vorst (2014) **Green transportation fuel production:** 3. Janic (2014) 4. Yang *et al.* (2014) 5. Raslavičius *et al.* (2014) **Transportation speed design:** 6. Wang and Meng (2012) 7. Maloni, Paul, and Gligor (2013). Psaraftis and Kontovas (2013) 8. Psaraftis and Kontovas (2014)

(continued overleaf)

Table 5.2 (Continued)

Decisions / Drivers	Products (number of papers: 42)	Facilities (number of papers: 19)	Transportation (number of papers: 31)
	Product recovery: 15. Ting, Feng, and Bin (2014) 16. Pal, Sana, and Chaudhuri (2013) 17. Johnson and McCarthy (2013) 18. Ziout, Azab, and Atwan (2014) 19. Ondemir and Gupta (2014)		**Transportation planning:** 9. Rosskopf, Lehner, and Gollnick (2014) 10. Díaz-Madroñero, Peidro, and Mula (2014) 11. Khoo and Teoh (2014) 12. Tsai et al. (2012) 13. Jeon, Amekudzi, and Guensler (2013)
	Procurement: 20. Walker et al. (2012) 21. McMurray et al. (2014) 22. Correia et al. (2013) 23. Kannana et al. (2013) 24. Su and Lin (2014)	**Facility planning:** 12. Muellera, Cannataa, and Herrmann (2014) 13. Somplák et al. (2013)	
Tactical (number of papers: 27)	**Manufacturing planning:** 25. Mirzapour Al-e-hashem, Baboli, and Sazvar (2013) 26. Ohara et al. (2014) 27. Liu et al. (2014) 28. Newman et al. (2012)		**Fleet management:** 14. Harris et al., 2011 15. Biellia, Biellib, and Rossic (2011) 16. Pan, Ballot, and Fontane (2011) 17. Bae, Sarkis, and Yoo (2011) 18. Stasko and Gao (2012)

Operational (number of papers: 23)			
Inventory planning: 29. Hua et al. (2011) 30. Bouchery et al. (2012) 31. Sazvar et al. (2014) 32. Konur (2014) 33. Konur and Schaefer (2014) 34. Arıkan and Jammernegg (2014) **Revenue-management:** 35. Lovrić et al. (2013) 36. Panda (2014) 37. Hsueh (2014) 38. Govindan and Popiuc (2014) **Product quality control:** 39. Giménez, Ares, and Ares (2012) 40. Etaio et al. (2011) 41. Bettayeb, Bassetto, and Sahnoun (2014) 42. Zhang et al. (2012)	**Control of order picking:** 14. Andriansyah et al. (2011) 15. Moellera (2011) 16. Zheng, Mohr, and Yoon (2014) **Work methods:** 17. Soon and Baines (2012) 18. Teixeira, Jabbour, and de Sousa Jabbour (2012) 19. Jabbour (2013)		**Maintenance:** 19. McKinnon et al., 2010 20. Vujanović et al. (2012) 21. Go, Kim, and Lee (2013)

(continued overleaf)

Table 5.2 (Continued)

Decisions / Drivers	Products (number of papers: 42)	Facilities (number of papers: 19)	Transportation (number of papers: 31)
			Vehicle routing: 22. Lin et al. (2014) 23. Erdogan and Miller-Hooks (2012) 24. Demir, Bektas, and Laporte (2014a) 25. Jabali, Van Woensel, and de Kok (2012) 26. Cirovic, Pamucar, and Bozanic (2014) 27. Bing et al. (2014) 28. Demir, Bektas, and Laporte (2014b)
			Speed control: 29. Yun et al. (2011) 30. Li et al. (2014) 31. Liu, Han, and Lu (2013)

5.4   Summary and Conclusions

In this chapter, we propose a hierarchical GSCM framework that aims, through an extensive literature review of academic research efforts from 2011 onwards, to assist supply chain stakeholders in: (i) providing an outline of the main decisions that affect the environmental performance of all three main physical drivers of a supply chain, namely products, facilities, and transportation, along with the quantitative models employed for quantifying this performance; and (ii) identifying how could these transcend the strategic, tactical, and operational decision phase. The chapter updates the work of Dekker, Bloemhof, and Mallidis (2012) and further extends it to consider additionally: (i) product design; (ii) manufacturing design and planning; (iii) transportation planning; (iv) product quality control; (v) equipment maintenance; and (vi) green work methods. Moreover, it provides a completely different review approach compared with that of Brandenburg *et al.* (2014), who provide an extensive sustainable supply chain management literature review, focusing on quantitative policies classified by the objective function, the employed research methodology, and the type of the sustainability focus of the supply chain. To this end, this chapter provides a critical synthesis of academic research efforts, where important observations are derived. As such, a gap is identified in the academic research areas of green facility design, planning, and equipment maintenance, while it is further documented that green product related research areas have gained significant popularity in recent last years.

Acknowledgments

This study has been conducted in the context of the GREEN-AgriChains project that is funded from the European Community's Seventh Framework Programme (FP7-REGPOT-2012-2013-1) under Grant Agreement No. 316167. This chapter reflects only the authors' views; The EU is not liable for any use that may be made of the information contained herein.

References

Abdallah, T., Diabat, A., and Rigter, J., 2013. Investigating the option of installing small scale PVs on facility rooftops in a green supply chain. *International Journal of Production Economics*, **146**, 465–477.

Andriansyah, R., de Koning, W., Jordan, R., Etman, L., and Rooda, J., 2011. A process algebra based simulation model of a miniload-workstation order picking system. *Computers in Industry*, **62**, 292–300.

Arıkan, E. and Jammernegg, W., 2014. The single period inventory model under dual sourcing and product carbon footprint constraint. *International Journal of Production Economics*, **62**, 1–9.

Bae, S., Sarkis, J., and Yoo, C., 2011. Greening transportation fleets: insights from a two-stage game theoretic model. *Transportation Research Part E: Logistics and Transportation Review*, **47**, 793–807.

Barlo, C. and Morgan, D., 2013. Polymer film packaging for food: an environmental assessment. *Resources, Conservation and Recycling*, **78**, 74–80.

Bettayeb, B., Bassetto, S., and Sahnoun, M., 2014. Quality control planning to prevent excessive scrap production. *Journal of Manufacturing Systems*, **33**, 400–411.

Biellia, M., Biellib, A., and Rossic, R., 2011. Trends in models and algorithms for fleet management. *Procedia Social and Behavioral Sciences*, **20**, 4–18.

Bing, X., de Keizer, M., Bloemhof, J., and van der Vorst, J., 2014. Vehicle routing for the eco-efficient collection of household plastic waste. *Waste Management*, **34**, 719–729.

Bouchery, Y., Ghaffari, A., Jemai, Z., and Dallery, Y., 2012. Including sustainability criteria into inventory models. *European Journal of Operational Research*, **222**, 229–240.

Bovea, M. and Belis, P., 2012. A taxonomy of ecodesign tools for integrating environmental requirements into the product design process. *Journal of Cleaner Production*, **20**, 61–71.

Brandenburg, M., Govindan, K., Sarkis, J., and Seuring, S., 2014. Quantitative models for sustainable supply chain management: developments and directions. *European Journal of Operational Research*, **233**, 299–312.

Cardwell, D. and Krauss, C., 2013. Trucking Industry Is Set to Expand Its Use of Natural Gas. The New York Times, http://www.nytimes.com/2013/04/23/business/energy-environment/natural-gas-use-in-long-haul-trucks-expected-to-rise.html?pagewanted=all&_r=0 (accessed June 22, 2015).

Chen, C. and Liu, L., 2014. Pricing and quality decisions and financial incentives for sustainable product design with recycled material content under price leadership. *International Journal of Production Economics*, **147**(Part C), 666–667.

Chen, C., Zhu, J., Yu, J., and Noori, H., 2012. A new methodology for evaluating sustainable product design performance. *European Journal of Operational Research*, **221**, 348–359.

Chu, C., Luh, Y., Li, T., and Chen, H., 2009. Economical Green Product design based on simplified computer-aided product structure variation. *Computers in Industry*, **60**, 485–500.

Cirovic, G., Pamucar, D., and Bozanic, D., 2014. Green logistic vehicle routing problem: routing light delivery vehicles in urban areas using a neuro-fuzzy model. *Expert Systems with Applications*, **41**, 4245–4258.

Correia, F., Howard, M., Hawkins, B., Pye, A., and Lamming, R., 2013. Low carbon procurement: an emerging agenda. *Journal of Purchasing & Supply Management*, **19**, 58–64.

Cyprus Ministry of Labour, 2006. A Guide to CSR. MLSI, Cyprus, http://www.mlsi.gov.cy/mlsi/sws/sws13.nsf/3465aea6a6052333c22570e1002eeae2/f8a05fe8440f5cb4c22570e100305444/$FILE/%CE%95%CF%84%CE%B1%CE%B9%CF%81%CE%B9%CE%BA%CE%AE%20%CE%9A%CE%BF%CE%B9%CE%BD%CF%89%CE%BD%CE%B9%CE%BA%CE%AE%20%CE%95%CF%85%CE%B8%CF%8D%CE%BD%CE%B7.pdf (accessed July 16, 2015).

Deif, A., 2011. A system model for green manufacturing. *Journal of Cleaner Production*, **19**, 1553–1559.

Dekker, R., Bloemhof, J., and Mallidis, I., 2012. Operations Research for Green Logistics – an overview of aspects issues and contributions. *European Journal of Operational Research*, **219**, 671–679.

Demir, A., Bektas, T., and Laporte, G., 2014a. The bi-objective pollution-routing problem. *European Journal of Operational Research*, **232**, 464–478.

Demir, E., Bektas, T., and Laporte, G., 2014b. A review of recent research on green road freight transportation. *European Journal of Operational Research*, **237**, 775–793.

Díaz-Madroñero, M., Peidro, D., and Mula, J., 2014. A fuzzy optimization approach for procurement transport operational planning in an automobile supply chain. *Applied Mathematical Modelling*, **38**, 5705–5725.

Duflou, J., Sutherland, J., Dornfeld, D., Herrmann, C., Jeswiet, J., Kara, S., Hauschild, M., and Kellens, K., 2012. Towards energy and resource efficient manufacturing: a processes and systems approach. *CIRP Annals - Manufacturing Technology*, **61**, 587–609.

Elhedhli, S. and Merrick, R., 2012. Green supply chain network design to reduce carbon emissions. *Transportation Research Part D*, **17**, 370–379.

Erdogan, S. and Miller-Hooks, E., 2012. A green vehicle routing problem. *Transportation Research Part E*, **48**, 100–114.

Etaio, I., Gil, P., Ojeda, M., Albisu, M., Salmerón, J., and Elortondo, F., 2011. Improvement of sensory quality control in PDO products: an example with txakoli white wine from Bizkaia. *Food Quality and Preference*, **23**, 138–147.

Fratocchi, L., Di Mauro, C., Barbieri, P., Nassimbeni, G., and Zanoni, A., 2014. When manufacturing moves back: concepts and questions. *Journal of Purchasing & Supply Management*, **20**, 54–59.

Giménez, A., Ares, F., and Ares, G., 2012. Sensory shelf-life estimation: a review of current methodological approaches. *Food Research International*, **49**, 311–325.

Go, H., Kim, J., and Lee, D., 2013. Operation and preventive maintenance scheduling for containerships: mathematical model and solution algorithm. *European Journal of Operational Research*, **229**, 626–636.

Govindan, K., Diabat, A., and Shankar, M., 2015. Analyzing the drivers of green manufacturing with fuzzy approach. *Journal of Cleaner Production*, **96**, 182–193.

Govindan, K. and Popiuc, M., 2014. Reverse supply chain coordination by revenue sharing contract: a case for the personal computers industry. *European Journal of Operational Research*, **233**, 326–336.

Green Packaging, 2011. New US Industry Forecasts for 2014 & 2019. http://recyclablepackaging.files.wordpress.com/2011/01/greenpackaging.pdf (accessed July 16, 2015).

Haralambides, H. and Thanopoulou, H. (2014). The Economic Crisis of 2008 and World Shipping: Unheeded Warnings. *SPOUDAI Journal of Economics and Business*, **64**, 5–13.

Harris, I., Mumford, C., and Naim, M., 2014. A hybrid multi-objective approach to capacitated facility location with flexible store allocation for green logistics modeling. *Transportation Research Part E*, **66**, 1–22.

Harris, I., Naim, M., Palmer, A., Potter, A., and Mumford, C., 2011. Assessing the impact of cost optimization based on infrastructure modelling on CO_2 emissions. *International Journal of Production Economics*, **131**, 313–321.

Hsueh, C., 2014. Improving corporate social responsibility in a supply chain through a new revenue sharing contract. *International Journal of Production Economics*, **151**, 214–222.

Hua, T., Guowei, T., Cheng, T., and Wang, S., 2011. Managing carbon footprints in inventory control. *International Journal of Production Economics*, **132**, 178–185.

Jabali, O., Van Woensel, T., and de Kok, A., 2012. Analysis of travel times and CO_2 emissions in time-dependent vehicle routing. *Production and Operations Management*, **21**, 1060–1074.

Jabbour, C., 2013. Environmental training in organisations: from a literature review to a framework for future research. *Resources, Conservation and Recycling*, **74**, 144–155.

Janic, M., 2014. Greening commercial air transportation by using liquid hydrogen (LH_2) as a fuel. *International Journal of Hydrogen Energy*, **39**, 16426–16441.

Jensen, P.Ø., Larsen, M., and Pedersen, T., 2013. The organizational design of offshoring: taking stock and moving forward. *Journal of International Management*, **19**, 315–323.

Jeon, C., Amekudzi, A., and Guensler, R., 2013. Sustainability assessment at the transportation planning level: performance measures and indexes. *Transport Policy*, **25**, 10–21.

Jin, M., Marulanda, N., and Down, I., 2013. The impact of carbon policies on supply chain design and logistics of a major retailer. *Journal of Cleaner Production*, **85**, 1–9.

Johnson, M. and McCarthy, I., 2013. Product recovery decisions within the context of extended producer responsibility. *Journal of Engineering and Technology Management*, **34**, 9–28.

Kannana, D., Khodaverdi, R., Olfat, L., Jafarian, A., and Diabat, A., 2013. Integrated fuzzy multi criteria decision making method and multi-objective programming approach for supplier selection and order allocation in a green supply chain. *Journal of Cleaner Production*, **47**, 355–367.

Khoo, H. and Teoh, L., 2014. A bi-objective dynamic programming approach for airline green fleet planning. *Transportation Research Part D*, **33**, 166–185.

Konur, D., 2014. Carbon constrained integrated inventory control and truckload transportation with heterogeneous freight trucks. *International Journal of Production Economics*, **153**, 268–279.

Konur, D. and Schaefer, B., 2014. Integrated inventory control and transportation decisions under carbon emissions regulations: LTL vs. TL carriers. *Transportation Research Part E*, **68**, 14–38.

Li, S., Yang, L., Li, K., and Gao, Z., 2014. Robust sampled-data cruise control scheduling of high speed train. *Transportation Research Part C*, **46**, 274–283.

Lin, C., Choy, K., Ho, G., Chung, S., and Lam, H., 2014. Survey of green vehicle routing problem: past and future trends. *Expert Systems with Applications*, **41**(Part 1), 1118–1138.

Liu, C., Dang, F., Li, W., Lian, J., Evans, S., and Yin, Y., 2014. Production planning of multi-stage multi-option seru production systems with sustainable measures. *Journal of Cleaner Production*. DOI: 10.1016/j.jclepro.2014.03.033.

Liu, W., Han, J., and Lu, X., 2013. A high speed railway control system based on the fuzzy control method. *Expert Systems with Applications*, **40**, pp. 6115–6124.

Lovrić, M., Li, T., VervestLovrić, P., Li, T., and Vervest, P., 2013. Sustainable revenue management: a smart card enabled agent-based modeling approach. *Decision Support Systems*, **54**, 1587–1601.

Mallidis, I., Dekker, R., and Vlachos, D., 2012. The impact of greening on supply chain network design and cost: a case for a developing region. *Journal of Transport Geography*, **22**, 118–128.

Maloni, M., Paul, J., and Gligor, D., 2013. Slow steaming impacts on ocean carriers and shippers. *Maritime Economics and Logistics*, **15**, 151–171.

McKinnon, A., Cullinane, S., Browne, M., and Whiteing, A., 2010. *Green Logistics: Improving the Environmental Sustainability of Logistics*. London: Kogan.

McMurray, A., Islam, M., Siwar, C., and Fien, J., 2014. Sustainable procurement in Malaysian organizations: practices, barriers and opportunities. *Journal of Purchasing & Supply Management*, **20**, 195–207.

Mirzapour Al-e-hashem, S., Baboli, A., and Sazvar, Z., 2013. A stochastic aggregate production planning model in a green supply chain: considering flexible lead times, nonlinear purchase and shortage cost functions. *European Journal of Operational Research*, **230**, 26–41.

Moellera, K., 2011. Increasing warehouse order picking performance by sequence optimization. *Procedia Social and Behavioral Sciences*, **20**, 177–185.

Moon, K., Youn, C., Chang, J., and Yeung, A., 2013. Energy-saving fashion: products design scenarios. *International Journal of Production Economics*, **146**, 392–401.

Muellera, F., Cannataa, A., and Herrmann, C., 2014. Procedure of modular Green Factory Planning to enhance collaboration and decision making. *Procedia CIRP*, **15**, 331–336.

Newman, S., Nassehi, A., Imani-Asrai, R., and Dhokia, V., 2012. Energy efficient process planning for CNC machining. *CIRP Journal of Manufacturing Science and Technology*, **5**, 127–136.

Ohara, K., Tsugeno, M., Imanari, H., and Sakiyama, Y., 2014. Process optimization for the manufacturing of sheets with estimated balance between product quality and energy consumption. *CIRP Annals - Manufacturing Technology*, **63**, 257–260.

Ondemir, O. and Gupta, S., 2014. Quality management in product recovery using the Internet of things: an optimization approach. *Computers in Industry*, **65**, 491–504.

Pal, B., Sana, S., and Chaudhuri, K., 2013. A stochastic inventory model with product recovery. *CIRP Journal of Manufacturing Science and Technology*, **6**, 120–127.

Palanivelu, P. and Dhawan, M., 2011. Green Logistics. White Paper Tata Consulting Systems, http://www.tcs.com/resources/white_papers/Pages/Green_Logistics.aspx (accessed July 9, 2015).

Pan, S., Ballot, E., and Fontane, F., 2011. The reduction of greenhouse gas emissions from freight transport by pooling supply chains. *International Journal of Production Economics*, **143**, 86–94.

Panda, S., 2014. Coordination of a socially responsible supply chain using revenue sharing contract. *Transportation Research Part E*, **67**, 92–104.

Pishvaee, M., Razmi, J., and Torabi, S., 2014. An accelerated Benders decomposition algorithm for sustainable supply chain network design under uncertainty: a case study of medical needle and syringe supply chain. *Transportation Research Part E*, **67**, 14–38.

Plehn, J., Zust, R., Kimura, F., Sproedt, A., and Schonsleben, P., 2012. A method for determining a functional unit to measure environmental performance in manufacturing systems. *CIRP Annals - Manufacturing Technology*, **61**, 415–418.

Psaraftis, H. and Kontovas, C., 2013. Speed models for energy-efficient maritime transportation: a taxonomy and survey. *Transportation Research Part C*, **26**, 331–351.

Psaraftis, H. and Kontovas, C., 2014. Ship speed optimization: concepts, models and combined speed-routing scenarios. *Transportation Research Part C*, **44**, 52–69.

Raslavičius, L., Semenov, V., Chernova, N., and Keršys, A., 2014. Producing transportation fuels from algae: in search of synergy. *Renewable and Sustainable Energy Reviews*, **40**, 133–142.

Regmi, M. and Hanaoka, S., 2009. *A Framework to Evaluate Carbon Emissions from Freight Transport and Policies to Reduce CO_2 Emissions through Mode Shift in Asia*. Tokyo Institute of Technology, Tokyo, http://www.ide.titech.ac.jp/~hanaoka/Modal_Shift_Paper__T_LOG__Final.pdf (accessed June 22, 2015).

Rosskopf, M., Lehner, S., and Gollnick, V., 2014. Economic-environmental trade-offs in long-term airline fleet. *Journal of Air Transport Management*, **34**, 109–115.

Russo, D. and Rizzi, C., 2014. Structural optimization strategies to design green products. *Computers in Industry*, **65**, 470–479.

Sazvar, Z., MirzapourAl-e-hashem, S., Baboli, A., and AkbariJokar, M., 2014. A bi-objective stochastic programming model for a centralized green supply chain with deteriorating products. *International Journal of Production Economics*, **150**, 140–154.

Schmeisser, B., 2013. A systematic review of literature on offshoring of value chain activities. *Journal of International Management*, **19**, 390–406.

Silva, D., Renó, G., Sevegnani, G., Sevegnani, T., and Truzzi, O., 2013. Comparison of disposable and returnable packaging: a case study of reverse logistics in Brazil. *Journal of Cleaner Production*, **47**, 377–387.

Somplák, R., Ferdan, T., Pavlas, M., and Popela, P., 2013. Waste-to-energy facility planning under uncertain circumstances. *Applied Thermal Engineering*, **61**, pp. 106–114.

Soon, J. and Baines, R., 2012. Food safety training and evaluation of handwashing intention among fresh produce farm workers. *Food Control*, **23**, 437–448.

Soysal, M., Bloemhof-Ruwaard, J., and Van der Vorst, J., 2014. Modelling food logistics networks with emission considerations: the case of an international beef supply chain. *International Journal of Production Economics*, **152**, 57–70.

Stasko, T. and Gao, H., 2012. Developing green fleet management strategies: repair/retrofit/replacement decisions under environmental regulation. *Transportation Research Part A*, **46**, 1216–1226.

Su, T. and Lin, Y., 2014. Fuzzy multi-objective procurement/production planning decision problems for recoverable manufacturing systems. *Journal of Manufacturing Systems*. DOI: 10.1016/j.jmsy.2014.07.007.

Tate, W., 2014. Offshoring and reshoring: U.S. insights and research challenges. *Journal of Purchasing & Supply Management*, **20**, 66–68.

Teixeira, A., Jabbour, C., and de Sousa Jabbour, A., 2012. Relationship between green management and environmental training in companies located in Brazil: a theoretical framework and case studies. *International Journal of Production Economics*, **140**, 318–329.

Ting, Y., Feng, W., and Bin, L., 2014. Quantification of end-of-life product condition to support product recovery decision. *Procedia CIRP*, **15**, 257–262.

Tsai, W., Lee, K., Liu, J., Lin, H., Chou, Y., and Lin, S., 2012. A mixed activity-based costing decision model for green airline fleet planning under the constraints of the European Union Emissions Trading Scheme. *Energy*, **39**, 218–226.

Vujanović, D., Momčilović, V., Bojović, N., and Papić, V., 2012. Evaluation of vehicle fleet maintenance management indicators by application of DEMATEL and ANP. *Expert Systems with Applications*, **39**, 10552–10563.

Walker, H., Miemczyk, J., Johnsen, T., and Spencer, R., 2012. Sustainable procurement: past, present and future. *Journal of Purchasing & Supply Management*, **18**(4), 201–206.

Wang, S. and Meng, Q., 2012. Sailing speed optimization for container ships in a liner shipping network. *Transportation Research Part E*, **48**, 701–714.

Yang, L., Ge, X., Wan, C., Yu, F., and Li, Y., 2014. Progress and perspectives in converting biogas to transportation fuels. *Renewable and Sustainable Energy Reviews*, **40**, 1133–1152.

Yangjian, J., Roger, J., Chen, L., and Wu, C., 2013. Green modular design for material efficiency: a leader–follower joint optimization model. *Journal of Cleaner Production*, **41**, 187–201.

Yun, B., Baohua, M., Fangming, Z., Yong, D., and Chengbing, D., 2011. Energy-efficient driving strategy for freight trains based on power consumption analysis. *Journal of Transportation Systems Engineering and Information Technology*, **9**, 43–50.

Zhang, G. and Zhao, Z., 2012. Green packaging management of logistics enterprises. *Physics Procedia*, **24**(Part B), 900–905.

Zhang, S., Wang, F., He, D., and Jia, R., 2012. Real-time product quality control for batch processes based on stacked least-squares support vector regression models. *Computers and Chemical Engineering*, **36**, 217–226.

Zhang, X., Liu, P., Li, Z., and Yu, H., 2013. Modeling the effects of low-carbon emission constraints on mode and route choices in transportation networks. *Procedia - Social and Behavioral Sciences*, **96**, 329–338.

Zheng, B., Mohr, C., and Yoon, S., 2014. A mixed-integer programming model for order picking schedule. In: Y. Guan and H. Liao, eds. *Proceedings of the 2014 Industrial and Systems Engineering Research Conference*, http://www.xcdsystem.com/iie2014/abstract/finalpapers/I1212.pdf (accessed June 22, 2015).

Zhou, M., Pan, Y., Chen, Z., Yang, W., and Li, B., 2012. Selection and evaluation of green production strategies: analytic and simulation models. *Journal of Cleaner Production*, **26**, 9–17.

Ziout, A., Azab, A., and Atwan, M., 2014. A holistic approach for decision on selection of end-of-life products recovery options. *Journal of Cleaner Production*, **65**, 497–516.

Zuo, J. and Zhao, Z., 2014. Green building research–current status and future agenda: a review. *Renewable and Sustainable Energy Reviews*, **30**, 271–281.

6

Safety and Traceability

Fabrizio Dabbene,[1] Paolo Gay,[2] and Cristina Tortia[2]

[1] IEIIT-CNR, Politecnico di Torino, Turin, Italy
[2] Department of Agricultural, Forest and Food Sciences,
University of Turin, Turin, Italy

6.1 Introduction

The most general and internationally recognized definition of food traceability is contained in the Codex Alimentarius, where traceability in the food sector is primarily defined as "the ability to follow the movement of a food through specified stage(s) of production, processing, and distribution" (Codex Alimentarius Commission, 2006). Here, it was also recognized that, at the international level, methods are not harmonized and are often complicated, thus leading to barriers to trade (Codex Alimentarius Commission, 2007). Each country, in issuing a traceability law, refers to the principles set out by the Codex Alimentarius Commission.

Several definitions of traceability exist in the technical literature, laws, and standards, and it is debated whether the term traceability is to be intended as covering all aspects of tracking other than legal or safety issues (logistics, management, process control, etc.) (e.g., Bosona and Gebresenbet, 2013; Karlsen *et al.*, 2013; Olsen and Borit, 2013).

Essentially, a food Traceability System (TS) consists of acquiring information on food products, which can be retrieved along the supply chain all the way to the consumer. In particular, the term *tracking* generally refers to the forward process of registering information about the item/lot that is being processed and transported

Supply Chain Management for Sustainable Food Networks, First Edition. Edited by Eleftherios Iakovou,
Dionysis Bochtis, Dimitrios Vlachos and Dimitrios Aidonis.

along the supply chain, while *tracing* is generally ascribed to the reverse path, when stored information is retrieved. If traceability is referred to processes which take place inside the boundaries of a single firm, it is called *internal traceability* or *process traceability*, while if the tracing is extended throughout the other stakeholders, it is referred to as *external traceability* or *supply-chain traceability* (Moe, 1998; Opara, 2003; Ene, 2013). Technically, the information can physically follow the delivered items flux, or the information and object fluxes are decoupled and the information is managed by means of information technology (IT) systems in collaborative distributed networks.

The term *traceability* is broadly used in different manufacturing sectors, as it represents a tool for achieving a number of different objectives. As complete traceability is not possible, decisions on the amount of information to be kept and transmitted through each of the successive nodes of the supply chain and the appropriate means to correctly and efficiently maintain traceability data are crucial (Golan *et al.*, 2004). The degree of complexity of the recording systems depends on the different traceability objectives. In many cases, due to the increasing health impact of contaminated food consumption, mandatory traceability has been established for many food products, also by extending tracking to the whole food supply chain, including animal feed and substances that could be present in food products [e.g., meat supply chain in the European Union (EU)].

The aim of this chapter is to provide an overview of the current objectives, methodologies, and solutions adopted in TSs.

The chapter is organized as follows: in Section 6.2 the main drivers for food traceability are discussed, with special attention to food safety issues. Section 6.3 is devoted to a comprehensive overview on international legislations and standards regulating food traceability in different countries. In Section 6.4 the design of a TS is considered, first by providing formal definitions of the main concepts involved in the design and management of a TS, then by overviewing the technological developments, and finally by describing how a TS can be effectively optimized to minimize the cost of a possible food safety crisis. A discussion of future trends and developments concludes the chapter.

6.2   Drivers for Food Traceability

In the public sector, traceability is mainly targeted toward enhancing food safety levels and managing safety alerts while, in the private sector, TSs are implemented also to improve competitiveness and profitability, thereby increasing the food supply chain value. To avoid the spread of diseases, the establishment of an efficient cooperation framework of public authorities and private stakeholders highly improves the traceability efficiency to control food safety and eventually withdraw contaminated food from the market. Safety is the major concern due to the constant increase in food related crises, which cause population hospitalizations and deaths and which are very frequent and underestimated due to the fact that the majority of food illnesses do not require hospitalization and are not detectable.

The food sector is periodically subject to scandals related to the consumption of food products dangerous to the health, which spread through the market and generate a national or international alert and recall actions. These food injuries are typically related to adulteration (milk with melamine, horse meat scandal, oil adulteration, etc.), presence of pathogenic microorganisms (*Listeria*, *Salmonella*, dangerous strains of *Escherichia coli*, etc.), food poisoning due to the presence of undesired chemicals (e.g., histamine, toxins of various origin, pesticide residues), viruses (*hepatitis A* in berries, *norovirus*), wrong composition, or insufficient labeling (e.g., about allergens).

The number of acknowledged food safety emergencies, which activated national and international alerts and recall procedures, is continuously updated in the bulletins of the various organizations for food alert as, for example, the reports of the Rapid Alert System for Food and Feed (RASFF) (see, e.g., European Union, 2012, 2013) or the web site of the US Food and Drug Administration (FDA) (Food and Drug Administration, 2014b). An example of a list of various crises and scandals can be also found in Da-Wen Sun (2008). Moreover, especially with perishable food products, inefficiency in supply chain management can cause food spoilage, which leads to safety and quality concerns as well as production waste.

National and supranational food safety authorities have the responsibility to protect citizens' health with maximum effectiveness. Legal requirements, control methods, and authorities that monitor unsafe food products force food business operators to identify and destroy potentially unsafe lots. Although enforcement and legal actions are needed, the self-regulation of aware business operators is also a strong prevention tool, but in this case the efficiency of the tracking and tracing method relies on strong agreements among the groups of companies.

To assist food producers to enhance product safety, hazard control methods, and plans to monitor risks in food processes were standardized, recognized worldwide, and further adopted in several laws. Based on control methods fit for the industry, which supplied food for the astronauts at NASA, the standard Hazard Analysis and Critical Control Point (HACCP) was established to control preventively microbiological, chemical, and physical hazards. Its principles were early recognized by the Codex Alimentarius (Codex Alimentarius Commission, 1969) and, starting from the 1990s, many governments mandated the use of HACCP in sectors of the food industry (European Union, 1993; Food and Drug Administration, 1995, 2001; European Commission, 2004a). The approach of HACCP is to enhance the consciousness of firms in analyzing their processes, defining the critical control points, the control monitoring, the human resource involved, the recordkeeping and management of the information, and the strategy and corrective actions to put in place in case of noncompliance.

To enhance product value, food producers mainly point at quality enhancement. Food quality is a very general concept, implying many expectations which can be different from consumer to consumer. Apart from the overall experience characteristics as sensorial aspects, which are directly perceived by consumers, other quality aspects influence a product's value to the consumer. Quality does not refer solely to the properties of the food itself, but also to the ways in which those properties have been achieved (Morris and Young, 2000). Some product properties, often referred to as credence attributes, cannot be perceived by the consumer without proper tracing, certification, and label declaration (Golan *et al.*, 2004). In particular, since health properties are not directly perceivable, they are listed among credence attributes or, more broadly, reflection traits, and the consumer attitude toward food purchase is based strongly on trust (Lee and Yun, 2015). Adding this type of information has the aim to differentiate food products, targeted for different groups of consumers. Among these, one can list also GMO, organic, "free-range" livestock, animal welfare, dolphin free, and so on.

As product origin is an important characteristic about which consumers need to be informed (Van Rijswijk *et al.*, 2008), the adoption of a geographical mandatory or voluntary certified TS [e.g., Protected Designation of Origin (PDO) and Protected Geographical Indication (PGI) labeling] can lead to a noticeable enhancement of revenues. The influence of product origin on quality is debated, but it could be linked to characteristics that are attributable to a precise geographical region and these could help the consumer to perceive attributes that are difficult to detect more easily, thus acting as a sort of consumer taste training.

Traceability is primarily viewed as a tool for food safety and quality, but it can also be used to prevent fraudulent or deceptive practices as well as the adulteration of food, which is an important challenge the food industry is facing. Research has developed modern techniques for food authentication (e.g., DNA analysis, isotopic analysis and chromatography, enzymatic analysis, electrophoresis) (Lees, 2003; Da-Wen Sun, 2008) while others are still in progress, with the purpose of detecting fraudulent activities. As these methods, which are often very costly, cannot be used routinely in TSs, they can be used by private firms to monitor traceability schemes and by public authorities and regulators to enforce legal actions against frauds.

Other certified labeling declarations could satisfy consumers concerned about social issues and ethical aspects regarding fair labor conditions (fair wage and trade) or the respect of religious requirements during production (kosher, halal products). With the growth of international food trade, the environmental impact of the food supply chain has become a growing concern. As agriculture and food manufacturers are often considered to contribute largely to pollution (e.g., by chemicals) or CO_2 emissions (e.g., livestock sector), transparency in supply chain sustainability is often required. In food supply chains, traceability is part of the sustainable supply chain management (SSCM), which is defined as the *management of material, information, and capital flows as well as cooperation among companies along the supply chain while taking into account the goals of all three dimensions of sustainable development, that is, economic, environmental, and social, which are derived from customer and stakeholder requirements* (Seuring and Müller, 2008). Labeling declarations about low use of energy and resources, the reduced emission of greenhouse gases and other pollutants, "carbon labeling," distance from production sites, as well as research methods used in SSCM [e.g., Life Cycle Assessment (LCA)] imply a high level of information sharing, and strong cooperation among all the supply chain stakeholders and supply chain management. This approach requires the involvement and the cooperation of all supply chain stakeholders including transport, logistic, and waste management.

If properly connected to the information systems adopted by single firms and food supply chains, TSs can contribute to enhance the efficiency of the supply chain.

Traceability should be used for the management of a single firm or of a supply chain, improving suppliers' control, monitoring stock levels and quality of ingredients and finished goods, enhancing the logistics of delivering and transport, and identifying critical points and bottlenecks. Bosona and Gebresenbet (2013) argue that traceability can be considered as a part of logistics management and that IT real-time systems applied both for logistics and food traceability as well as monitoring should be integrated in a collaborative environment, which favors efficient logistics recall processes.

While in huge firms balancing costs and benefits of traceability is difficult and time consuming, in short supply chains the improvement is rapidly achieved and perceived and the added value is immediately assessed. Recently, the market of small food producers who are changing their selling strategies using, for example, e-commerce has been increasing and new methods are being adopted to manage logistics and data sharing.

The selection of the type and amount of data about the food products that influences the consumer and fulfills the consumer's requirements is based on the modern buyer's fast buying attitude of spending very little time in acquiring information. Nevertheless, the potential information source of access has increased because of the availability of other communication channels that are different from traditional paper labels (web site, email communications).

The use of Internet-based communication has also emerged as a key enabler in driving supply chain integration in supply chain transactions among business operators (business-to-business, B2B). Firms and traders can use the Internet to take advantage of operating in a collaborative network by tracing in enlarged boundaries and strengthening supply chain traceability. These networks of trading partners are very effective in responding quickly to sudden negative events such as recalls, out of stock, delivery problems, and so on.

6.3 Traceability: Legislations and Standards

6.3.1 International Legislation

In Europe, EC General Food Law Regulation 178/2002 (European Commission, 2002), in force since 2005, requires the establishment of a TS for all food products. The General Food Law clearly states that the traceability details are to also be extended to each ingredient of the food product, defining traceability as "the ability to trace and follow a food, feed, food-producing animal or substance intended to be or expected to be incorporated into a food or feed, through all stages of production, processing, and distribution." The General Food Law is based on the *one-step-forward* and *one-step-back* traceability scheme and does not state any specific method or technique that food operators have to follow. Therefore, in the absence of other more restrictive laws related to a specific food product or of national laws issued by member states, some details, such as for instance lot size, are not defined. The requirement for traceability is limited to ensuring that businesses are at least able to identify the immediate supplier of the considered product and the immediate subsequent recipient, with the exemption of retailers to final consumers. The General Food Law (art. 33) established the European Food Safety Authority (EFSA) and the RASFF for food alert notifications from member states (on the basis of art. 50, 51, and 52). However, when an alert arises, legal and sanitary actions to be put in place to face the emergency are left to each member state resulting in non-homogeneous action, causing delays in the time lapse from alert to recall.

In recent years, further regulations have constantly been issued on food and feed safety. Traceability on GMOs (European Commission, 2003a,b) and allergens (European Commission, 2003c), was regulated just after the General Food Law, together with other mandatory rules about food hygiene (European Commission, 2004a,b,c). More recently, EC regulation 931/2011 (European Commission, 2011a), requires, among other things, more detailed information about food description and quantity, and unit-level traceability and identification (lot, batch, consignment) on products of animal origin, even if in the law it has been officially recognized that "business operators do not generally possess the information needed to ensure that their systems identifying the handling or storage of foods is adequate, in particular in the sector of food of animal origin." This reflects the actual difficulties encountered in regulating and managing efficiently food traceability data.

Moreover, a recent EU Regulation 1169/2011 has been issued, specifically targeted at food information for consumers (European Commission, 2011b). The aim is to enable the consumer to "identify and make appropriate use of a food and to make choices that suit their individual dietary needs." This law sets principles and rules for labeling all types of food and giving information even with other means than labels (e.g., in distance selling). Food business operators are responsible for inadequate declarations. The law entered into force in December 2014, apart for nutrition declarations that will become mandatory from December 2016.

Following art. 26 of Regulation 1169/2011, the EU has also issued Regulation 1337/2013 (European Commission, 2013) which sets out the obligations regarding information on labeling of meat products: fresh, chilled, and frozen swine, sheep or goats, and poultry meat. This Regulation imposes reporting on the label the name of the Member State or third Country where the animal was reared and, in a separate indication, slaughtered. A set of precise labeling rules has been established in all the possible animals' age options when moved through premises or slaughterhouses, and rearing involved in the supply chain from farm to fork. The Regulation applies to packed meat and meat products with mandatory specifications about the complete traceability of lots by means of batch codes. In the text of the law the additional costs to stakeholders are recognized following its provision and it states that "a balance needs to be struck between the need of the consumers to be informed and the additional cost for operators and national authorities, which finally has an impact on the final price of the product" (art. 2).

In the US, compulsory traceability has been only recently introduced in the food sector, and food safety was previously assured mainly by private companies in order to guarantee good quality for the consumer. Following this approach, TSs are integrated by a firm's internal information management after a careful assessment of the benefits and costs, adopting strategies for enhancing the brand's reputation and adding value. This type of traceability management, which can be managed with a high degree of collaboration between logistics and planning systems, is often very effective as it is free from governmental restriction boundaries.

Traceability first became mandatory only as a reaction against bioterrorism (United States, 2002). The Food Safety Modernization Act (United States, 2011) was signed on January 2011, in which a system of preventive controls and inspections was enforced as a response to violations (recalls) on domestic as well as on foreign US food products.

Nowadays in the US, the release of new laws has accelerated and the US FDA is also adopting rules for foreign food importers for human and animal consumption. A new regulation, the Foreign Supplier Verification Program (FSVP) (Food and Drug Administration, 2014a), concerns a considerable number of firms exporting food and feed in the US. Following this provision, a detailed list of non-compliant food lots is published on the FDA web site, classified in three risk levels, and can be freely accessed in real-time by cell phones.

Mandatory Country of Origin Labeling (COOL) is still strongly debated among public and private US and other countries stakeholders, as it directly affects product revenues. In the US, under the COOL labeling system, retailers since 2009 have had to provide their customers with information about the origin of various food products, including fruit, vegetables, fish, shellfish, and meats (United States Department of Agriculture, 2009). While the US legislators' intention is to provide, by means of retailer declarations, food provenance information to respond to consumer needs, the opponents of mandatory COOL argue that there is no evidence that consumers want such labeling and that the cost of COOL labeling is not worthwhile. Some meat producers have stated that scientific principles, and not geographical position, should be considered the safety criterion for food products, and that the COOL program should not replace any other established regulatory food safety or traceability programs. For these reasons, in 2009 Canada and Mexico challenged COOL at the World Trade Organization (WTO), arguing that COOL was a trade barrier for the meat industry on both sides of the border (Jurenas and Greene, 2013; World Trade Organization, 2014).

In Japan, food TSs can broadly be classified into mandatory systems and voluntary systems. For some products (beef, since 2004 and rice, since 2009) TSs are mandatory and imposed by corresponding laws, whereas for other products traceability is only encouraged on a voluntary basis. In the beef sector, a full TS is required on domestic beef on the basis of the individual animal's ID, with the adoption of high-tech IT systems (accessed by the consumer by cell phone). Japan's meat TSs are very efficient and broad since, as in the EU, they are mandatory, controlled by public and private auditors and post-slaughter extended. Voluntary traceability is strongly encouraged and the share of voluntarily traced food products is increasing (Shaosheng and Zhou, 2014). Moreover, Japan is a strong food importer and exporters have to comply with high quality standards. In particular, Japan has established the JAS system which was introduced in 1950 by the Agricultural and Forestry Standard Law (Ministry of Agriculture, Forestry and Fisheries of Japan, 1950), and assumed its current status in 1970 with the addition of the quality labeling standards system. The system is comprehensive incorporating the Japanese Agricultural Laws and the Japanese Agricultural Standards, which certify voluntary traceability and quality labeling.

Australia and New Zealand, since the National Food Authority Act (1991), have decided to adopt joint standards on traceability collected in the Food Standards Australia New Zealand (FSANZ) Code. The Standards in the FSANZ Code became legislative instruments under the Legislative Instruments Act 2003 (Australian Government, 2003). The compliance with the Code is monitored by the local authorities of the Australian states and territories and New Zealand. Some divergences between the two countries still exist (e.g., COOL).

In developing countries, the fulfillment of the required traceability standards is an important requisite for foreign market access. Standardization, transparency, certification, fair work (respect of children and women), and wage are important quality and safety issues in exporting food products and they have become a social challenge in rural areas. In specific supply chains where exports are very high (cocoa, coffee, meat from South America), high standards of traceability and quality certification have been achieved, while in other sectors (e.g., fresh produce) the target is far from having been reached.

6.3.2   Standards

Different commercial standards have been issued by organizations and associations to set traceability requirements, facilitating traceability data sharing, and adopting product identification standards for commercial purposes. Organizations such as ISO, GS1, GlobalGAP, and British Retail Consortium (BRC), deliver guidelines and requirements for traceability, defining the principles as well as effective TS designs and tests. These commercial standards are adopted in different contexts and required, for example, to access a given market or to comply with stakeholder requests.

The ISO 9000 series standard is not specifically addressed to food traceability, but concerns Quality Management Systems in undefined production environments, while ISO 9001 (International Organization for Standardization, 2000) defines a model for quality management and quality assurance. In ISO 9001:2008 the concept of product identification is introduced, requiring that "where appropriate, the organization shall identify the product by suitable means throughout product realization and where traceability is the requirement, the organization shall control the unique identification of the product and maintain records" and that "preservation shall also apply to the constituent parts of a product" (International Organization for Standardization, International Organization for Standardization 9001:2008, 2008). ISO 22000:2005 deals with the requirements for food safety management systems and addresses the establishment and application of TSs "that enable the identification of product lots and their relation to batches of raw materials, processing and delivering records" (International Organization for Standardization, International Organization for Standardization 22000:2005, 2005).

ISO 22005:2007 introduces principles and basic requirements for the design and the implementation of a food (and feed) TS. Even if it does not specify how this should be achieved, it introduces the requirement that organizations involved in a food supply chain have to define information that should be, at each stage, obtained and collected from the supplier and then provided to customers, in addition to product and processing history data (International Organization for Standardization, International Organization for Standardization 22005/2007, 2007). To this extent, a number of ISO Standards [e.g., ISO/International Electrotechnical Commission (IEC) 15961, 15962, 24791, 15459,15418, and 15434] have been delivered to regulate data encoding on radio frequency identification (RFID) devices and their interoperability with barcode-based systems (see Chartier and Van Den Akker, 2008 for a complete report delivered by the Global RFID Interoperability Forum for Standards).

Finally, GS1 is constantly emanating guidelines to apply traceability in different food sectors such as, for example, meat, fruit and vegetables, and seafood.

6.4   Design of Traceability Systems

6.4.1   Definitions of Traceability Related Concepts

An important feature of TSs for improving the performances and the safety level of a food supply chain is the ability to monitor the location and the composition of each lot in the production and supply chains. This information can also be used to define new management objectives and the relative actions to be undertaken for their fulfillment. To this end, many new concepts and definitions have been recently introduced in the literature to quantify the level of accuracy, and, more generally, the performances gained by the TS. Indeed, the definition and the evaluation of the performance of a TS represent a fundamental step forward in developing traceability-oriented management policies.

The level of traceability was first considered by Golan *et al.* (2004) and then by McEntire *et al.* (2010) by introducing four quantities: *depth* (how far upstream or downstream in the food supply chain the TS traces the lot/unit correctly), *breadth* (amount of attributes connected to each traceable unit), *precision* (the degree of assurance with which the system can pinpoint a particular product's movement or characteristic), and *access* (the speed with which tracking and tracing information can be communicated to supply chain members and the speed with which the required information can be disseminated to public health officials during food-related emergencies).

Breadth quantifies the amount of information related to the traced food unit. The information flow can be coupled to the physical flow also in aggregated form or it can be physically distributed and accessed remotely at different levels of detail (Trienekens and Beulens, 2001; Bechini *et al.*, 2008) and even contracted independently. Depth varies with the type of attributes and interests in the different production stages and marketing agreements. Together with unit size, traceability depth level has been deeply discussed from an economic as well as a safety perspective. Asioli, Boecker, and Canavari (2014) proposed a method to quantify breadth, depth, and precision and a regression model for the estimation of the relationship between these parameters and TS costs and benefits.

Another important aspect is the definition of the product unit that is singularly traced through the food supply chain. Moe (1998), following the terminology first introduced by Kim, Fox, and Gruninger (1995), proposed the concept of a *traceable resource unit* (TRU) for batch processes as a "unique unit, meaning that no other unit can have exactly the same, or comparable, characteristics from the point of view of traceability." This concept has been more recently formalized in ISO Standard 22005/2007 (International Organization for Standardization, International Organization for Standardization 22005/2007, 2007), where the lot is defined as a "set of units of a product which have been produced and/or processed or packaged under similar circumstances."

This concept was further elaborated by Bollen, Riden, and Cox (2007) who proposed the concept of an *identifiable unit* (IU). This represents the product unit which has to be uniquely identifiable through any system in which it is processed. The *granularity* of the TS, for which many definitions have been proposed in the literature, derives from the size of the IUs. Granularity, defined by Karlsen *et al.* (2012) as a quantity "determined by the size of a traceable unit and the number of the smallest traceable units necessary to make up the traceable unit at a specific granularity level," is determined by the size and number of batches. A finer granularity allows for adding even more detailed product, and for acting at a more detailed and range-limited level in the case of a possible recall. The optimal granularity level in a production or supply chain is very difficult to design and tune, but it is determining in a possible recall action. Nowadays, the granularity at which the involved products are traced in most parts of current supply chains does not come from the results of a formal analysis and optimization study, but it is often the consequence of a combination of short-term convenience, tradition, and use of available facilities.

Bollen, Riden, and Cox (2007) provided a formal definition of the *precision* of a TS, which can be evaluated as the ratio between the IU sizes at two points in the supply chain. Precision depends on the number and the nature of the transformations in which IUs are involved, and on the extent, nature, and accuracy of the recorded data. Whenever an IU is split up, the separated parts keep the identification of the parent IU, while if some IUs are joined, the identification of the IU is different from the identification of the parent IUs. Hence, precision reflects the degree of assurance with which a TS can identify a particular food product movement or characteristic (Golan *et al.*, 2004). The measure of *purity* has been defined by Riden and Bollen (2007) as the (composition) percentage of an output lot when sourced from a single raw material input lot. In other words, for a given lot, purity expresses the percentage of the input lot making the largest contribution to its composition.

Possible systematic information loss, as for instance when information about the composition or process conditions is not properly linked to the product and systematically recorded, is the cause of degradation in the performance of a TS. Karlsen, Donnelly, and Olsen (2010) defined the *critical traceability point* (CTP) as the point in the food supply chain where this systematic loss occurs. The identification and mapping of CTPs is typically performed by qualitative methods (direct observation, structured interviews, and document analysis), and leads to the definition of a CTP analysis plan (Karlsen and Olsen, 2011). Donnelly, Karlsen, and Olsen (2009) and Karlsen, Donnelly, and Olsen (2011) have described some applications of CTP mapping and validation.

Whatever measures are adopted for the characterization of a TS, it is crucial to define monitoring and validating schemes to evaluate the effectiveness of the system. In fact, whenever possible the TS should be validated by external methods (e.g., physicochemical, genetic, or microbiological) able to identify and discriminate products (see e.g., Peres *et al.*, 2007; Aceto *et al.*, 2013; Galimberti *et al.*, 2013). As reported in ISO 22005:2007 (section 5.1 General design considerations), the proper functioning of traceability procedures has to be periodically checked by setting suitable procedures. Examples were proposed by Randrup *et al.* (2008) and by Mgonja, Luning, and Van der Vorst (2013) who considered the effects of simulated occurrences of food safety hazards in Nordic fish supply chains. To become a reliable risk management tool, a TS should also record the necessary data to determine process mass balances (Lavelli, 2013).

6.4.2   Current Technologies for Traceability

Automation in data collection enhances the precision and the reliability of identification of the traced unit and new and improved technologies and devices are constantly being brought to market. Recorded information on produce can be retrieved automatically by Personal Digital Assistant (PDA) scanning barcodes, Quick Response (QR) codes, data matrix or RFID, and accessing a database. This has increased the potential for adding information during the product's physical flow, also avoiding the excess of written information on traditional labels, which have a low level of usability.

The ability to identify objects wirelessly and without contact, even at item level, by RFID systems has already improved speed, accuracy, and efficiency in tracking objects in different sectors. The adoption of information and communications technology systems allows the collection of traceability data on shared repositories, which can be accessed remotely by stakeholders at each node along the supply chain. RFID was thought to be a revolutionary solution that could replace traditional barcodes in automatic tracking from production to consumer. However, after a period of enthusiasm following the introduction of the technique, with examples of RFID tracking being adopted by major retailers such as Walmart in the US and Tesco in the UK, which required their suppliers to put RFID tags on their products, some obstacles delayed the wide spread use of RFID. This was due to both technical as well as economic problems. Tag costs, which were expected to decrease very steeply, did not reach the break-even level to justify tag use in identification of low value items, since the cost of the copper coils contained in the low-frequency (LF) and high-frequency (HF) tags constituted the lower tag price limit. Despite this, these frequency ranges have been successfully adopted in many sectors (e.g., animal identification, ticketing, access control, anti-theft, libraries).

Systems operating in the ultra-high-frequency (UHF) band, which can identify objects within a wide reading zone (2–2.5 m) and allow dynamic identification of several objects by means of powerful anti-collision algorithms, have spread rapidly, also facilitated by the favorable cost of the aluminum used for UHF tags.

In each supply chain node, the correct identifier must be chosen according to different factors such as food composition, environmental conditions, operability during processing, and transport. Throughout the whole supply chain, food items could be identified by means of different types of coded tags and even disassembled and reassembled in different packaging with multiple or single item identification. Along the supply chain, the identifier can even be changed automatically by transferring, adding, and removing information.

A high level of interoperability in traceability data exchange improves the reliability and efficiency of the systems. The adoption of open standards implemented in the systems by the different chain actors (controllers, producers, food processing and storage firms, service providers, internal and external consultants, etc.) can increase tracking security and efficiency. At present, lack of data standardization in food supply chains limits the boundaries of TSs at international level, decreasing interoperability, and profitability.

The standardization of a common language for data interchange involves the public (e.g., territorial and sanitary databases) and private sectors (B2B initiatives). Where public databases exist, data are homogeneous, well consolidated and validated; nevertheless, some obstacles in using public databases for traceability are encountered regarding accessibility, privacy level, and updating frequency. Another important issue is the lack of uniformity of national databases even inside the EU. Private databases are generally more usable for management and marketing purposes, but are more difficult to certify and control.

The main obstacle in introducing UHF in food product traceability is the technical difficulty of obtaining exhaustive and reliable food item identification in critical production environments characterized by high amounts of water and metal. Success in object identification by RFID in manufacturing and logistics systems depends on the correct choice and coupling of the transmitting and receiving devices since the reliability of an RFID system depends both on the tag and the antenna radiation pattern and the resulting electromagnetic coupling when transponders are attached to the objects. RFID tags can be applied on the processing line to single food items (e.g., cheese; Barge *et al.*, 2014) or to the primary packaging (food contained in plastic films or boxes), batch of items (boxes, pallets), live animals, and meat (Barge *et al.*, 2013).

RFID passive tag performance and reading range depend on the alteration of the electromagnetic field due to the environment near the object (reflections, absorption, and demodulation) and on the realized gain of the tag antenna when attached to food items which can influence correct tag response. The limits and benefits of RFID system integration in the food industry can be evaluated through the analysis of the tag response conducted experimentally in a controlled environment and by identifying the best combinations of tag and reader type, positioning, and mutual orientation.

In chilled food products, rapid chilling and temperature control along the chain are required to prevent microbial growth which reduce spoilage and enhance shelf life, thus preventing food-related illnesses. Cold chain monitoring can be applied in parallel to TSs with other active RFID sensors which can acquire both room and food core temperature during storage and transport.

6.4.3   Performance Optimization

Motivated by the increasing frequency of food safety crises, with the consequent necessity of recalling very large quantities of products, increasing attention has been given to designing and optimizing TSs able to reduce the amounts and costs of product to be recalled. In this regard, efforts have recently been made to formally define the performance index of a TS in terms of its recall cost (RC). The RC has been introduced to measure how efficiently a TS manages a food recall action. This efficiency is high when a TS is able to precisely track lots of products, reducing unnecessary mixings, and contaminations, during whole production and distribution phases, avoiding – in the case – a massive spreading of the unwanted products in the supply chain and, finally, on the market. A low RC corresponds to potential reduced economic impact on the company in the case of a recall action – which in some recent food scandals caused market collapse, but also an implicit guarantee for the safety of the consumer. In fact, the lighter the amount of product to be recalled, the easier, faster, and more successful will be the recall action.

Indeed, as was pointed out by Fritz and Schiefer (2009), the recall of a product follows two main phases: first, potentially deficient lots of raw materials have to be identified in a backward manner starting from the damaged product; and secondly, a forward phase has to be performed in order to identify the products that could have been potentially affected and thus have to be withdrawn.

The RC depends on three main factors: (i) the size of the batches that have been individually tracked and managed by the TS; (ii) the way the batches of the different materials have been processed and mixed to obtain the final product; and (iii) the level of segregation adopted by the company to manage and maintain different batches of product separately.

In particular, Resende-Filho and Buhr (2010) consider notification costs, retrieval logistic costs, and lost sales as direct costs associated with the recall action. They assume that these cost components are directly proportional to the amount of product to be recalled, that is

$$RC = \alpha P_r Q_R$$

where P_r denotes the retail value of the product, Q_R the quantity of product to be recalled and where $\alpha > 1$ is a coefficient accounting for notification, logistics, and so on. Similarly, in Fritz and Schiefer (2009), the overall cost of a TS is expressed as the sum

$$C(overall) = RC + C(\text{tt}) + C(\text{e}) + C(\text{q})$$

where $C(tt)$, $C(e)$, and $C(q)$ denote, respectively, the system cost, and the costs induced by the possible reductions in efficiency and in quality caused by the adoption of the tracking and tracing systems.

Apparently, the first attempt to provide a precise formalization of RCs was made in Dupuy, Botta-Genoulaz, and Guinet (2005) by formally introducing the batch dispersion cost (BDC). In particular, the downward and upward dispersion indices are introduced: the downward dispersion of a lot denotes the number of batches of finished product that contain part of that lot, and the upward dispersion of a finished product lot is determined by the number of raw material lots involved in the production of that lot. Then, the total batch dispersion of a TS is defined as the sum of downward and upward dispersion indices of all raw materials. In practice, this index corresponds to the number of active paths (links) between raw materials and finished products. A similar index was introduced in Rong and Grunow (2010) where the distribution phase was described by the chain dispersion measurement

$$D_b = \frac{N_b(N_b - 1)}{2}$$

where N_b denotes the number of retailers served by lot b. Note that D_b depends on the number of links, but it increases quadratically with it.

Recently, Dabbene and Gay (2011) further elaborated on this framework and introduced a different measurement: the worst-case recall cost (WCRC) index which corresponds to the maximum *quantity* of product to be recalled when a batch of raw material is found to be unsafe. Similarly, they defined the average recall cost (ARC) index as the average mass of product to be recalled when one of the entering materials is found to be inappropriate The frameworks proposed by Dupuy, Botta-Genoulaz, and Guinet (2005) and Dabbene and Gay (2011) are based on modeling the food supply chain in terms of an interconnected graph, in which the nodes represent the different lots of raw materials, and the arrows represent the mixing operations that lead to the final products. This formalism captures the essential characteristics of the production process from a traceability viewpoint.

To summarize the frameworks considered in these papers, and in particular to better explain the BDC, WCRC, and ARC indices, an illustrative example is presented in Figure 6.1. The scheme depicts, in a simplified way, the production line of an industrial bakery, which produces different fresh or baked goods (ranging from bread to pasta and cakes). Starting from raw materials (flour, eggs, sugar, and butter), different mixes and doughs are produced and then processed to obtain the different products. In the example, three different lots of flour are available (a, b, c). The supply-chain manager has the possibility of deciding the production by acting on parameter α. The numbers on the arrows represent the quantity (mass) of materials involved in the mixing. Notice that, by changing the value of α, the three costs can differ significantly. In particular, while batch dispersion (which corresponds to the total number of links from raw materials to final product) does not change, the WCRC is rather different. This means that different solutions, which lead to the same BDC index, may result in very different amounts of product to be recalled in the case of a food safety crisis.

The introduction of the performance measures discussed in this section constitutes a key step toward designing efficient TSs, by determining appropriate batch sizes and mixing rules for each step in the production and supply chains. This is obtained

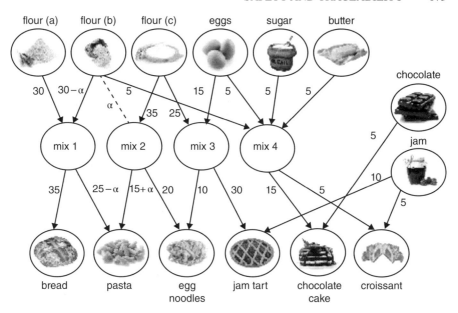

flour (a)   flour (b)   flour (c)   eggs   sugar   butter

chocolate

30   30−α   5   15   5   5   5

α

35   25

5

jam

mix 1   mix 2   mix 3   mix 4

35   25−α   15+α   20   10   30   15   5

10

5

bread   pasta   egg noodles   jam tart   chocolate cake   croissant

Figure 6.1   Example of a simple traceability system from an industrial bakery. In the illustrative example, we compute the different traceability indices for different values of the parameter α in the range 0–25. In particular, we consider the two extreme cases of α = 0 and 25. Note that, in both cases, the batch dispersion cost of the above system is equal to 21 (i.e., there exist 21 distinct paths from the raw materials to the produced goods). Conversely, the worst-case recall cost (WCRC) for α = 25 is equal to 35, and it is relative to raw material flour (b). This means that if flour (b) is found defective or contaminated, a total of 135 kg of final products will have to be recalled (in practice, everything except jam tarts would be recalled). Conversely, the WCRC for α = 0 is equal to 110, and is relative to raw material flour (c). That is, if flour (c) is found defective, only 110 kg of final product (corresponding to pasta, egg noodles, and jam tarts) will have to be recalled. The average recall costs are 65 for α = 0 and 68.75 for α = 25, respectively.

by properly designing the chain so that these indices are optimized, and consequently the performance of the TS is enhanced. A first step in this direction is done in Dupuy, Botta-Genoulaz, and Guinet (2005), where the mixing rules (that is the values of the different flows of materials in the graph) are optimized so that batch dispersion measurement is minimized. To this end, the problem is formulated in the form of a mixed integer linear program (MILP). A solution based on genetic algorithms (GAs) for the same problem is proposed by Tamayo, Monteiro, and Sauer (2009). The GA-based solution is generally suboptimal, but larger sized problems can be handled by this approach.

Motivated by the consideration that the BDC measurement, even if somehow related to the final quantity to be recalled since it aims at reducing the mixing of different batches, may lead to solutions with high RCs (as shown in the example in

Figure 6.1), Dabbene and Gay (2011) proposed direct minimization of the *WCRC* and *ARC* indices. Also in this case, the ensuing optimization problem can be formulated in MILP terms.

The concept of batch dispersion has also been adopted in Donnelly, Karlsen, and Olsen (2009) to study traceability problems arising in the lamb meat industry. Detailed material and information flow diagrams are provided, leading to the specification of joining and splitting points, and to the ensuing identification of CTPs. A different approach was proposed in Rong and Grunow (2010), where a joint production and distribution model is discussed. This model also takes into account some simplified food product degradation dynamics, and proposed a MILP-based optimization of lot sizing and routing.

Other approaches can be found in Wang, Li, and O'Brien (2009), where an optimization procedure integrating operational and traceability objectives is proposed, in Saltini and Akkerman (2012), in which the potential impact of the improvements of a chocolate TS on production efficiency and product recall is evaluated by comparing different scenarios, and in Thakur and Donnelly (2010) and Thakur, Wang, and Hurburgh (2010), who consider the case where the blending of different batches is necessary to achieve some desired characteristics, such as sensory properties, moisture content and test weight, and propose a multi-objective optimization model aimed at minimizing the number of storage bins and the total cost of blending and shipping grain products.

6.5   Future Trends

Food traceability is a growing area of research, as discussed in the previous sections, and it is continuously evolving in order to cope with increasing industry and market demands in a global context. For a detailed overview of this area, we refer the interested reader to the survey in Dabbene, Gay, and Tortia (2014). In this section, we briefly overview some problems which represent the frontier of future research and still require definitive solutions.

As pointed out in Dabbene, Gay, and Tortia (2014), "to be really effective, a traceability system should be conceived and implemented at the entire supply chain level, going beyond the basic principle of one step back-one step forward traceability adopted to comply with EC Regulation 178/2002 (European Commission, 2002), where every actor in the chain handles only the data coming from his supplier and those sent to his client." Indeed, the lack of widely accepted standards constitutes a key problem encountered by many companies, as it hinders the possibility of sharing information at supply chain level. Hence, large benefits would ensue both from the engagement of all the involved stakeholders, and from the improvement of the TS in the entire supply chain, by increasing the depth of the TS (see, e.g., Saltini and Akkerman, 2012), and by implementing a fast and efficient data exchange (Anica-Popa, 2012) based on an inter-organizational communication and sharing information system between all organizations across the food supply chain.

The benefits of such a system include the reduction in the time necessary for identifying all the movements and food processes a product undergoes along the chain, the detection and elimination of possible CTPs (Karlsen and Olsen, 2011), and the possibility of adopting more sophisticated managing rules taking into account the whole product history, including the distribution phase. Indeed, integrated production and distribution planning represent a promising approach involving not only traceability issues but, as discussed, for example, by Kainuma and Tawara (2006), Amorim, Günther, and Almada-Lobo (2012), and Zarei *et al.* (2011), general management policies.

A second important aspect worth investigating is connected to the consequences of uncertainty in the information provided by the TS. This issue was, for example, pointed out by Riden and Bollen (2007), who underlined the fact that the information from the TS, which is always assumed to be exact, could in reality be affected by underlying and uncontrollable stochastic phenomena. In industrial practice, absolute certainty is typically obtained by introducing over-bounds on the size of the lots. This leads to conservative TSs that, as a consequence, exhibit poor performances. An improvement of indices such as WCRC can be achieved by admitting tolerances expressed in probabilistic terms. This corresponds to admitting events at very low misclassification probability, in accordance with EC Regulation 1829/2003 (European Commission, 2003a), where segregation of genetically modified (GM) and non-GM products is regulated by means of a tolerance-based criterion.

Another issue, which is attracting increasing interest, concerns the mixing and processing of raw bulk products. Indeed, many ingredients used in the food industry are either liquids (vegetable oils, milk), powders (flour, cocoa), or crystals (sugar, salt). Clearly, the ideal solution for such products would becomplete segregation, with cleaning of the silos between different product batches (Cocucci, Ferrari, and Martello, 2002). However, these procedures, besides being usually rather costly, are not possible in the case of continuous production systems in which products are refined gradually and with minimal interruptions through a series of operations (Dennis and Meredith, 2000; Skoglund and Dejmek, 2007).

The traceability of flowing products, in continuous processing conditions, was considered by Bollen, Riden, and Cox (2007) and by Riden and Bollen (2007), for the specific case of apples processed in a pack house, and by Skoglund and Dejmek (2007), in which the changeover of lots of a liquid product in a pipe is modeled by introducing the concept of fuzzy traceability. A dynamic simulation solution is proposed to trace the ensuing *virtual batches*. Recently, Comba *et al.* (2013) further elaborated on these ideas by introducing the *composition distance*. This criterion allows to formally define the homogeneity of a lot from the viewpoint of raw material composition, in accordance with the definition of a lot given in ISO Standard 22005/2007, and also with the already mentioned current regulation for the management and traceability of GM products (European Commission 2003a,b), which states that a product can be labeled as GM-free if its percentage of GM content is less than 0.9%. Comba *et al.* (2013) introduced a compartmental model aimed at monitoring the composition of homogenous lots of products (referred to as cohorts) and their flow inside the production line, thus allowing tracking of the composition, in terms of lots of raw material, of any portion of product processed in the plant. This also bears similarities to the approach introduced in Comba, Belforte, and Gay (2011) for determining precise thermal conditions of fluid products processed in mixed continuous–discontinuous flow conditions.

Another approach is based on the development of specific markers, based on RFID technology, to be mixed with granular products in order to monitor the composition during flows. This method was first introduced by Kvarnström, Bergquist, and Vännman (2011) specifically for iron pellets, but the same can be applied to agricultural granular products. Interesting ideas are contained in Lee *et al.* (2010), Liang *et al.* (2012), and Liang *et al.* (2013), where it is proposed to insert pill-sized food-grade tracer particles into the grains during harvest carrying identity information. These tracers are composed of materials, such as cellulose or sugar, which can be safely eaten, and are labeled using food-grade ink.

At present, these first approaches for bulk products, which define general frameworks for TS implementation, still need additional research efforts to gain the robustness level required in food safety applications.

To conclude, current TSs are in many cases not able to handle fast dynamics in supply chains. Even if complying with art. 18 of EU Regulation 178/2002, it is a fact that many industries are not fully prepared to promptly start recall procedures after the primary signals of potential injuries and then to quickly trace back their product along the supply chain. To evaluate the speed of the TS in responding to information requests regarding the traded items, Mgonja, Luning, and Van der Vorst (2013) introduced the concept of rapidity of response. Beside the fact that a delay in the recall can be perceived by the consumer as company negligence, it can increase the number of possible injuries and even deaths (Magno, 2012). The recall process requires some time to effectively take place, and this introduces delays that have to be taken into account in the planning strategy. In fact, this generates an implicit relationship between the rapidity in removing the involved products and the measure of their dispersion. The earlier the contaminated product is removed from the production line, the smaller its dispersion, which becomes a function of time, will be.

6.6   Conclusions

The integration of the TS with supply chain management processes constitutes a key feature that could change the way traceability is currently perceived by the food industry.

Indeed, many food companies frequently consider traceability as a hindering task, whose costs overcome the potential benefits. Moreover, the increasingly stringent food safety, quality and commercial standards, and law requirements, as well as a growing demand for food characterized by a certain identity (GM, non-GM, ethical, organic, low carbon footprint, subject to religious constraints, etc.), motivate the development of efficient, well performing, and comprehensive TSs.

The efficiency and performance of TSs can be improved by adopting suitable management policies able to cope with aspects such as product recall dynamics, information retrieval, and tracking robustness. Indeed, even if traceability by itself cannot directly improve the quality and safety of food products, it constitutes a key element in a global production and distribution control scheme and, if coupled with other tools such as production planning, logistics, and HACCP, could improve the whole supply chain performance.

Nowadays, these objectives can be achieved profitably by exploiting the increasing availability of novel technologies such as active and passive RFID, embedded sensors, and localization devices that allow automatic sensing and identification, and the growing diffusion of computationally efficient simulation and optimization models.

References

Aceto, M., Robotti, E., Oddone, M., Baldizzone, M., Bonifacino, G., Bezzo, G., Di Stefano, R., Gosetti, F., Mazzucco, E., Manfredi, M., and Marengo, E., 2013. A traceability study on the Moscato wine chain. *Food Chemistry*, **138**, 1914–1922.

Amorim, P., Günther, H.O., and Almada-Lobo, B., 2012, Multi-objective integrated production and distribution planning of perishable products. *International Journal of Production Economics*, **138**, 89–101.

Anica-Popa, I., 2012. Food traceability systems and information sharing in food supply chain. *Management & Marketing*, **7**, 749–758.

Asioli, D., Boecker, A., and Canavari, M., 2014. On the linkages between traceability levels and expected and actual traceability costs and benefits in the Italian fishery supply chain. *Food Control*, **46**, 10–17.

Australian Government, 2003. Legislative Instruments Act 2003, No. 139, 2003 as amended, ComLaw, July 15, 2014, The Government of the Commonwealth of Australia, Canberra.

Barge, P., Gay, P., Merlino, V., and Tortia, C., 2013. Radio frequency identification technologies for livestock management and meat supply chain traceability. *Canadian Journal of Animal Science*, **93**, 1–11.

Barge, P., Gay, P., Merlino, V., and Tortia, C., 2014. Item-level Radio-Frequency IDentification for the traceability of food products: application on a dairy product. *Journal of Food Engineering*, **125**, 119–130.

Bechini, A., Cimino, M., Marcelloni, F., and Tomasi, A., 2008. Patterns and technologies for enabling supply chain traceability through collaborative e-business. *Information and Software Technology*, **50**, 342–359.

Bollen, A., Riden, C., and Cox, N., 2007. Agricultural supply system traceability, part I: role of packing procedures and effects of fruit mixing. *Biosystems Engineering*, **98**, 391–400.

Bosona, T. and Gebresenbet, G., 2013. Food traceability as an integral part of logistics management in food and agricultural supply chain. *Food Control*, **33**, 32–48.

Chartier, P. and Van Den Akker, G., 2008. GRIFS Global RFID Forum for Standards. D1.3 RFID Standardisation State of the Art Report Version 3. GRIFS, Brussels, http://www.iotstandards. org/sites/default/files/GRIFS%20D1.5%20RFID%20Standardisation%20State%20of%20 the%20art_revision%203.pdf (accessed July 17, 2015).

Cocucci, M., Ferrari, E., and Martello, S., 2002. Milk traceability: from farm to distribution. *Italian Food and Beverage Technology*, **29**, 15–20.

Codex Alimentarius Commission, 1969. *Codex Commission of Food Hygiene, General Principles of Food Hygiene*, CAC/RPC 1-1969. Rome: Food and Agriculture Organization.

Codex Alimentarius Commission, 2006. *Principles of Traceability/Product Tracing as a Tool within Food Inspection and Certification System*, CAC/GL 60-2006. Rome: Food and Agriculture Organization.

Codex Alimentarius Commission, 2007. *Joint FAO/WHO Food Standards Programme Codex Committee on Food Import and Export Inspection and Certification Systems*, CX/FICS 07/16/7. Rome: Food and Agriculture Organization.

Comba, L., Belforte, G., Dabbene, F., and Gay, P., 2013. Methods for traceability in food production processes involving bulk products. *Biosystems Engineering*, **116**, 51–63.

Comba, L., Belforte, G., and Gay, P., 2011. Modelling techniques for the control of thermal exchanges in mixed continuouse–discontinuous flow food plants. *Journal of Food Engineering*, **106**, 177–187.

Dabbene, F. and Gay, P., 2011. Food traceability systems: performance evaluation and optimization. *Computers and Electronics in Agriculture*, **75**, 139–146.

Dabbene, F., Gay, P., and Tortia, C., 2014. Traceability issues in food supply chain management: a review. *Biosystems Engineering*, **120**, 65–80.

Dennis, D. and Meredith, J., 2000. An empirical analysis of process industry transformation systems. *Management Science*, **46**, 1085–1099.

Donnelly, K.A.-M., Karlsen, K.M., and Olsen, P., 2009. The importance of transformations for traceability – A case study of lamb and lamb products. *Meat Science*, **83**, 68–73.

Dupuy, C., Botta-Genoulaz, V., and Guinet, A., 2005. Batch dispersion model to optimise traceability in food industry. *Journal of Food Engineering*, **70**, 333–339.

Ene C., 2013. The relevance of traceability in the food chain. *Economics of Agriculture*, **60**, 217–436.

European Commission, 2002. Regulation (EC) No. 178/2002 of the European Parliament and of the Council of 28 January 2002 laying down the general principles and requirements of food law, establish the European Food Safety Authority and laying down procedures in matters of food safety. *Official Journal of the European Union*, **L31**, 1–24.

European Commission, 2003a. Regulation (EC) No. 1829/2003 of the European Parliament and of the Council of 22 September 2003 on genetically modified food and feed. *Official Journal of the European Union*, **L268**, 1–23.

European Commission, 2003b. Regulation (EC) No. 1830/2003 of the European Parliament and of the Council of 22 September 2003 concerning the traceability and labelling of genetically modified organisms and the traceability of food and feed products produced from genetically modified organisms and amending Directive 2001/18/EC. *Official Journal of the European Union*, **L268**, 24–28.

European Commission, 2003c. Directive 2003/89/EC of the European Parliament and the Council of 10 Novemebr 2003 amending Directive 2001/13/EC as regards indication of the ingredients in foodstuff. *Official Journal of the European Union*, **L308**, 15–18.

European Commission, 2004a. Regulation (EC) No. 852/2004 of the European Parliament and of the Council of 29 April 2004 laying down specific hygiene rules on the hygiene of foodstuffs. *Official Journal of the European Union*, **L226**, 3–21.

European Commission, 2004b. Regulation (EC) No. 853/2004 of the European Parliament and of the Council of 29 April 2004 laying down specific hygiene rules on the hygiene of foodstuffs. *Official Journal of the European Union*, **L226**, 22–82.

European Commission, 2004c. Regulation (EC) No. 854/2004 of the European Parliament and of the Council of 29 April 2004 laying down specific hygiene rules on the hygiene of foodstuffs. *Official Journal of the European Union*, **L226**, 83–127.

European Commission, 2011a. Regulation (EU) of 19 September 2011 implementing Regulation 931/2011 on the traceability requirements set by Regulation (EC) No. 178/2002 of the European Parliament and of the Council for Food of Animal Origin. *Official Journal of the European Union*, **L242**, 2–3.

European Commission, 2011b. Regulation (EU) 1169/2011 of the European Parliament and of the Council of 25 October on the provision of food information to consumers, amending Regulations (EC) No. 1924/2006 and (EC) No. 1925/2006 of the European Parliament and of the Council, and repealing Commission Directive 87/250/EEC, Council Directive 90/496/EEC, Commission Directive 1999/10/EC, Directive 2000/13/EC of the European Parliament and of the Council, Commission Directives 2002/67/EC and 2008/5/EC and Commission Regulation (EC) No. 608/2004. *Official Journal of the European Union*, **L304**, 18–63.

European Commission, 2013. Commission implementing Regulation (EC) No. 1337/2013 of the European Parliament and of the Council 13 December laying down rules for the application of Regulation (EU) No. 1169/2011 of the European Parliament and of the Council as regards the indication of the country of origin or place of provenance for fresh, chilled and frozen meat of swine, sheep, goats and poultry. *Official Journal of the European Union*, **L335**, 19–22.

European Union, 1993. Council Directive 93/43/EEC of 14 June 1993 on the hygiene of foodstuffs. *Official Journal of the European Communities*, **L175**, 1–11.

European Union, 2012. *The Rapid Alert System for Food and Feed (RASFF) 2012 Annual Report*. Brussels: Publication Office of the European Union.

European Union, 2013. *The Rapid Alert System for Food and Feed (RASFF) 2013 Annual Report*. Brussels: Publication Office of the European Union.

Food and Drug Administration, 1995. Procedures for the Safe and Sanitary Processing and Importing of Fish and Fishery Products. *US Federal Register*, vol. **60**, p. 242.

Food and Drug Administration, 2001. Hazard Analysis and Critical Control Point (HAACP): Procedures for the Safe and Sanitary Processing and Importing of Juice. *US Federal Register*, vol. **66**, p. 13.

Food and Drug Administration, 2014a. Foreign Supplier Verification Programs for Importers of Food for Humans and Animals. *Federal Register*, vol. **79**, p. 188.

Food and Drug Administration, 2014b. Outbreak Investigations. CORE Investigations,s http://www.fda.gov/Food/RecallsOutbreaksEmergencies/Outbreaks/ucm272351.htm (accessed November 20, 2014).

Fritz, M. and Schiefer, G., 2009. Tracking, tracing, and business process interests in food commodities: a multi-level decision complexity. *International Journal of Production Economics*, **117**, 317–329.

Galimberti, A., De Mattia, F., Losa, A., Bruni, I., Federici, S., Casiraghi, M., Martellos, S., and Labra, M., 2013. DNA barcoding as a new tool for food traceability. *Food Research International*, **50**, 55–63.

Golan, E., Krissoff, B., Kuchler, F., Calvin, L., Nelson, K., and Price, G., 2004. *Traceability in the US Food Supply: Economic Theory and Industry Studies*, Agricultural Economic Report No. 380 Economic Research Service. Washington, DC: United States Department of Agriculture.

International Organization for Standardization 9000:2000, 2000. *Quality Management Standards*. ISO, Geneva.

International Organization for Standardization 22000:2005, 2005. *Food Safety Management System and Requirements for Any Organization in the Food Chain*. ISO, Geneva.

International Organization for Standardization 22005/2007, 2007. *Traceability in the Feed and Food Chain – General Principles and Guidance for System Design and Development.* ISO, Geneva.

International Organization for Standardization 9001:2008, 2008. *Quality Management Systems and Requirements.* ISO, Geneva.

Jurenas, R. and Greene, J.L., 2013. Country-of-Origin Labeling for Foods and the WTO Trade Dispute on Meat Labeling. CRS report for Congress 16 September 2013, https://www.fas.org/sgp/crs/misc/RS22955.pdf (accessed July 17, 2015).

Kainuma, Y. and Tawara, N., 2006. A multiple attribute utility theory approach to lean and green supply chain management. *International Journal of Production Economics*, **101**, 99–108.

Karlsen, K.M., Donnelly, K.A.-M., and Olsen, P., 2010. Implementing traceability: pratical challenges at a mineral water bottling plant. *British Food Journal*, **112**, 187–197.

Karlsen, K.M., Donnelly, K.A.-M., and Olsen, P., 2011, Granularity and its importance for traceability in a farmed salmon supply chain. *Journal of Food Engineering*, **102**, 1–8.

Karlsen, K.M., Dreyer, B., Olsen, P., and Elevoll, E., 2012. Granularity and its role in implementation of seafood traceability. *Journal of Food Engineering*, **112**, 78–85.

Karlsen, K.M., Dreyer, B., Olsen, P., and Elevoll, E., 2013. Literature review: does a common theoretical framework to implement food traceability exist? *Food Control*, **32**, 409–417.

Karlsen, K.M. and Olsen, P., 2011. Validity of method for analysing critical traceability points. *Food Control*, **22**, 1209–1215.

Kim, H., Fox, M., and Gruninger, M., 1995. An ontology of quality for enterprise modelling. In: *Proceedings of the Fourth Workshop on Enabling Technologies: Infrastructure for Collaborative Enterprises*, Berkeley Springs, WV, April 20–22, 1995. Piscataway, NJ: IEEE, pp. 105–116.

Kvarnström, B., Bergquist, B., and Vännman, K., 2011. RFID to improve traceability in continuous granular flows – an experimental case study. *Quality Engineering*, **23**, 343–357.

Lavelli, V., 2013, High-warranty traceability system in the poultry meat supply chain: a medium-sized enterprise case study. *Food Control*, **33**, 148–156.

Lee, K., Armstrong, P.R., Thomassonn, J.A., Sui, R., Casada, M., and Herrman, T.J., 2010. Development and characterization of food-grade tracers for the global grain tracing and recall system. *Journal of Agricultural and Food Chemistry*, **58**, 10945–10957.

Lee, H.-J. and Yun, Z.-S., 2015. Consumers' perceptions of organic food attributes and cognitive and affective attitudes as determinants of their purchase intentions toward organic food. *Food Quality and Preference*, **39**, 259–267.

Lees, M., 2003. *Food Authenticity and Traceability.* Sawston: Woodhead Publishing.

Liang, K., Thomasson, J.A., Lee, K., Shen, M., Ge, Y., and Herrman, T.J., 2012. Printing data matrix code on food-grade tracers for grain traceability. *Biosystems Engineering*, **113**, 395–401.

Liang, K., Thomasson, J.A., Shen, M.X., Armstrong, P.R., Ge, Y., Lee, K.M., and Herrman, T.J., 2013. Ruggedness of 2D code printed on grain tracers for implementing a prospective grain traceability system to the bulk grain delivery system. *Food Control*, **33**, 359–365.

Magno, F., 2012. Managing product recalls: the effects of time responsible vs. opportunistic recall management and blame on consumers' attitudes. *Procedia – Social and Behavioral Sciences*, **58**, 1309–1315.

McEntire, J.C., Arens, S., Bernstein, M., Bugusu, B., Busta, F. F., Cole, M., Davis, A., Fisher, W., Geisert, S., Jensen, H., Kenah, B., Lloyd, B., Mejia, C., Miller, B., Mills, R., Newsome, R., Osho, K., Prince, G., Scholl, S., Sutton, D., Welt, B., and Ohlhorst, S., 2010. Traceability (product tracing) in food systems: an IFT (Institute of Food Technology) report submitted to the FDA (Food and Drug Administration), volume 1: technical aspects and recommendations. *Comprehensive Reviews in Food Science and Food Safety*, **9**, 92–158.

Mgonja, J.T., Luning. P., and Van der Vorst, J.G.A.J., 2013. Diagnostic model for assessing traceability system performance in fish processing plants. *Journal of Food Engineering*, **118**, 188–197.

Ministry of Agriculture, Forestry and Fisheries of Japan, 1950. Concerning Standardization and Proper Labeling of Agricultural and Forestry Products, 1950, Law No. 175, Ministry of Agriculture, Forestry and Fisheries of Japan, Tokyo.

Moe, T., 1998. Perspectives on traceability in food manufacture. *Trends in Food Science and Technology*, **9**, 211–214.

Morris, C. and Young, C., 2000. 'Seed to shelf', 'teat to table', 'barley to beer' and 'womb to tomb': discourses of food quality and quality assurance schemes in the UK. *Journal of Rural Studies*, **16**, 103–115.

Olsen, P. and Borit, M., 2013. How to define traceability. *Trends in Food Science and Technology*, **29**, 142–150.

Opara, L.U., 2003. Traceability in agriculture and food supply chain: a review of basic concepts, technological implications, and future prospects. *Food, Agriculture and Environment*, **1**, 101–106.

Peres, B., Barlet, N., Loiseau, G., and Montet, D., 2007. Review of the current methods of analytical traceability allowing determination of the origin of foodstuffs. *Food Control*, **18**, 228–235.

Randrup, M., Storøy, J., Lievonen, S., Margeirsson, S., Árnason, S., Møller, S., and Frederiksen M.T., 2008. Simulated recalls of fish products in five Nordic countries. *Food Control*, **19**, 1064–1069.

Resende-Filho, M. and Buhr, B.L., 2010. Economics of Traceability for Mitigation of Food Recall Costs. MPRA Paper, http://mpra.ub.uni-muenchen.de/27677/ (accessed June 24, 2015).

Riden, C. and Bollen, A., 2007. Agricultural supply system traceability, part II: implications of packhouse processing transformations. *Biosystems Engineering*, **98**, 401–410.

van Rijswijk W., Frewer L.J., Menozzi D., and Faioli, G., 2008. Consumer perceptions of traceability: a cross-national comparison of the associated benefits. *Food Quality and Preference*, **19**, 452–464.

Rong, A. and Grunow, M., 2010. A methodology for controlling dispersion in food production and distribution. *OR Spectrum*, **32**, 957–978.

Saltini, R. and Akkerman, R., 2012. Testing improvements in the chocolate traceability system: impact on product recalls and production efficiency. *Food Control*, **23**, 221–226.

Seuring, S. and Müller, M., 2008. From a literature review to a conceptual framework for sustainable supply chain management. *Journal of Cleaner Production*, **16**, 1699–1710.

Shaosheng, J. and Zhou, L., 2014. Consumer interest in information provided by food traceability systems in Japan. *Food Quality and Preference*, **36**, 144–152.

Skoglund, T. and Dejmek, P., 2007. Fuzzy traceability: a process simulation derived extension of the traceability concept in continuous food processing. *Food and Bioproducts Processing*, **85**, 354–359.

Sun, D.-W., 2008. *Modern Techniques for Food Authentication*. Waltham, MA: Academic Press.

Tamayo, S., Monteiro, T., and Sauer, N., 2009. Deliveries optimization by exploiting production traceability information. *Engineering Applications of Artificial Intelligence*, **22**, 557–568.

Thakur, M. and Donnelly, K.M., 2010. Modeling traceability information in soybean value chains. *Journal of Food Engineering*, **99**, 98–105.

Thakur, M., Wang, L., and Hurburgh, C., 2010. A multi-objective optimization approach to balancing cost and traceability in bulk grain handling. *Journal of Food Engineering*, **101**, 193–200.

Trienekens, J. and Beulens, A., 2001. The implications of EU food safety legislation and consumer demands on supply chain information systems. In: *Food and Agribusiness Management Association: Proceedings of the 11th Annual World Food and Agribusiness Forum*, Sydney, Australia, June 25–28, 2001. Washington, DC: IFAMA.

United States, 2002. Public Health Security and Bioterrorism Preparedness and Response Act of 2002, 12/6/2002, Public Law 107–188, United States Congress, Washington, DC.

United States, 2011. Public Law 111–353 Food Safety Modernization Act, 4/1/2011, 111th Congress, 124 Stat 3885, United States Congress, Washington, DC.

United States Department of Agriculture, 2009. Final Rule (FR) on Mandatory Country of Origin Labeling of Beef, Pork, Lamb, Chicken, Goat Meat, Wild and Farm-Raised Fish and Shellfish, Perishable Agricultural Commodities, Peanuts, Pecans, Ginseng, and Macadamia Nuts. Amending FR – Federal Register, 1/15/2009. Federal Register, vol. 78, p. 101.

Wang, X., Li, D., and O'Brien, C., 2009. Optimisation of traceability and operations planning: an integrated model for perishable food production. *International Journal of Production Research*, **47**, 2865–2886.

World Trade Organization, 2014. United States – Certain Country of Origin labelling (COOL) Requirements, WT dispute DS384/RW; DS386/RW, http://www.wto.org/english/tratop_e/dispu_e/384_386rw_e.pdf (accessed June 24, 2015).

Zarei, M., Fakhrzad, M.B., and Jamali Paghaleh, M., 2011. Food supply chain leanness using a developed QFD model. *Journal of Food Engineering*, **102**, 25–33.

7

Information Technology for Food Supply Chains

Katerina Pramatari

Department of Management Science and Technology, ELTRUN e-Business Research Centre, Athens University of Economics and Business, 47A Evelpidon and 33 Lefkados Street, Athens, Greece

7.1   Introduction

Information technology (IT) has been a key enabler in the management of the supply chain, and over the last decades we have witnessed a significant industry investment and increased focus on the exploitation of this enabling role. Very often the driver behind IT deployments in the supply chain environment has been the increase of operational efficiency and decrease of costs. Reduced inventory levels, operational benefits and reduced cycle times are just some of the benefits reported as a result of IT implementation in the supply chain (Tseng, Wu and Nguyen, 2011). In many cases though, the role of IT has been extended to provide advanced information provision capabilities that can further support coordinated decision-making, demand forecasting and higher efficiency in supply chain activities (Wu *et al.*, 2006).

Building on the experience of more than two decades, many aspects of IT implementation in the supply chain have now matured and there is a common ground of knowledge shared among involved stakeholders regarding the anticipated costs and benefits of such implementations. For example, there is quite a good understanding of how the activities related to distribution and warehousing processes are supported

Supply Chain Management for Sustainable Food Networks, First Edition. Edited by Eleftherios Iakovou, Dionysis Bochtis, Dimitrios Vlachos and Dimitrios Aidonis.
© 2016 John Wiley & Sons, Ltd. Published 2016 by John Wiley & Sons, Ltd.

by a distribution management system (DMS) and a warehouse management system (WMS) or how inventory management, production and replenishment processes are handled through manufacturing resource planning (MRP) and enterprise resource planning (ERP) systems. While the industry feels quite confident today about the cost–risk benefits and the way to move forward with such implementations, this cannot be said for the many new challenges that have only emerged in the last few years and make the role of IT in the supply chain environment more demanding and more critical than ever before.

The globalisation of markets and food supply chains, the changing consumer trends, the scarcity of resources, the impact on global warming and the emergence of new technologies are some of the dynamics that shape a new challenging environment for the IT and supply chain management professional. The globalisation of markets and food supply chains impose increased requirements for food traceability, while competition at a global scale asks for higher operational efficiency than ever before. Changing consumer trends, such as online and omni-channel purchasing, smart delivery options, smaller and more frequent buys, urbanisation, and so on, put further pressure on supply chain efficiency. At the same time, the scarcity of resources and impact of supply chain organisations on global warming require that the above are achieved with the lowest possible cost and use of resources. When considering food products, we need to add to the above special logistics requirements associated with food security and safety and very short lead times to ensure freshness, in order to get the full picture.

The above challenges present many new requirements for the use of IT in the supply chain environment, asking for enhanced information capabilities, the need for automation, information sharing, and advanced decision support. At the same time, the emergence of new technologies and information-rich environments [e.g. Internet-of-Things, sensors, radio frequency identification (RFID), big data, etc.] open up new possibilities and provide novel capabilities and tools for addressing these requirements. In the following sections, a first attempt is made to map the latest IT capabilities with emerging needs and contribute to this discussion through initial empirical evidence. The enabling role of IT is initially discussed in the supply chain context and a high-level IT architecture is proposed. Three contemporary trends and challenges for food supply chains are then selected, namely RFID-enabled supply chain management, carbon footprint monitoring in the supply chain, and urban shared food logistics, and the enabling role of IT through the proposed architecture is discussed. The focus of this book on sustainability and the current market trends justify this selection. This chapter is concluded with an overall discussion of the main barriers and drivers behind IT adoption to address the contemporary challenges of food supply chains.

7.2   Information Technology Architecture in a Supply Chain Context

In order to map IT capabilities with current supply chain challenges, we distinguish between different uses of IT and the respective scope and impact of implementation. Subramani (2004), when discussing how suppliers benefit from IT use in supply chain relationships, differentiates between the use of IT for exploitation, where the goal is on improving, applying and incrementally refining firm capabilities, and the use of IT for exploration, where the goal is on creating new capabilities and devising novel solutions to current problems. The former is mostly associated with business processes and leads to clearly definable benefits (e.g. cost reduction, process efficiency, etc.) while the latter is mostly related to domain knowledge and the benefits are soft and difficult to evaluate in advance.

Along the same lines, we differentiate between the enabling role of IT for process improvement and the use of IT to provide advanced information capabilities and management support. Sometimes the objective of an IT implementation in the supply chain is limited to process improvement, whereas in other cases process automation is seen as the first step of a series of incremental stages with the vision to ultimately achieve an information-rich environment to support management decisions and advanced services. There are also cases when an investment in IT aims from the beginning at improving the information rather than the process flow, although this is not very common. Advanced information capabilities may refer to new types of information or new views relying on the combination of information coming from different sources. These advanced information capabilities can provide enhanced domain knowledge to supply chain partners and inform critical management decisions. Moreover, the process improvement and advanced information capabilities can be extended at the lower end of the supply chain to empower the end-consumer, which is commonly done nowadays via a web or mobile application interface.

The data to support either process improvements or advanced information capabilities and management decisions rely on other internal information systems (ISs) or information sharing and collaboration with supply chain partners. The scope and granularity of data and respective information further depend on the capabilities of the employed IT infrastructure.

Figure 7.1 presents a graphical representation of a proposed IT architecture in the supply chain context. Each layer of the architecture builds on the capabilities provided by the lower layer to offer enhanced capabilities. In that respect, the choices made at the IT infrastructure layer already define, to a certain extent, the quality and granularity of the information that can be available to the upper layers. For example, if barcode technology is used at the IT infrastructure layer for product identification, then this imposes certain limitations on the level of process automation or type of information that is available to the higher layers of the architecture. On the other hand, if unique instance identification is enabled by an RFID infrastructure, then the higher layers can benefit from this and support, for example a traceability service offered to end-consumers for each unique product instance that would not have been possible in the previous case. Of course, such a service would also rely on the appropriate internal IT/IS capabilities and information sharing with supply chain partners, as well as on the appropriate process control and management of information.

In the following sections, the enabling role of IT based on this proposed architecture is demonstrated in the context of three different areas of interest for food supply chains, namely RFID-enabled supply chain management, carbon footprint monitoring, and urban shared logistics.

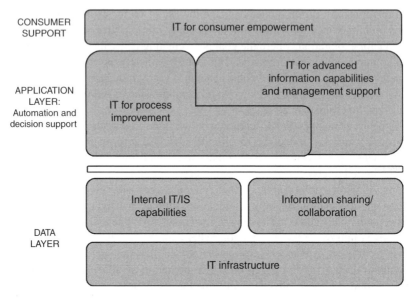

Figure 7.1   Information technology architecture in a supply chain context.

7.3   RFID-Enabled Supply Chain Management

The emergence of new technologies, such as RFID, is expected to revolutionise many of the supply chain operations by reducing costs, improving service levels, and offering new possibilities for identifying unique product instances. The advanced data capture capabilities of RFID technology coupled with unique product identification and real-time information coming from different data sources, such as environmental sensors, define a new and rich information environment that opens up new horizons for efficient management of supply chain processes and decision support.

While efforts and discussions around the deployment of RFID in the supply chain environment have been around for over a decade now, it is only in the last few years that we have witnessed a more rational approach to RFID adoption in the supply chain. Initial efforts were greatly driven by external factors, such as the pressure imposed by a leading retailer (e.g. Walmart or Metro) on its supply chain partners, especially at times when the technology was not mature enough and various technical challenges had yet to be addressed. This fact created negative industry perceptions overall and the capabilities of RFID technology were treated with scepticism for several years afterwards. However, looking at current RFID deployments, one can claim that we have entered the rationality phase, when investments are mostly driven by internal motives and cost–benefit assessment.

A recent survey conducted among European enterprises (Pramatari and Dimakopoulou, 2014) shows that there is clearly an increasing trend of adopting RFID technology in the supply chain, with RFID projects related to traceability, asset tracking,

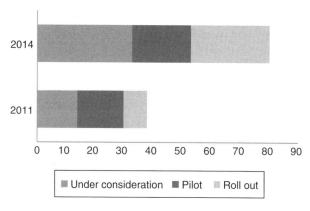

Figure 7.2   Responses related to RFID deployment for traceability (Pramatari and Dimakopoulou, 2014).

inventory audit and inbound/outbound logistics found at the top of the list. Figure 7.2 shows the difference in the responses to the same survey conducted in 2014 and 2011 among respondents declaring an interest in traceability. In 2011 less than 40% of the respondents declared any interest in deploying RFID for traceability, with less than 5% of the companies declaring they had such a project in roll-out phase. In 2014, almost 80% of the respondents declared an interest and about 30% declared they already had such a project in roll-out phase, showing a significant increase in RFID adoption. A similar trend is found in projects related to asset tracking, inventory audit and inbound/outbound logistics. It is also interesting to note that only 13% of the respondents that have adopted RFID technology said that this came as a request by a major trading partner.

For many companies, RFID is seen as an infrastructure that can support a family of applications and they take this into account when considering investment decisions (60% of the respondents of the same survey). In the food supply chain, RFID can potentially empower a broad spectrum of applications, ranging from upstream warehouse and distribution management down to retail-outlet operations, including shelf management, promotions management and innovative consumer services, as well as applications spanning the whole supply chain, such as product traceability (Pramatari, Doukidis and Kourouthanassis, 2005).

More specifically, RFID is a technology that uses radio waves to automatically identify objects. The identification is done by storing a serial number, and perhaps other information, on a microchip that is attached to an antenna. This bundle is called an RFID tag. The antenna enables the chip to transmit the identification information to a reader. The reader converts the radio waves reflected back from the RFID tag into digital information that can be passed on to an enterprise IS.

Traditionally, the food supply chain uses barcodes as the main identifier for cases, pallets and products. However, many in the industry are already looking to the business case of RFID as the 'next generation of barcode' through its ability to identify products automatically not requiring line-of-sight and store much more information, thus enabling mass serialised identification of every single product instance in the supply chain. The Electronic Product Code (EPC) is the standard adopted in this case.

In the various application domains, the contribution of RFID can be sought in the following areas:

a. the automation of existing processes, leading to time/cost savings and more efficient operations;

b. the enablement of new or transformed business processes and innovative consumer services, such as monitoring of product shelf availability or traceability of unique product instances;

c. the improvement achieved in different dimensions of information quality, such as accuracy, timeliness, and so on;

d. the formation of new types of information, leading to a more precise representation of the physical environment, for example a product's exact position in the store, a specific product's production and distribution history, and so on.

If RFID technology is only exploited internally by a network leader looking solely at internal benefits, for example a big retailer trying to improve store operations, then suppliers deal with RFID technology as another unfortunate strategic necessity (Barua and Lee, 1997). As already mentioned, this trend had a negative impact on RFID market acceptance and adoption rates. Subramani (2004) argues that suppliers benefit from IT use in supply chain relationships when they use IT either in order to gain higher *business-process specificity* or in order to gain *higher domain-knowledge specificity*. We could say the first two points listed above are associated with business-process specificity while the latter two are associated with domain-knowledge specificity. From this perspective, the question that arises is how to enable collaborative processes and decision-making exploiting the aforementioned RFID capabilities, so that not only network leaders–retailers but also suppliers can benefit from employing RFID both in improving process management and in gaining domain knowledge.

In order to address this question, the data layer of the proposed architecture (including the respective IT infrastructure, the internal IT/IS capabilities and the information sharing with supply chain partners) should consider the following requirements:

- the immense amount of data that needs to be processed in real time; already in cases where products are still identified at product-type level through barcodes, the handling of information in real-time for decision-support purposes is quite a technical challenge;

- the need to ensure synchronised product information between supply chain partners; although the sector has adopted barcode technology as a standard to identify products, the information is maintained at different levels in either the

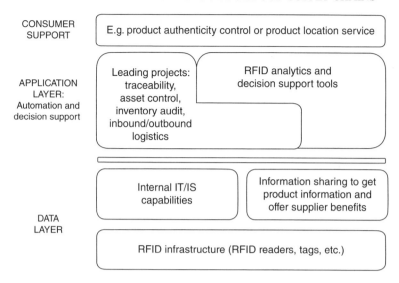

Figure 7.3   IT architecture for RFID-enabled supply chain management.

retailers' or the manufacturers' systems causing serious integrity issues when data exchange and synchronisation is required;

• the many different business relationships that need to be supported and the different collaboration scenarios that may be applicable in each supply chain relationship; each retailer may collaborate with hundreds of suppliers and vice versa;

• the need to support seamless information sharing and collaborative decision-support through automated and secure interorganisational system links.

Figure 7.3 presents how the proposed IT architecture is used where RFID technology is employed to support supply chain management. As already mentioned, the same RFID infrastructure combined with internal IT capabilities and other systems (such as ERP and WMS) is often used to support different processes, such as traceability, asset control, inventory audit and inbound/outbound logistics. It is then when the justification of investments in RFID technology is easier to achieve, as the costs of the same RFID infrastructure are compared with the benefits gained from many different process improvements. As the industry moves more and more today towards rolling-out RFID in many supply chain processes, the discussion of exploiting the acquired information to extract domain knowledge and support advanced decisions slowly opens up. The same applies to the exploitation of RFID technology for supporting innovative consumer services, especially in sectors that deploy RFID at item level, such as fashion retailing.

In fashion retailing, RFID is widely adopted today mainly for inventory control purposes and management of backroom operations. Investments in RFID in this sector are easily justified based on the high value of garments relative to the cost of an RFID tag. In such sectors, the industry has already moved to the next stage of empowering consumers with innovative services associated with the handling of unique product instances in the retail store environment (e.g. www.seramis-project. eu). In sectors where the difference in value between the item to be tagged and the RFID tag is not significant (e.g. the food supply chain), RFID is either deployed at case or pallet level or is focussed on products of high value or addresses the need to preserve a product's unique identity. In such sectors, the exploitation of RFID technology to support innovative consumer services is expected to be further delayed.

Especially relevant to food products and traceability, RFID technology provides great opportunities for effective and efficient traceability system design. Based on automated data capture, traceability information can be obtained at significantly reduced labour costs and with small changes in the enterprises' business processes. Furthermore, RFID technology combined with the appropriate information infrastructure can enable end-to-end traceability in the supply chain at small costs, affordable to small and medium enterprises that comprise the majority of food companies. Kelepouris, Pramatari and Doukidis (2007) identify the requirements related to traceability that the respective data layer should meet, in order to enable both process improvements related to enabling traceability in the supply chain as well as advanced information capabilities and management support. These refer to:

- *The Level of Identification*: This determines traceability accuracy and resolution. Identification may take place at item, case, pallet or even batch/lot level. As we decrease identification resolution, both traceability accuracy and information management costs decrease. Apparently, there is a trade-off between traceability resolution and the related costs that a firm must suffer. For most food product categories, the most balanced choice seems to be the identification at case or pallet level, as traceability resolution is maintained at good levels and the costs are affordable.

- *The Product Transformation Stage*: Two different types of stages are distinguished. The first refers to stages where the production process is a sequence of activities transforming a listing of raw materials, parts, intermediaries and subassemblies into one particular end-product. The second refers to stages where the product is not subject to any transformations, such as distribution or retail. Clearly, the needs and complexity of traceability are much higher and difficult to address in the first type of stages.

- *Operations and Capacity Units*: A production process is a network of manufacturing steps, which have been aggregated into operations for the purpose of manufacturing control. These operations take place in actual capacity units. Furthermore, the operations include specific variables that affect the quality of the outcome of each operation. Traceability requires the recording of both the variables and their values under which the operation took place. Moreover, the capacity unit in which the operation was held should be recorded.

- *Location Information*: Apart from the information regarding product composition, information regarding product location throughout the chain should also be recorded. This aspect of information refers to physical traceability. In order to support this kind of information, item observations should be recorded across the chain. Each observation should consist of a triplet of arguments: location, timestamp and item identity.

Kelepouris, Pramatari and Doukidis (2007) discuss the aforementioned traceability requirements and suggest how RFID technology can efficiently address them. Compared with the other application areas of RFID, traceability clearly puts the most pressure of all on information sharing and collaboration with supply chain partners, in order to extend the scope of traceability, either upwards or downwards in the supply chain.

7.4   Carbon Footprint Monitoring in the Supply Chain

Another major challenge that companies in the food supply chain clearly face today is related to environmental impacts. Climate change, legal compliance, rising energy prices and customers' increasing ecological awareness are exerting strong pressures (Bunse *et al.*, 2011) and have raised an imperative need for wise management of energy resources and industrial environmental performance. To effectively meet this requirement, sustainability practices have come to the societal and governmental forefront (Watson, Boudreau and Chen, 2010) and organisations are consciously interested in analysing and improving the environmental impact of their products and processes.

Firms, under the pressure of their stakeholders, have implemented various environmental strategies in order to manage the interface between their business and the natural environment (Aragon-Correa and Sharma, 2003). A firm's environmental strategy may materialise through various environmental practices, such as product and process innovations for pollution prevention, use of life cycle analysis, acquisition of clean technology/equipment, shipments consolidation, selection of cleaner transportation methods, as well as through the implementation of ISs (Jenkin, Webster and McShane, 2011). The deployment of ISs to cope with environmental challenges is a new area of development and only lately have we seen an increasing number of publications addressing this topic (e.g. Melville, 2010; Watson, Boudreau and Chen, 2010; Elliot, 2011). Some initial efforts have examined the role of IT/IS as a tool to help organisations implement more sustainable business processes (Watson, Boudreau and Chen, 2010) or as the basis for new ways of delivering products with less energy and carbon emissions (Bunse *et al.*, 2011). Jenkin, Webster and McShane (2011) first make the link between environmental motives and the development of an environmental IS strategy and propose a framework to guide future research in the field. At the basis of all these efforts lies the capability to monitor and control energy consumption and carbon emissions, in order to further drive energy efficiency and

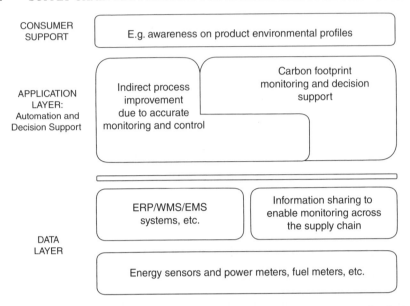

Figure 7.4   IT architecture for carbon footprint monitoring in the supply chain.

lower environmental impacts. As the common principle that 'what cannot be measured, cannot be managed' is also applied in this case, issues related to measurement and monitoring of a firm's environmental performance become significantly important.

In the following paragraphs we discuss how IT can support carbon footprint monitoring in the supply chain, building on the proposed architecture presented earlier in this chapter. Usually companies engaged in such efforts put emphasis on two separate but related metrics: (i) energy consumption, as a key cost driver and environmental sustainability metric behind many supply chain operations and (ii) carbon footprint or carbon dioxide equivalent, as the most universal and widely adopted environmental performance indicator, used also for reporting and comparison purposes.

This is a case where the focus of IT deployment is clearly on offering advanced information capabilities and management support. Supply chain processes may be improved as a result of the acquired domain knowledge, but the driver behind this improvement may not be the IT per se but an innovative logistics practice, for example shared logistics. This is depicted in Figure 7.4, demonstrating the IT architecture supporting energy and carbon footprint monitoring.

If an accurate and comparable metric of carbon footprint is acquired through the appropriate IT architecture, then this information could further be used to inform consumers through innovative services, offering, for example product environmental impact information at the point-of-sales. However, the industry is not there yet to agree on a common approach to monitor carbon footprint across all different product categories so that these are comparable in the eyes of a consumer. While approaches like Life Cycle Assessment (LCA) have been widely adopted and databases of Life Cycle Inventories (LCIs) have been constructed, there is still way to reach industry consensus on how to present carbon footprint information related to a specific product to the end-consumer. In the following, we thus leave aside this perspective and examine what are the capabilities that the underlying layers of the architecture should provide in order to be able to monitor carbon footprint in the supply chain and support respective decisions.

In this context, at the lower level of the architecture, the infrastructure layer comprises energy sensors and power meters or fuel consumption meters that can automatically provide real-time information regarding energy consumption. While more and more organisations start deploying such infrastructures at their premises and in their fleets of vehicles, so as to acquire more accurate information, if such infrastructures are not available then energy consumption measurements usually rely on electricity bills and other energy-related costs/invoices providing an aggregated picture.

In order for this information to be associated with the products flowing through the supply chain, integration of the energy consumption data flows with information coming from other sources is required. More specifically, at the data layer, an IS supporting carbon footprint monitoring in the supply chain should be able to integrate the following information flows:

- Contextual information flow, including information on machines, processes, products, facilities, supply chain partnerships, and so on, which supports the interpretability of the transactional information.

- Transactional information flow, including information on the transactions that take place in the supply chain such as production processes, ordering, distribution and inventory management.

- Environmental information flow, including information on energy consumption that is either measured by energy sensors or retrieved by existing Building Management Systems and fuel consumption information referring usually to vehicle fuel refills or to actual fuel consumption monitored through sensors and metering devices installed on vehicles.

- Product environmental information flow, including information on the environmental profile of products, as provided by external sources, such as LCIs.

It easily follows that the above information flows come from different existing systems (such as ERP, WMSs, energy sensors, building management systems, environmental management systems, etc.). Moreover, if this information comes from internal ISs then the scope of carbon footprint monitoring is limited to a company's boundaries. If this needs to be extended, so that information about a product's environmental profile is acquired as it moves across the supply chain, then information sharing with supply chain partners is also required. This is usually the case when other supply chain partners are involved in a product's distribution, warehousing and other supply chain activities.

The concept of monitoring carbon footprint across the supply chain, based on the integration of the different data flows as presented above, has been validated in the context of the e-SAVE research project (www.e-save.eu). More specifically, this concept has been applied in three different food supply chain field settings in Germany, Italy and Greece with the involvement of an international food manufacturer and a supermarket retailer in order to support:

- The monitoring of energy consumption, carbon-footprint and other environmental performance indicators of operations and/or products through analysis, reporting and collaborative business intelligence.

- Day-to-day operations and supply chain management decisions (such as inventory management, logistics and distribution planning, vehicle rooting, etc.).

- Strategic supply chain planning through simulation, giving the possibility for managers to simulate alternative supply chain configurations and evaluate them using both energy efficiency as well as cost and benefit <termDef>Key Performance Indicator</termDef> (<abbrev>KPI</abbrev>s).

- Green consumer services, informing consumers about the environmental footprint of products in the form of a dynamic energy-efficiency label, for example through smart-phone applications, in order to stimulate green consumer demand.

The main findings from these field tests can be summarised in the following three points:

- There is still significant room for improvement in the supply chain and distribution processes in terms of environmental performance; the respective opportunities can be identified by considering a combined view of both environmental and operational efficiency metrics.

- Collaboration with supply chain partners is critical in acquiring a holistic view of the supply chain operations but also for improving environmental performance metrics.

- The difficulty of integrating different information flows coming from disparate systems and supply chain partners into meaningful performance indicators should not be disregarded.

7.5   Urban Shared Food Logistics

Collaborative logistics, usually also called logistics pooling or collaborative distribution, are more than ever at the forefront of corporate priorities and have led to the emergence of new practices in managing supply chain flows that enhance a firm's performance and add value (Vanovermeire and Sörensen, 2014). Especially the urban environment presents many new opportunities for shared food logistics. In Europe, around 75% of the population lives in urban areas and it is foreseen to increase up to 80% by 2020 (EEA, 2010). At the same time, looking through the urban environmental sustainability assessment carried out by cities and local authorities, the major impacts are generated by food transport, as confirmed by various 'food miles' studies (Pirog and Benjamin, 2005; Coley, Howard and Winter, 2008). Managing freight transport operations, especially in the urban environment, is a complex task. This is due to the fact that urban transport of goods has specific characteristics such as: multiple operators, delivery vehicles of varying capacity, different organisations [Less than Truckload (LTL)/Full Truckload (FTL), delivery and collection], a large part of own account transport, and a significant part of subcontracting. On top of that, an expansive last mile movement worsened by congestion and increasing constraints in road sharing, new traffic regulations and planning by local authorities make matters worse. Thus, it is not surprising that urban freight distribution is accompanied by numerous negative environmental effects (e.g. traffic congestion, noise, gas emissions, accident fatalities, etc.) and significant transport costs.

Lately, a number of solutions and methods have been developed in an urban context to provide support both to public authorities and transport stakeholders. Such solutions include Urban Consolidation Centres (UCCs), off-peak/night deliveries, cargo trams, cargo bikes, tolls in city centres, and so on. Another solution that has been adopted over recent years is the use of logistics resource sharing. Especially in the food supply chain, the shared logistics practice has drawn a lot of attention among manufacturers, logistics service providers and retailers. While collaborative logistics solutions have been developed and implemented for several years now, recent changes have increased the pace of this development. In a recent industry survey conducted by ECR France with the participation of 74 companies representing manufacturers/distributors and logistics service providers, 62% of the respondents reported that they are already using collaborative logistics solutions and 83% of them stated that they plan to engage in a new collaborative logistics project within the next 2 years (Efficient Consumer Response France, 2013). The high interest in collaborative logistics practices is attributed to today's economic environment with virtually no increase in sales volumes, but sharp increases in the cost of energy, transport and taxes, with more just-in-time deliveries and more players aimed at reducing stock in inventory. On top of that, we can identify three major consumer trends that increase the transport burden in urban contexts and call for further collaborative logistics practices in the food retail sector:

a. An increasing number of consumers prefer frequent, smaller-basket visits to neighbourhood-based convenience stores instead of monthly shopping trips to big hypermarkets.

b. Consumers turn more and more towards quality food (e.g. organic food) and production origins (or site-specific farming or preference for local producers).

c. There is a significant increase in online shopping, especially in big city centres, where consumers often request home delivery out of normal working hours.

Table 7.1   Top consumer trends leading to increased distribution flows in urban areas.

Consumer trend	Impact on distribution flows in urban areas
(a) Increasing move of consumers from monthly shopping trips to big hypermarkets to frequent smaller-basket visits to neighbourhood-based convenience stores	Increased distribution flows from manufacturers of packaged and fresh food to convenience stores in urban areas, either directly or through retail distribution centres
(b) Consumers turn towards quality of food (e.g. organic food) and production origins (or site-specific farming)	Increased product flows from local producers to retail stores in urban areas as well as to consumers directly
(c) Significant increase in online shopping, especially in big city centres, where consumers often request home delivery out of normal working hours	Increasing number of previously non-existent direct-to-consumer distribution flows, from either retail stores or distribution hubs of online retailers

Table 7.1 summarises how these new consumer trends contribute to the significant increase in food distribution flows in urban areas, resulting in subsequent increase in transportation cost, traffic congestion and environmental burden.

Complementing the above, there is currently a tendency of big retailers pushing back to manufacturers the cost of logistics operations, as a response to the economic crisis. This further decreases the degree of logistics synergies and economies of scale achieved through centralised deliveries to retail outlets. Thus, we find more and more small, convenience and speciality stores in urban areas that need frequent direct store deliveries of food products as well as more and more consumers requesting direct home delivery.

The above emerging trends create a new distribution landscape that calls for collaboration and consolidation of flows (as depicted in Figure 7.5) in order to control cost, traffic and environmental burden in urban areas.

Moreover, food logistics have specific distribution requirements that are quite strict compared with the distribution of other goods, such as traceability requirements, temperature, smell and vehicle-condition limitations, restrictions in the combination of goods, and so on. This fact necessitates the need to address food logistics as a separate field compared with the distribution of other goods in urban areas and imposes specific limitations and new parameters in any collaboration effort.

While the above demonstrate the need for consolidation of the increasing distribution flows in urban areas, there is still a long way to go in this direction. In the same study conducted by ECR France as mentioned above (Efficient Consumer Response France, 2013), the three top reasons identified as the main barriers

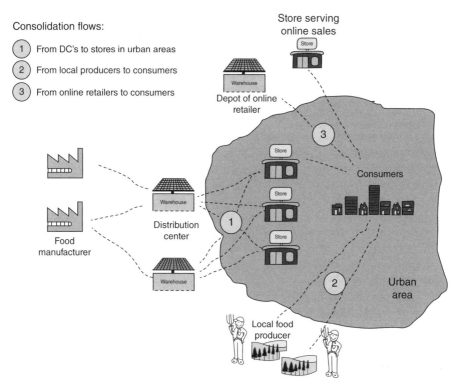

Consolidation flows:

① From DC's to stores in urban areas

② From local producers to consumers

③ From online retailers to consumers

Figure 7.5   Trends for flows consolidation in urban food logistics. DCs, Distribution Centres.

behind the adoption of shared logistics are: difficulty in identifying a suitable partner (44%); difficulty in sharing information (31%); and not being convinced about the benefits (25%).

In this context, IT can again act as a key enabler, supporting the identification of opportunities, evaluation of alternatives and management of shared logistics practices. Figure 7.6 demonstrates how the proposed IT architecture is employed to support shared logistics. The emphasis in this case is on information sharing and collaboration as the key prerequisite to identify the right matching partners for distribution flow consolidation.

Information sharing with interested parties allows a firm to determine compatibility of flow consolidation, depending on the type of products (e.g. ambient temperature, positive or negative refrigeration, seasonality, etc.), degree of identical delivery locations, times of pick-up and delivery, frequency of deliveries, vehicle load rates, and so on. On the other hand, information sharing is one of the sensitive points involved in the strategy behind collaborative logistics projects and if not appropriately addressed can be a barrier to collaborative logistics projects. The fear of information loss can be alleviated by providing the appropriate functionality through

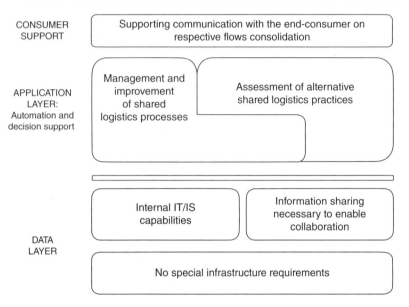

CONSUMER SUPPORT

APPLICATION LAYER: Automation and decision support

DATA LAYER

Figure 7.6   IT architecture for shared logistics.

the design and development of a platform to be operated by an independent third party and protecting firms from unnecessary information disclosure. A collaboration platform established to support logistics pooling strategies could, for example enable companies to:

- Identify opportunities for consolidating transportation flows with the objective to reduce the environmental impacts and the transportation cost. Based on the provided data, the platform could identify and propose specific logistics sharing matches between different food suppliers serving the same retail outlets, or between different local food producers or between different online food retailers supplying the same area.

- Measure the impact of the identified practices and strategies, via comparative analysis, financial assessment and simulation experimentation, and evaluating alternative policies based on KPIs addressing adverse economic, environmental and social effects.

- Share information with their supply chain partners and managing the respective logistics sharing partnerships.

Moreover, previous studies have identified the data-intensive nature of collaborative logistics, the need for data flow synchronisation, the need for more sophisticated ISs to support decision-making and the identification of the specificities of the various cases as some of the main issues that should be tackled (e.g. Ramanathan, 2014; Gonzalez-Feliu and Salanova, 2012).

Overall, in food supply chains, shared logistics practices bring forward novel processes requiring the use of IT to enable them. The use of IT in this context builds on new and existing internal ISs as well as on information sharing and collaboration with supply chain partners to enable process automation and decision support. The information that is generated through shared logistics practices can be further explored to generate new domain knowledge and support management decisions. However, the industry is still currently evaluating shared logistics opportunities rather than broadly using systems to support them.

7.6   Discussion

This chapter is concluded with an overall discussion of the main barriers and drivers behind IT adoption to address the contemporary challenges of food supply chains. As revealed from the cases presented above, IT is seen as a key enabler for many of the challenges that food supply chains have to face now and in the near future. In two of these cases, new information infrastructures are put in place in order to enable the automatic capturing of information related to either product movements (e.g. RFID) or other data (e.g. energy consumption). The justification of investments in such infrastructures is a main barrier behind their adoption, and this has for several years hindered the deployment of RFID in the supply chain. This barrier is alleviated when the cost of the infrastructure can be shared among different applications and process improvements that render adequate benefits to counterbalance this cost. Especially for food products, traceability is an additional application area that provides further benefits for such investments. It is, thus, not surprising that many companies have rolled-out RFID projects across many different application domains in parallel, including traceability, asset tracking, inventory audit and inbound/outbound logistics. On the other hand, investments in energy consumption monitoring infrastructures (e.g. electricity power and fuel consumption meters) follow a slower pace of adoption, with the reduction of energy consumption being the main argument behind their justification. In this area, the deployment of IT infrastructures just to enable more accurate monitoring and reporting of environmental impacts cannot be easily justified yet. The difficulty to measure the benefits accruing from more accurate environmental reporting and decision support, paired with the difficulty to integrate all the different information flows required to support it (energy consumption, product flows, etc.), indicate that the adoption of such systems by the industry will be rather slow. The pace of adoption may change though, if external pressures are strengthened, such as stricter regulation or need to report environmental impacts to consumers. From respective field experimentation (e.g. in the course of the e-SAVE project, www.e-save.eu), it has been shown that more accurate reporting of environmental impacts usually demonstrates lower values of environmental KPIs for companies that have employed environmental strategies. In other words, relying on aggregate figures and market averages in order to estimate the environmental impact of products, works against companies with environmental focus and in favour of less environmentally focused ones. Thus, unless the information about a product's environmental impact gains higher importance, companies will not feel the need to adopt information systems to support more accurate environmental reporting and decision support.

In all three cases, the integration of information flows from different internal information systems and new information infrastructures has been a key challenge. The information quality and granularity level of the data to be combined as well as the timeliness of the information raise many interesting conditions and issues to be tackled. Various validation and interpretation mechanisms are often required to combine the data from different sources and translate them into meaningful information that can be used by applications at the higher levels of the architecture.

This exercise becomes even more difficult when part of the information comes from external information systems of supply chain partners. This has been a condition extending internal possibilities in the first two cases, but has been an absolute necessity in the latter case of shared logistics. From all three cases it is shown that information sharing and collaboration among supply chain partners will surely expand in the future and offer competitive advantage to the companies that cope with its complexity. We thus anticipate new forms of supply chain collaboration to be slowly established among supply chain partners, extending the current practices of the Continuous Replenishment Program (CRP), Vendor Managed Inventory (VMI) or Collaborative Planning, Forecasting and Replenishment (CPFR) (Pramatari, 2007). IT has played an enabling role in all these collaboration practices and there is a clear evolution path in the capabilities and sophistication of the underlying IT infrastructure supporting former versus later forms of collaboration.

The sophistication, level of automation and coordination of information flows at the data layer determine to a great extent the capabilities provided at the upper layers of the architecture for process improvement and advanced decision support. The main focus of the industry today is clearly on applications centred around process automation and improvement, where the benefits can be easily quantified to justify the respective investments. This leaves big room for improvement and further developments in areas exploiting the new types of generated information to enhance domain knowledge and offer advanced decision support. Given the increasing amounts of generated data, big data technologies will need to be employed to meet the requirements for advanced business analytics in this field.

Last but not least, the first signs are already there of firms exploiting information stemming from their supply chain operations to empower consumers and gain a competitive advantage on this front. Given the fact that almost every consumer today has power through the use of a smart phone or tablet device and can start to interact with the Internet of Things infrastructure, creates a new competitive landscape and offers room for service innovation and consumer engagement. Thus, it is anticipated that in the near future there will be more and more food supply chain companies innovating in this domain.

7.7   Conclusions

This chapter has presented three different opportunities and challenges that companies in the food supply chain face today and has discussed the role that IT can play in these different cases through a proposed IT architecture. From this discussion and the evidence provided by the three cases, some interesting findings related to the enabling role of IT in the supply chain environment are discussed. The work presented in this chapter and the conclusions drawn are relevant to practitioners and academics having either technical or business orientation. Technical people can gain evidence from watching the application of different technologies in a specific application field. On the other hand, practitioners in the food supply chain but also in other supply chain areas can assess the risk and capabilities associated with the deployment of IT in the supply chain environment and better anticipate future developments.

The main findings associated with the adoption of advanced IT/IS infrastructures in the supply chain, as supported by the presented cases, are summarized as follows:

- Quantitative benefits related to process improvements are much easier to justify and appear to be today the main driver behind the adoption of advanced IT infrastructures in the supply chain (e.g. RFID infrastructures).

- The domain knowledge that can be acquired from the new information and data flows provided by new IT infrastructures is largely underexplored. There seems to be a significant time lag between the deployment of IT for automation and process improvements and the exploitation of the advanced information capabilities provided by such infrastructures for decision support.

- Established IT/IS systems in the supply chain environment, such as ERP, WMS, and so on, set specific limitations related to information quality and level of data granularity and their integration with new sources of data, such as RFID data streams, traceability information or energy consumption data, is not a trivial task. This represents a main barrier to adoption or diminishes the value of such investments.

- The integration of information flows becomes a much more difficult exercise and respective barrier behind the adoption, when information sharing with supply chain partners is involved. We thus anticipate that supply chain practices relying on information sharing with supply chain partners, such as shared logistics, need to first address this barrier before they can be widely supported by IT.

Based on the evidence provided by the three cases presented in this chapter, it can be anticipated that advanced IT/IS infrastructures enabling internal process improvement are the first to be adopted and there are already significant steps made in this direction by the industry (e.g. RFID adoption in supply chain processes); the internal exploitation of advanced information for decision support is a subsequent next phase. When information sharing with supply chain partners is required, cases that can demonstrate significant gains from process improvement, such as shared logistics, are expected to be the main drivers to push companies in this direction. Unless a much broader sharing of information is witnessed among supply chain partners than is happening today, it is difficult to anticipate the gain of domain knowledge and advanced business analytics and decision support relying on supply chain partner information. Last but not least, the provision of advanced information capabilities to the end-consumer, who is the last node in the supply chain, is seen today as a key challenge. This can be a source of competitive advantage for companies that can achieve a smooth information flow along the supply chain and support new services and the provision of integrated information to the end-consumer.

However, the three cases discussed provide qualitative findings that would be useful to validate through further empirical evidence and quantitative work, and also through application in different field settings. Interesting findings may also be gained by not only looking at the technological aspects of these three cases, but at several other aspects relating to IT deployment and supply chain collaboration, such as the governance rules and relations between supply chain partners, the organisational issues associated with existing and new practices, the motives for adoption as well as the pertinent costs and benefits for the various partners. These are interesting issues for academics and practitioners alike and call for further research in this area.

References

Aragon-Correa, J.A. and Sharma, S., 2003. A contingent resource-based view of proactive corporate environmental strategy. *The Academy of Management Review*, **28**, 71–88.

Barua, A. and Lee, B., 1997. An economic analysis of the introduction of an electronic data interchange system. *Information Systems Research*, **8**, 398–422.

Bunse, K., Vodicka, M., Schönsleben, P., Brülhart, M. and Ernst, F.O., 2011. Integrating energy efficiency performance in production management–gap analysis between industrial needs and scientific literature. *Journal of Cleaner Production*, **19**, 667–679.

Coley, D., Howard, M. and Winter M., 2008. Local food, food miles and carbon emissions: a comparison of farm shop. *Food Policy*, **34**, 150–155.

EEA, 2010. *The European Environment – Urban Environment*. Copenhagen: European Environment Agency.

Efficient Consumer Response France, 2013. Collaborative Logistics Barometer. ECR France, Paris, https://docs.google.com/file/d/0B-rSIEhOXi6pUU9xdFJvRzhaaHc/edit (accessed 24 June 2015).

Elliot, S., 2011. Transdisciplinary perspectives on environmental sustainability: a resource base and framework for IT-enabled business transformation. *MIS Quarterly*, **35**, 197–236.

Gonzalez-Feliu, J. and Salanova, J.-M., 2012. Defining and evaluating collaborative urban freight distribution systems. *Procedia – Social and Behavioral Sciences*, **39**, 172–183.

Jenkin, T.A., Webster, J. and McShane, L., 2011. An agenda for 'Green' information technology and systems research. *Information and Organization*, **21**, 17–40.

Kelepouris, T., Pramatari, K. and Doukidis, G.I., 2007. RFID-enabled traceability in the food supply chain. *Industrial Management and Data Systems*, **107**, 183–200.

Melville, N.P., 2010. Information systems innovation for environmental sustainability. *MIS Quarterly*, **31**, 1–21.

Pirog, R. and Benjamin, A., 2005. *Calculating Food Miles for a Multiple Ingredient Food Product*. Ames, IA: Leopold Center for Sustainable Agriculture, Iowa State University.

Pramatari, K.C., 2007. Collaborative supply chain practices and evolving technological approaches. *Supply Chain Management – An International Journal*, **12**, 210–220.

Pramatari, K.C. and Dimakopoulou, A., 2014. Assessing the impact of RFID projects: European survey results. International RFID Congress 2014, Marseille, France, 7–8 October 2014.

Pramatari, K.C., Doukidis, G.I. and Kourouthanassis, P., 2005. Towards 'smarter' supply and demand-chain collaboration practices enabled by RFID technology. In: P. Vervest, E. Van Heck, K. Preiss and L.F. Pau (eds) *Smart Business Networks*. New York: Springer Verlag, pp 197–210.

Ramanathan, U., 2014. Performance of supply chain collaboration – A simulation study. *Expert Systems with Applications*, **41**, 210–220.

Subramani, M., 2004. How do suppliers benefit from information technology use in supply chain relationships. *MIS Quarterly*, **28**, 45–73.

Tseng, M.-L., Wu, K.-J. and Nguyen, T.T., 2011. Information technology in supply chain management: a case study. *Procedia – Social and Behavioral Sciences*, **25**, 257–272.

Vanovermeire, C. and Sörensen, K., 2014. Measuring and rewarding flexibility in collaborative distribution, including two-partner coalitions. *European Journal of Operational Research*, **239**, 157–165.

Watson, R.T., Boudreau, M.C. and Chen, A.J., 2010. Information systems and environmentally sustainable development: energy informatics and new directions for the IS community. *MIS Quarterly*, **34**, 23–38.

Wu, F., Yeniyurt, S., Kim, D. and Cavusgil, S.T., 2006. The impact of information technology on supply chain capabilities and firm performance: a resource-based view. *Industrial Marketing Management*, **35**, 493–504.

8

Carbon Footprint Management for Food Supply Chains: an Integrated Decision Support System

Agorasti Toka,[1] S. C. Lenny Koh,[2] and Victor Guang Shi[3]

[1]*Laboratory of Quantitative Analysis, Logistics and Supply Chain Management, Department of Mechanical Engineering, Aristotle University of Thessaloniki, Thessaloniki, Greece*
[2]*Advanced Resource Efficiency Centre (AREC), University of Sheffield, Sheffield, UK*
[3]*Advanced Manufacturing Research Centre (AMRC), University of Sheffield, Sheffield, UK*

8.1 Introduction

Food consumption has been increasing continuously as a result of population growth, economic development, and improvement of standards of living throughout the world, thus stimulating higher food production and development of extensive globalized food supply chains (FSCs) (Clay, 2011; Foley *et al.*, 2011). Interestingly, world food exports expanded by 12% in 2010, with the European Union (EU) increasing its exports by 5% and remaining a major exporter of food products with a value of more

Supply Chain Management for Sustainable Food Networks, First Edition. Edited by Eleftherios Iakovou, Dionysis Bochtis, Dimitrios Vlachos and Dimitrios Aidonis.

than US$450 billion per year (World Trade Organization, 2011). At the same time, the food industry emerges as one of the most carbon-intensive in the world, posing challenges to policy-makers, companies, and consumers to reduce the environmental impact of contemporary food supply networks, while securing more affordable, reliable, and safe food for society.

In general, FSCs contribute between 15% and 28% to the total greenhouse gas (GHG) emissions in developed countries, with the largest portion of the total food system emissions being attributed to agricultural production (Vermeulen, Campbell, and Ingram, 2012; CCAFS, 2014). Interestingly, direct impacts of agriculture are estimated to represent 10–12% of global emissions (5100–6100 $MtCO_2e$), while this figure rises up to 30% when additional emissions stemming from fuel use, fertilizer production and agriculturally induced land use change are included (Smith *et al.*, 2007; Bellarby *et al.*, 2008). According to the Food and Agriculture Organization (FAO), emissions from agriculture, forestry, and fisheries have nearly doubled over the past 50 years and are estimated to increase by one-third by 2050 unless drastic efforts to reduce them are undertaken (FAO, 2014).

From the policy-makers' perspective, managing the environmental impact of food systems is of primary importance and needs to be addressed by governments, in order to comply with the enacted international environmental regulations that require substantial reductions in GHG emissions in the long run (European Commission, 2012; United Nations, 2012). On the other hand, from the food industry's perspective, achieving high environmental standards, through improving the sustainability of processes and products, appears to provide competitive advantages to companies and further stimulates the development of mutually beneficial partnerships with suppliers (Smith, 2008). To that effect, several measures of the total amount of GHG emissions produced during the life cycle of a product have been employed in the food sector lately. Carbon Footprint (CF), which is considered in general as a part of Life Cycle Analysis (Wiedmann and Minx, 2008), has been recognized as a widely well-accepted methodology both by academics and practitioners alike (International Trade Centre, 2012; Jensen, 2012).

Several CF standardization initiatives have been developed, including PAS 2050 developed by the British Standards Institution (BSI, 2011, 2012), the GHG Protocol (WRI and WBSCD, 2011), and the ISO 14067 standard (ISO, 2013). The food sector in particular is highly regulated and has been subject to continuous introduction of environmental European and international legislation. Notably, the ENVIFOOD Protocol indicates that a solid life cycle-based methodology for CF assessment could assist the stakeholders of FSC networks in quantifying successfully the environmental impact of food products and proceed with a sustainability oriented management process (Food SCP RT, 2013). Moreover, several standardization initiatives have been developed for specific sectors, such as the dairy industry (IDF, 2010) and the beverage industry (BIER, 2010).

The food industry represents a dynamic environment, where customers have a growing demand for a sustainable food supply and high awareness of how food is produced and offered (Beske, Land, and Seuring, 2014). Quality, safety, and environmental conformity are included among the top-rated criteria for purchase decisions by consumers, who are often even willing to pay more for "green" products (Wognum *et al.*, 2011). Major companies in the food sector have already responded to these changing consumer demand patterns by adopting several sustainable practices, such as Unilever (2009), Nestlé (2012), Walmart (2012), Tesco (2013), Kellogg's (2014), and General Mills (2014). Indicatively, Walmart, one of the largest multinational retail corporations globally, accomplished a GHG emissions reduction of 20% at its stores and distribution centers in 2012 (compared with 2005), Sainsbury has adopted advanced energy technologies for refrigeration uses (The Economist Intelligence Unit, 2010) and Kellogg's (2014) has committed to reduce energy and GHG emissions by an additional 15% (per metric tonne of food produced) from 2015, while increasing the use of low carbon energy in plants by 50%. Several types of CF labeling schemes with varying types of labels are currently in use worldwide, such as the Carbon Trust (2013) and the Casino (Groupe Casino, 2013) labels, used by British and French retailers, respectively.

In this chapter, Carbon Footprint Management (CFM) for sustainable food supply networks is addressed. More specifically, a comprehensive analysis of the CF of contemporary global food networks is provided and an integrated methodological approach is proposed. The proposed Decision Support System (DSS) could be employed by all involved stakeholders, such as potential investors, involved regulators, and decision-makers, as a strategic decision-making tool for assessing the environmental impact of FSCs, as well as for investigating the impact of several sustainable practices toward decarbonizing food networks. To that effect, first the environmental impact of food supply networks is comprehensively discussed (Section 8.2). The most significant carbon hotspots that may arise across the entire FSC echelons are highlighted and critical CFM challenges are presented. Following, a holistic and practical evidence-based framework is proposed to aid the decision-making process for the decarbonization of FSCs (Section 8.3). More specifically, the need for development of new tools for tackling CFM issues within complex food systems is addressed, while the DSS known as the supply chain environmental analysis tool (SCEnAT) is presented analytically. Following that, the application of SCEnAT is demonstrated through an illustrative case study motivated by a real-world UK wheat industry, while interesting managerial insights are presented and practice-oriented low carbon interventions are discussed (Section 8.4). Finally, a summary and conclusions are provided (Section 8.5).

8.2   The Carbon Footprint of Food Supply Chains

8.2.1   Background

FSC networks produce GHG emissions within all stages of the food life cycle, from the pre-farming processes through to farming, production, distribution, retailing, consumption, and waste disposal. The main carbon-related activities that take place throughout FSCs may be classified in general into three categories: (i) pre-farm processes, including manufacturing and distribution of inputs to the farm, such as seed, animal feed, fertilizers, pesticides, growth substrates, pharmaceuticals, machinery, buildings, and other capital goods; (ii) on-farm processes, referring to agricultural production of crops, livestock, fisheries, and other products, along with the associated carbon-intensive or chemical-based farming activities; and (iii) post-farm processes, including primary and secondary processing, packaging, storage, refrigeration, transport, wholesale, retail and catering activities, distribution, domestic food processing and consumption, and waste disposal.

The value of the CF varies considerably across different activities of a FSC, and important variances are also identified among different systems or countries. More specifically, the post-production stages tend to have a greater role in high-income countries, while in other cases specific economic subsectors emerge as important, such as the high GHG contribution from fertilizer manufacture in China (Vermeulen, Campbell, and Ingram, 2012). A graphical qualitative illustration of the average GHG emissions at different stages in the FSC has been provided by Garnett (2011) and is presented in Figure 8.1.

More specifically, the left half of the illustrated pie chart denotes the on-farm and pre-farm emissions comprised of nitrous oxide (N_2O) from soil and livestock processes, methane (CH_4) emitted from ruminant digestion, rice cultivation, and anaerobic soils. A lower contribution of carbon dioxide (CO_2) emissions is also attributed to additional activities, such as fossil-fueled machinery, manufacture of fertilizers, and burning of waste. The pale gray part on the left illustrates the additional CO_2 emissions that arise from agriculturally induced land use change. Finally, the right half of the pie

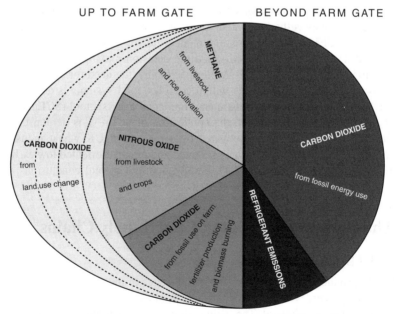

Proportions for illustrative purposes only

Figure 8.1   Food chain impacts and the distribution of the different gases. Adapted from Garnett, 2011.

chart represents the CO_2 emissions related to post-farm activities, which are attributed mainly to fossil fuel energy use, as well as to emissions associated with refrigeration processes.

In the following paragraphs the environmental impact of all FSC processes is addressed, while key insights from state-of-the-art and up-to-date literature are provided in order to reveal the contemporary CFM challenges that need to be tackled both by scientific researchers and associated decision-makers in the food sector.

8.2.2   Carbon Footprint of Pre-farm Processes

The intensification of agriculture during the last decade has increased dramatically the use of several inputs such as seeds, animal feed, fertilizers, pesticides, growth substrates, and pharmaceuticals for disease control (Tilman *et al.*, 2011). Fertilizers are identified as the major carbon carriers since they have increased by over 800% in about 45 years, while they are found to be utilized less in developed countries as a result of the present strict environmental legislation. Fertilizer production accounts for approximately 1.2% of the world's total GHGs (Wood and Cowie, 2004), which are attributed mainly to energy-demanding production processes, as well as to N_2O emissions associated with the manufacturing techniques of nitrate fertilizers employed.

However, there is a great variety in energy requirements and associated GHG emissions for the production of several types of fertilizers. In the case where the origin of the fertilizer is known, the uncertainty in the GHG emissions from production has a range of ±30% (Röös, Sundberg, and Hansson, 2011). Notably, this uncertainty is considered to be small compared with the uncertainty of other type of emissions, for example, soil emissions. The CF modeling of organic fertilizers has also been recently addressed (Meier *et al.*, 2012). The crop yield and the amount of nitrogen fertilizer applied are parameters with a major influence on the CF of roots, cereals, and open-air vegetables (Röös, 2013).

Additional GHG emissions incur from the transport of fertilizers to the farms, as well as from their final application through machinery. Indicatively, about 284–575 $MtCO_2e$ were emitted by fertilizer production in 2007, while global estimates of emissions from crop protection provide a wide range of 3–140 $MtCO_2e$ per year (Bellarby *et al.*, 2008), even though GHG emissions from agricultural pesticide manufacture and use is scarce (Dijkman, Birkved, and Hauschild, 2012). As far as the GHG emissions from the production of feed for livestock and aquaculture is concerned, these stem either from fossil fuel inputs through cultivation, transport, and processing of feed, or indirectly, through land change.

8.2.3   Carbon Footprint of On-farm Processes

The CF of on-farm processes is associated with the agricultural production of crops, livestock, and fisheries, as well as with all the primary and secondary farming activities. It refers mainly to land use change emissions, soil emissions, and emissions from energy use on fields, greenhouses, animal houses, and so on. In addition,

GHG emissions from energy consumption in agriculture arise mainly from combustion of fuels used for field machinery, on-farm transportation, and crop processing and storage (drying, heating, lighting, ventilation, etc.).

Agriculture makes the greatest contribution to the total food system emissions, that is, 7300–12 700 $MtCO_2e$ each year, accounting for 80–86% of food systems emissions (Vermeulen, Campbell, and Ingram, 2012). Direct emissions from agricultural processes account for 50–70% of agricultural emissions (United States Environmental Protection Agency, 2012). These are dominated by emissions of CH_4 and N_2O, which arise naturally from activities such as fertilization or keeping large numbers of ruminants. Livestock production is recognized to contribute significantly to GHG emissions, mainly through CO_2, CH_4, and N_2O emissions (Steinfeld et al., 2006). Fisheries and aquaculture make only minor contributions to GHG emissions, but are linked to other types of environmental impact (Ziegler et al., 2012).

On the other hand, estimates of the GHG emissions of land use changes are probably the most uncertain when assessing the environmental impact of FSCs, since their actual volume depends on several factors (Ramankutty et al., 2007; Ponsioen and Blonk, 2012). Despite this uncertainty, land use change is considered as a major contributor to global GHG emissions. Deforestation and land use change account for 30–50% of agricultural emissions (Vermeulen, Campbell, and Ingram, 2012), even though the relative contribution of deforestation and degradation to anthropogenic carbon emissions has been declining during recent years (Houghton, 2012).

Soil emissions are the major source of GHGs in agriculture and are affected by several factors such as soil properties and climate conditions, as well as the type of crop and farming system. However, measuring N_2O emissions from large fields is expensive and challenging, and they are characterized by considerable variation in space and time (Röös, 2013). Few studies to date have included carbon emissions arising from soil organic matter changes in CF calculations of food products [see Bosco et al. (2013) for a review]. Indicatively, in Röös, Sundberg, and Hansson (2010) and Röös, Sundberg, and Hansson (2011) emissions due to changes in soil carbon are included in the CF assessment of potatoes and pasta.

Interestingly, several measures for reducing the CF of the agricultural production processes have been reported in recent literature, related mostly to the enhancement of carbon removal, the optimization of nutrient use, productivity improvement, exploitation of outputs, and use of alternative energy sources (Johnson et al., 2007; Smith et al., 2007; Neufeldt and Schäfer, 2008; Garnett, 2011). Regarding livestock production, certain practices toward GHG mitigation have been suggested, such as improving efficiency in animal production, reducing CH_4 emissions and emissions from manure management, sequestering of soil carbon, changing human consumption of animal-source food, and taxing mechanisms (Garnett, 2009; Gill, Smith, and Wilkinson, 2010; Wirsenius, Hedenus, and Mohlin, 2011).

8.2.4   Carbon Footprint of Post-farm Processes

Post-farm processes include primary and secondary food processing in industrial units, logistics operations (transport/distribution, storage, refrigeration, packaging), wholesale, retail and catering activities, domestic food processing and consumption, and waste disposal. In general, the proportion of emissions from post-farm stages of FSCs is larger in higher income countries than in lower income countries (Garnett, 2011). The increased complexity of food production at the post-farm stages of FSCs severely hampers the tracing of GHG emissions and the calculation of the food CF (McKinnon, 2010). Traditional logistics management aims at two key targets, namely cost efficiency and improved responsiveness, along with the design of the logistics network. However, intrinsic characteristics of food products and processes, such as product perishability and sustainability issues, require extension of the key logistical aims with quality and environmental considerations. This necessity leads to the need for decision support tools that integrate economic considerations with quality and environmental protection in FSCs (Soysal *et al.*, 2012).

8.2.4.1 Processing

GHG emissions from food processing include CO_2 (from combustion in cookers, boilers, and furnaces), CH_4, and N_2O (from wastewater systems) emissions. Processing of sugar, palm oil, starch, and corn drives most of the total GHG emissions caused by global food processing. Indicatively, corn wet milling is a significantly energy-intensive process, requiring 15% of total US food industry energy (Vermeulen, Campbell, and Ingram, 2012). Moreover, Ugaya and Majijn (2009) address the impact of cleaner technologies in food processing, indicating the large contribution of the crop drying process to the CF of high-demand food products (e.g., coffee, tea, rice).

8.2.4.2 Transportation

Food transportation is attributed with a direct contribution to GHG emissions of FSCs. However, large-scale studies of national food systems show that transportation accounts for less than 15% of the energy use and GHG emissions of food products (Weber and Matthews, 2008). The environmental impact of transportation activities is widely debated in the literature. Weber and Matthews (2008) reveal that although food is transported over long distances in general (6760 km life cycle supply chain on average), the GHG emissions associated with food are dominated by the production phase, contributing 83% of the average US household's footprint (i.e., 8.1 tCO_2e/year) for food consumption.

Moreover, there has been discussion comparing the environmental impact of regional and national-scale FSCs in recent literature (e.g., Wakeland, Cholette, and Venkat, 2012; Brodt *et al.*, 2013). The concept of "food miles," a measure of how far food travels between its production point and the final consumer, receives considerable attention in the scientific and business environment (Coley, Howard, and Winter, 2009). The focus on increased food miles due to a boost of international trade in the food sector has led many environmental advocates, retailers, and others to urge the localization of global food supply networks. However, this trend is questioned due to the variant production practices employed in different regions or the increased storage needed to "buy locally" through all seasons (van der Werf *et al.*, 2014). Interestingly, many researchers propose modeling approaches toward the design of sustainable food supply networks and distribution systems, while providing useful insights from specific food sectors [e.g., Akkerman, Yang, and Grunow (2009) for meal elements, van der Vorst, Tromp, and van der Zee (2009) for pineapples, Oglethorpe (2010) for pork, and Validi, Bhattacharya, and Byrne (2014) and Bosona and Gebresenbet (2011) for dairy products].

8.2.4.3 Refrigeration

Refrigeration processes are attributed to GHG emissions from operating energy use, from refrigerant manufacturing and the associated losses. Pelletier *et al.* (2011) report that refrigeration is a major energy-intensive component of the food chain, while James and James (2010) estimate that the cold chain accounts for approximately 1% of total global GHG emissions and this figure is expected to increase. Notably, leakage of chlorofluorocarbon (CFC) refrigerants accounted for 30% of direct GHG emissions from supermarkets in the UK in 2009. However, according to Wu, Hu, and Mo (2013), CO_2 emissions derived from the energy consumption associated with refrigeration processes are much larger than those due to GHG leakage. Practices intended to reduce refrigeration emissions include improved energy efficiency, advanced and novel technologies, and measures to eliminate refrigerant losses. Tassou, De-Lille, and Ge (2009) address the development and application of alternative technologies to refrigeration systems that have the potential to reduce the overall energy consumption and environmental impacts. Fitzgerald *et al.* (2011) have studied the energy use of integral refrigerated containers in maritime transportation, and they argue that uncertainties in transport and refrigeration data can make significant differences to the final results for CF calculations.

8.2.4.4 Packaging

The environmental impact of food packaging is regularly studied in the literature. Garnett (2011) reveals that packaging accounts for 7% of UK food-related GHG emissions. Moreover, Accorsi *et al.* (2014) provide a literature review that considers recent studies on sustainability in design and selection of packaging. The environmental impact of packaging has recently been subject to intensive investigation by the research community (Siracusa *et al.*, 2014; Zampori and Dotelli, 2014). Indicatively, Levi *et al.* (2011) have conducted a comparative Life Cycle Assessment (LCA) on the disposable and reusable packaging for the distribution of Italian fruit and vegetables. Barthel *et al.* (2007) have studied the sustainability of packaging systems for fruit and vegetable transport in Europe, in which single-use wooden and cardboard boxes are compared with multi-use plastic ones. In general, all beverage packaging options appear to have a lesser environmental impact when they are recycled rather than disposed of in landfills or incineration plants (Pasqualino, Meneses, and Castells, 2011). In other cases, such as tomato packaging, the use of lighter primary packaging material results in significant environmental benefits (Manfredi and Vignali, 2014).

Interestingly, state-of-the-art food packaging design needs to combine sustainable material use and recycling possibilities, as well as reduced food losses (Williams and Wikström, 2011). Moreover, eco-design of packaging cannot be considered only in terms of the materials employed, since the contribution of the consumers' behavior emerges as a determinant criterion in the design of food packaging (Bertoluci, Leroy, and Olsson, 2014). Indicatively, the environmental impact of a conventional wooden pallet in comparison with a fully recyclable plastic bulk packaging system used to transport empty yogurt containers has also been investigated by Lee and Xu (2004). Notably, the authors conclude that the plastic packaging has a lower environmental impact than the wooden pallet, mainly due to its lower weight, recyclable parts, and long service life.

8.2.4.5   Food Retailing

Energy consumption of food retail activities contributes significantly to GHG emissions, depending on the practices and the equipment used in retail outlets for food storage, preparation, preservation, and display. The final activities of food transport, use, and disposal remain to a large extent beyond the control of the producer, as they depend mostly on the consumer's behavior (Röös, 2013). Tassou et al. (2011) reveal that the total annual emissions associated with major retail food outlets in the UK amount to ~4 $MtCO_2e$, while the major portion of the electrical energy consumption, ranging from 25 to 60%, is attributed to refrigeration. Therefore, retailers are requested to renovate their environmental and energy management systems, with a focus on reduction of refrigerant related emissions, as well as on adoption of renewable energy sources (European Retail Forum for Sustainability, 2011).

8.2.4.6   Food Consumption

Preparing food contributes to GHG emissions through the energy use associated particularly with cooking and refrigeration. Garnett (2011) argues that catering activities account for 6% of direct UK food chain emissions, and home-related food cooking, storage, and preparation account for 9%. Another study conducted in the Finnish environment reveals that a Finnish household produces on average 170 kg of CO_2e per capita, of which 50% is derived from food preservation, 27% from food transportation, and 23% from food preparation (Kauppinen et al., 2010). Concerning consumer practices, actions to address the CF of food include the purchase of green products, as well as the adoption of advanced energy-efficient refrigerators, energy saving cooking practices, and less use of electrical devices. However, even though CF product labeling has increased the awareness of the environmental impact of food among FSC stakeholders, the impact on consumers' purchasing behavior remains below expectations (Röös, 2013).

8.2.4.7   Food Waste Disposal

Notably, approximately one-third of the food produced for human consumption, that is, 1.3 billion t per year, is lost or wasted (Gustavsson, Cederberg, and Sonesson, 2011). The main causes of food waste across the entire FSC include lack of consumer environmental awareness, stock management inefficiencies, retail marketing strategies, overproduction, product and packaging damage during food manufacturing and distribution, and inadequate storage capacity throughout the whole food chain. Food waste in developing countries is dominated by high losses at the post-harvest and processing stages due to financial, managerial, and technical limitations in difficult climatic conditions (Gustavsson, Cederberg, and Sonesson, 2011; Venkat, 2011), while in developed countries it is strongly associated with consumer waste. There are essential factors affecting post-consumer food waste worldwide that involve public awareness campaigns and raising awareness among consumers of the importance of food waste reduction, as well as government interventions, and food industry initiatives (Parfitt, Barthel, and Macnaughton, 2010). Other recommendations for reducing food waste within the FSC include better coordination between manufacturers and retailers, and development of packaging technology to extend product life-spans (Garnett, 2009, 2011).

8.2.5   Critical Challenges of CFM

CFM for FSCs is challenged by high uncertainty that stems from the increased variability in management practices, the case-dependent conditions, the great diversity in types of food products, as well as the variable accuracy of the existing CF calculation methods. In general, livestock products appear to have a considerably larger CF than crop-based agricultural products, although high CF values for plant-based foods have also been identified for a few products that are produced in heated greenhouses, transported by air, or produced in low-yielding systems (Stoessel *et al.*, 2012). Different food groups exhibit a large range in GHG intensity; on average, red meat is around 150% more GHG intensive than chicken or fish (Weber and Matthews, 2008).

However, the CF of FSCs may vary considerably even for the same type of product, depending on the production system employed, the logistics processes developed, the food flows, as well as on the applied methodologies for the CF assessment and carbon calculation issues. Indicatively, significant complexity is related to the system boundaries and the CF calculation of several agricultural systems that may produce the same products but use different amounts of land (e.g., McLaren, 2010), or in the case where there is joint production of food products (e.g., Flysjö *et al.*, 2011). The system boundary of the food CF commonly ends at the farm ("cradle-to-farm gate") or at the retail outlet ("cradle-to-retail") or at the plate ("cradle-to-plate"). The human digestion and management of human waste is seldom included, although related methods have been proposed (Muñoz, Milà i Canals, and Clift, 2008).

Today's globalized FSCs typically consist of a large variety of different supply chain partners such as retailers, wholesalers, distributors, various traders, processors, marketers/storage, farmers, or farm suppliers that increase complexity and decrease transparency (Roth *et al.*, 2008). Given geographical restrictions for agricultural production and the dispersion of consumer market destinations, food flows span great geographical distances (Grimm, Hofstetter, and Sarkis, 2014). Although business relationships even in multi-tier FSCs are mostly dyadic, effective CFM requires that businesses have a holistic understanding of their supply chains and partners have a shared awareness of their emissions within supply chain collaborative networks (Elhedhli and Merrick, 2012; Mena, Humphries, and Choi, 2013). Therefore, the development of state-of-the-art DSSs that provide food businesses with integrated CFM tools is crucial in order to overcome both the uncertainty of data and the complexity of the developed networks.

8.3   The Supply Chain Environmental Analysis Tools

8.3.1   The Need for New Tools

The global agricultural landscape is changing, while technology has led innovations of genetics, plant breeding, and agro-machineries to grow and manufacture food cost efficiently (Thompson and Scoones, 2009). The liberalization of agricultural market and global consolidation and expansion of agricultural services quietly revolutionize the conventional agricultural practices. New business models have evolved from production-centered practices to creating high-value service-oriented agricultural product offerings. For example, farmers in many developed countries are equipped with the latest GPS-linked system that lets the tractor do the driving itself, and implement within 30 cm of its last pass through the field. Thus, it virtually eliminates overlaps, thereby saving fuel, chemicals, and wear and tear, while reducing operator fatigue (Lane, 2008).

Despite the benefits delivered through advances in technology, the industrialization of modern agriculture is associated with environmental issues, such as tight control of crops and their genetics, and soil fertility via chemical fertilization and irrigation, affecting both biological and cultural diversity (Pretty, 2008; Thompson and Scoones, 2009). Leading technological innovation is not always compatible with the social, environmental, and political situation in developing countries; for example, duration of land tenure right and size can prohibit a farmer's capital investment to increase productivity. Consequently, sustainability research of an agri-food ecosystem lies behind innovations in the commercial system. In addition, the diversity of stakeholders in multiple scales increases the complexity of the agri-food chain, starting with a firm's involvement in biotechnology, extending through agrochemical inputs and production, and ending with highly processed food (Bonnano *et al.*, 1995). Firms from the agri-food sector are developing increasingly a variety of different alliances with other players in the system, forming new food system "clusters" (Hendrickson and Heffernan, 2002).

Sustainable agri-food ecosystems aim to ensure that all people have consistent physical, social, and economic access to sufficient, safe, and nutritious food that meets their dietary needs and food preferences for an active and healthy life (FAO, 2002). Furthermore, contemporary agri-food ecosystems require an integrated approach to FSCs, environmental responsibility, consumer behavior, and human health. Therefore decision-making for the development of sustainable agri-food ecosystems needs to be informed by supply chain sustainability research, which includes total visibility of the entire supply chains, both upstream and downstream.

8.3.2   Description of Methodological Approach

SCEnAT is a robust supply chain life cycle analytical modeling asset developed by Koh with support from her team (Koh *et al.*, 2012). It deploys the most advanced Hybrid LCA and Input–Output methodology to quantify the carbon impact across the supply chain from production to consumption. SCEnAT has already been used by leading industries, including advanced manufacturing, construction, materials, energy, food, and other sectors to assist decision-making. Indicatively, it has been employed in cases where interventions could be made in order to maximize return on investment and enjoy wider social and environmental benefits. The backbone of the SCEnAT methodology is Koh's notion of creating a balanced green supply chain system for a robust decision support process (map, calculate, intervene, evaluate, and decide cycle): mapping entire supply chains, tracing each product component from cradle to grave, revealing materials, and energy usage at each level of the supply chain, manufacturing processes, and logistical operations within and between each company, recycling, and waste management. SCEnAT scenario modeling provides advanced analytics to simulate changes and interventions in any modeled supply chains.

The current methodology behind SCEnAT has been coded based on the current state-of-the-art methodology in carbon accounting known as the Hybrid LCA methodology (Koh *et al.*, 2012). The Hybrid LCA method integrates two LCA methodologies (i.e., Traditional or Process LCA and Environmental Input–Output LCA) in order to make use of their respective advantages. The traditional approach to LCA is based on a cradle-to-grave analysis of all supply chain inputs (i.e., from raw material extraction through materials processing, manufacture, distribution, use, repair and maintenance, and disposal or recycling, etc.), hence process methodology (Guinée, 2002), allowing for input specific processes to be analyzed. The main limitation of the pure LCA approach is the truncation of the LCA system boundary (Lenzen, 2000).

The Environmental Input–Output LCA model, on the other hand, takes an economy-wide perspective of the environmental assessment (Wiedmann *et al.*, 2011). The whole economy is considered as the system boundary so that every input and associated impacts from any given sector of the economy that contributes to the supply chain can be estimated. Therefore, this offers the advantage of the extended system boundary. In relation to supply chain data input, SCEnAT takes primary data from organizations and secondary data from organizations or databases such as Ecoinvent, Global Trade Analysis Project (GTAP), National Occupational Standards (NOS), and national and regional input–output data. As illustrated in Table 8.1, the SCEnAT approach is designed to complement the system boundary issues, offering an integrated approach by combining LCA with environmental input and output analysis. According to Acquaye *et al.* (2011) the principle of the hybrid LCA methodology is that the more accurate Process LCA is prioritized, but missing or unaccounted inputs are estimated using the Environmental Input–Output LCA.

Table 8.1   Comparison of the existing carbon management tools.

Features	Type 1 offering	Type 2 offering	Type 3 offering	SCEnAT offering
Carbon calculation	Yes	Yes	Yes	Yes
SC mapping	N/A	N/A	Not clear	Yes
Carbon hotspot identification	N/A	N/A	Not clear	Yes
Methodology	Emissions inventory and formula-based calculations	Very Basic LCA	Comprehensive LCA	Process LCA + I/O analysis nested LCA
Industrial activity coverage	Sectorial focus	More than one sector	More than one sector	Entire economy
Behavioral flexibility within the approach	Little or no flexibility	Some flexibility	Good flexibility	Complete flexibility
Availability of option for interventions	No	No	Yes	Yes
Level of interventions	N/A	N/A	Three	Four
Impact estimation	No	No	Yes	Yes
Types of impacts	N/A	N/A	Environmental and economic	Social, economic, environmental
Optimization of carbon footprints	N/A	N/A	Yes	Yes
In-built database	Yes	Not clear	Yes	Yes
Case studies	No	No	Yes	Yes
Comprehensive system knowledge base	No	No	No	Yes

I/O, input/output; N/A, not available; SC, supply chain.
Adapted from Koh et al. (2012).

8.4   An Illustrative Case Study

8.4.1   Scenario Description

Sustainable FSC solutions depend on incremental and highly coordinated under-standing spread throughout multiple disciplines in the ecosystems, rather than a single technological innovation. For example, many Sub-Saharan countries do not have sufficient infrastructures such as enough road and scalable market to make and transport fertilizers (World Bank, 2007). This study applies the SCEnAT cloud-based sustainable supply chain analytical tool that increases the visibility of emissions reduction potential and allows firms and supply chain partners to collaborate on low carbon interventions (Koh *et al.*, 2012).

As a demonstration of how this can be done, we have considered the wheat supply chain and interventions to reduce GHG emissions as an exemplar metric of environ-mental impact. A supply chain perspective is achieved by examining relationships among multiple stakeholders; multiple specialized activities from growth to produc-tion and consumption, enabling each one to have a holistic understanding of the supply chain and ensuring partners within a supply chain have a shared understanding of their environmental impact (Acquaye *et al.*, 2011). This can establish a platform for understanding beyond narrow advancement within a single organization/discipline; rather, it embraces the co-evolution of technological, social, environmental, and economic dimensions of a sustainable FSC.

In Figure 8.2 a map of the wheat supply chain is presented that shows the stake-holders involved and the processes they carry out. Using SCEnAT, we were allowed to quantify the carbon emissions on both direct and indirect GHG impact. In this example, direct product and process specific data were captured from Ecoinvent and include validated secondary data on agricultural production processes, fertilizers, pesticides, seed, and transport. Indirect impact measures input across the whole supply chain from aggregated (18) sectors of the economy, which are usually unreported in carbon assessment. The scope of the supply chain system boundaries in this study include:

- Raw material supply-agriculture services including agricultural machinery, agricultural chemicals, bio-techs, farmer;

- Processors, distributors, and traders;

- Retailers.

SCEnAT is used to facilitate environmental collaboration through consistent interpretation of an environmental hotspot. Wheat supply chain systems rely on the interconnectedness of many complex business interrelationships; mismatched understanding of any one dimension can increase unintended consequences. According to Koh *et al.* (2012), the benefits of agriculture technology innovation have been unevenly distributed. Complexities of agricultural landscape and poor access to finance, seeds, and fertilizers undermine the green revolution in many

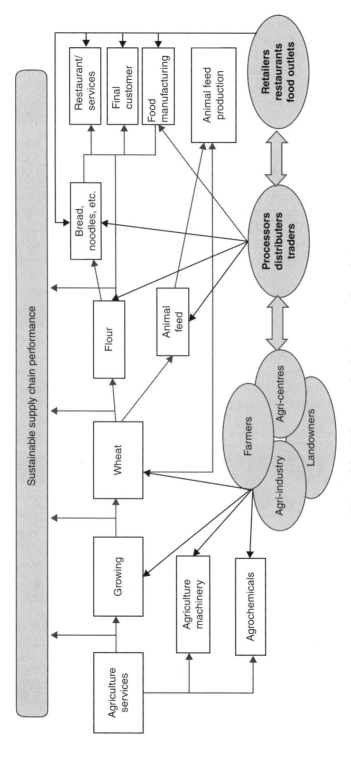

Figure 8.2 Wheat supply chain system boundaries.

developing countries. In order to realize sustainable change, innovators must take into account the entire system. For example, relevant technological changes in the supply chain (Paarlberg, 2006), understanding farming practices and identifying farmers needs and wants (Johnson and Christensen, 2008), the social side effects of mechanization replacing manual labor (Baines and Lightfoot, 2013), and policy structures such as land tenure, intellectual property rights, and agri-center research funding can similarly enable, encourage, and constrain agriculture (Baulcombe *et al.*, 2009).

8.4.2   Application and Analysis of SCEnAT Tools

In the following paragraph, the application of SCEnAT tools in the context of a wheat supply chain is presented. Figure 8.3 demonstrates a simple visual illustration of a carbon mapping result. Highlighted areas show ammonium nitrate and urea use contribute 37.1 and 11.6% of total CO_2e emissions. In addition to adverse GHG impact, ammonium nitrate is also a major cause of eutrophication due to nitrogen and phosphorus loss to ground and surface water. Interventions can be targeted to reduce fertilizer use through advanced practices in precision farming and digitized mechanization. However, access to finance and credit are major inhibitors, leading agri-service manufacturers, and suppliers to adopt new business models such as servitization to allow lowered outright capital investment by delivering advanced technologies in services instead of ownerships (Shelton, 2013).

The SCEnAT carbon map also shows that a wide variety of chemicals are used as pesticide. Although the total contribution to CO_2e from use of pesticide was less significant, the use of pesticide and insecticides is known to be a major cause of

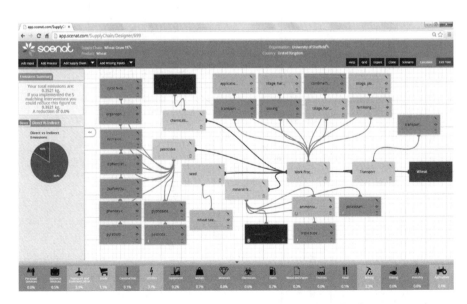

Figure 8.3   SCEnAT carbon map of wheat supply chain.

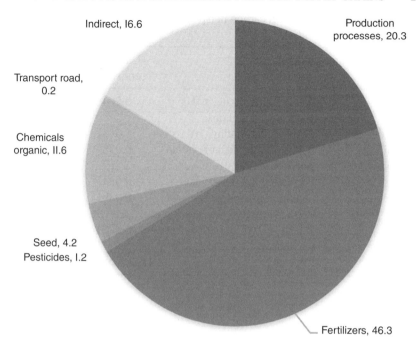

Figure 8.4   Process life cycle emissions (%) of wheat supply chain.

human health problems, the loss of beneficial insects, and water pollution (Baines and Lightfoot, 2013). The process impacts associated with direct supply chain inputs contributed 83.4% of the total emission. Indirect GHG emissions contributed 16.6% of total life-cycle emission.

Figure 8.4 shows the process life cycle emission of a wheat supply chain, both modeled and generated by SCEnAT. The production process carbon hotspot constitutes 20.3% of total wheat growing CO_2e emissions, where combined with harvesting contribute the highest emission (Figure 8.3), whilst agrochemical use (including fertilizers 46.3%, pesticides 1.2%, seed 4.2%, and organic chemicals 11.6%) accounts for 63.3% of total wheat growing CO_2e emissions.

SCEnAT has also been applied to model various scenarios of wheat supply chains, including the mass wheat supply chain scenario (The USA model), and conventional wheat supply chain analysis (The European model).

8.4.2.1   Scenario 1: Mass Wheat Supply Chain Carbon Hotspots in the USA (The USA Model)

Table 8.2 shows fertilizer use in ammonium nitrate contributed 15.2% of the total CO_2e emissions. Work processes using combine harvesting contributed 12.9% of the LCA CO_2e emissions. Interventions can be targeted in this category, by adopting practices such as advanced precision agriculture for pesticides and fertilizers,

Table 8.2   Direct and indirect carbon hotspots in the mass wheat supply chain in the USA (The USA model).

Process category	Input name	Carbon emissions	Emission (%)
Fertilizer	Ammonium nitrate, as N, at regional storehouse	0.0668	15.20
Work processes	Combine harvesting	0.0567	12.90
Inorganic chemicals	Ammonia, liquid, at regional storehouse	0.0339	7.70
Work processes	Fertilizing, by broadcaster	0.0335	7.60
Fertilizer	Diammonium phosphate, as P_2O_5, at regional storehouse	0.0291	6.60
Seed	Wheat seed IP, at regional storehouse	0.0264	6.00
Work processes	Irrigating	0.0255	5.80

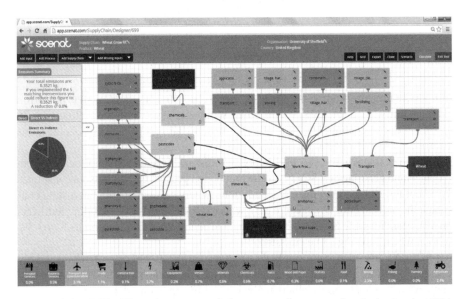

Figure 8.5   SCEnAT carbon map of the mass wheat supply chain in the USA (The USA model).

digitalized mechanization, and robotics to lower GHG emissions and other adverse environmental impacts (Baulcombe *et al.*, 2009). Figure 8.5 shows process LCA impact contributed 86.1% of the total CO_2e emissions, and indirect impact from aggregated economy contributed 13.9% of the total CO_2e emissions.

8.4.2.2 Scenario 2: Conventional Wheat Supply Chain Carbon Hotspots in Europe (The European Model)

This scenario is built around a wheat supply chain from production at farm gate to final consumption at feed mill. The functional unit adopted is 1 kg of wheat grain. The location is based in Switzerland, and the processes involve extensive production of wheat grain, transport, and resource use at the storehouse, and transport by road to the feed mill and processing at the feed mill.

Table 8.3 shows fertilizer use in ammonium nitrate and calcium ammonium nitrate together contributed 19% of the lifecycle CO_2e emissions. Indirect impacts from utilities (6.4%) and transportation (5.3%) have a significant contribution to total CO_2e in the wheat supply chain. Interventions can be targeted on utilities and transportation, and leading practices involve a fully integrated supply chain management in large farms, and proximity between smaller farms and warehousing/processing centers to reduce the need for transportation.

8.4.3 Implication of SCEnAT Tools in a Wheat Supply Chain

The visualized carbon mapping result from SCEnAT enables a clearer guide for sustainable agri-food chains that require the involvement of multiple actors. Sustainability demands multi-community research collaborations and information sharing occurs not just within the firm and its supply chains but also from the outside world such as the government, mass consumer, and agriculture research centers. Local farmers hold vital knowledge and have an understanding of soils, climate conditions, and agriculture practices (Chesbrough, 2006). They must also serve the complex needs of their consumers, in how crops keep, how they taste, and how they cook; these elements need to be incorporated into the process of designing more productive farming systems (Thompson and Scoones, 2009). According to Baulcombe *et al.* (2009), working through farmer organizations can channel effective environmental collaboration between formal science and local understanding.

Table 8.3   Direct and indirect carbon hotspots in the conventional wheat supply chain in Europe (The European model).

Process category	Input name	Carbon emissions	Emission (%)
Fertilizer	Ammonium nitrate, as N	0.085	19.00
Fertilizer	Calcium ammonium nitrate, as N, at regional storehouse	0.0427	9.50
Indirect	Utilities (indirect)	0.0287	6.40
Work processes	Combine harvesting	0.027	6.00
Indirect	Transport and communication (indirect)	0.0238	5.30

Consumers of the food chains are moving from physical retail to online shopping; they are more likely to be more expressive and have desire to share in the virtual space. This offers game changing opportunities for agri-food chain organizations to address these needs, by incorporating healthy, nutritious, and green values into their customers' "total experience." SCEnAT can unlock communication of environmental values, not just by computer systems, but by using tablets, iPhones, and Android, supported by cloud technologies that allow persistent customer digital engagement without any physical restrictions. According to Pretty (2008), businesses are standardizing information architecture; application programming interfaces (APIs) enable companies to take and give data not just to the companies' own application but also to third parties, and cloud computing is the virtual space where information is stored and disseminated. SCEnAT can be applied to add an additional information layer of environmental collaboration to engage with the consumer on social networks to communicate their views with instant accessibility (Shi *et al.*, 2012).

8.5   Summary and Conclusions

The environmental impact of food supply networks is significant, from the post- and on-farming processes through to production, logistics, and retailing operations, consumption, and waste disposal, with the largest portion of the total food system emissions attributed to agricultural production. To that effect, CFM for FSCs emerges as an application area of critical importance in the context of an integrated sustainable FSC management that needs to be addressed by stakeholders within all stages of the food life cycle. CFM is challenged by high uncertainty, since the CF may vary considerably even for the same type of product, depending on the production system employed, the developed logistics processes, the food flows, as well as on the accuracy of the existing CF calculating methods. At the same time, contemporary FSCs usually consist of a large variety of supply chain stakeholders located worldwide and thus, network complexity and decreased transparency hamper the shared awareness of emissions within supply chain networks. Hence, the development of innovative DSSs that provide food businesses with integrated CFM tools is crucial in order to overcome both the uncertainty of data and the complexity of the developed networks.

The SCEnAT tool creates an integrated robust methodology that can provide a pathway to establish dialogs and DSSs to implement effective sustainability oriented intervention. Therefore, it provides a more transparent understanding of sustainability both within the firms and across supply chain partners. Managers can also adopt the SCEnAT tool to harness the productive energy of its user communities through built-in intervention learning systems. SCEnAT users can access innovative solutions that come from outside of the organization, while it allows co-creation with external partners and connection with the outside world, where knowledge is abundant.

Our research elevates discussions of sustainable FSC complexity through examining relationships among multiple stakeholders, with multiple specialized activities from growth to production and consumption. A simplified carbon mapping illustration allows businesses to establish a holistic understanding of their supply chain wide environmental hotspots and ensure supply chain collaborative networks have a shared understanding of their emissions. Therefore, a platform for understanding beyond narrow advancement within a single organization/discipline should be established, in order to embrace the co-evolution of technological, social, environmental, and economic dimensions on a sustainable FSC.

References

Accorsi, R., Cascini, A., Cholette, S., Manzini, R., and Mora, C., 2014. Economic and environmental assessment of reusable plastic containers: a food catering supply chain case study. *International Journal of Production Economics*, **152**, 88–101.

Acquaye, A.A., Wiedmann, T., Feng, K., Crawford, R.H., Barrett, J., Kuylenstierna, J., Duffy, A.P., Koh, S.C.L. and McQueen-Mason, S., 2011. Identification of 'carbon hot-spots' and quantification of GHG intensities in the biodiesel supply chain using hybrid LCA and structural path analysis. *Environmental Science & Technology*, **45**, 2471–2478.

Akkerman, R., Yang, W., and Grunow, M., 2009. MILP approaches to sustainable production and distribution of meal elements. CIE 2009 International Conference on Computers and Industrial Engineering, Troyes, France, July 6–9, 2009, pp. 973–978.

Baines, T. and Lightfoot, H., 2013. *Made to Serve: What It Takes for a Manufacturer to Compete through Servitization and Product-Service Systems*. Hoboken, NJ: John Wiley & Sons, Inc.

Barthel, L., Albrecht, S., Deimling, S., and Baitz, M., 2007. The Sustainability of Packaging Systems for Fruit and Vegetable Transport in Europe Based on Life Cycle Analysis, EuPC, Brussels, http://www.plasticsconverters.eu/uploads/Final-Report-English-070226.pdf (accessed July 3, 2014).

Baulcombe, D., Crute, I., Davies, B., Dunwell, J., Gale, M., Jones, J., Pretty, J., Sutherland, W., Toulmin, C., and Green, N., 2009. *Reaping the Benefits: Science and the Sustainable Intensification of Global Agriculture*. London: The Royal Society.

Bellarby, J., Foereid, B., Hastings, A., and Smith, P., 2008. Cool Farming: Climate Impacts of Agriculture and Mitigation Potential, Greenpeace International, Amsterdam, http://www.greenpeace.org/international/en/publications/reports/cool-farming-full-report/ (accessed June 23, 2015).

Bertoluci, G., Leroy, Y., and Olsson, A., 2014. Exploring the environmental impacts of olive packaging solutions for the European food market. *Journal of Cleaner Production*, **64**, 234–243.

Beske, P., Land, A., and Seuring, S., 2014. Sustainable supply chain management practices and dynamic capabilities in the food industry: a critical analysis of the literature. *International Journal of Production Economics*, **152**, 131–143.

BIER, 2010. Beverage Industry Sector Guidance for Greenhouse Gas Emissions Reporting, Beverage Industry Environmental Roundtable (January 2010 Version 2.0), http://www.bieroundtable.com/pressreleaseGHGguidance.html (accessed July 4, 2014).

Bonnano, A., Busch, L., Friedland, W., Gouveia, L., and Mingione, E., 1995. *From Columbus to ConAgra: The Globalization of Agriculture and Food*. Lawrence, KS: University Press of Kansas.

Bosco, S., Di Bene A., Galli, M., Remorini, D., Massai, R., and Bonari, E., 2013. Soil organic matter accounting in the carbon footprint analysis of the wine chain. *International Journal of Life Cycle Assessment*, **18**, 973–989.

Bosona, T. and Gebresenbet, G., 2011. Cluster building and logistics network integration of local food supply chain. *Biosystems Engineering*, **108**, 293–302.

British Standards Institution PAS 2050:2011, 2011. *Specification for the Assessment of the Life Cycle Greenhouse Gas Emissions of Goods and Services*, BSI, London.

British Standards Institution PAS 2050-1:2012, 2012. *Assessment of Life Cycle Greenhouse Gas Emissions from Horticultural Products. Supplementary Requirements for the Cradle to Gate Stages of GHG Assessments of Horticultural Products Undertaken in Accordance with PAS 2050*, BSI, London.

Brodt, S., Kramer, K.J., Kendall, A., and Feenstra, G., 2013. Comparing environmental impacts of regional and national-scale food supply chains: a case study of processed tomatoes. *Food Policy*, **42**, 106–114.

Carbon Trust, 2013. Carbon Footprinting, http://www.carbontrust.com/ (accessed June 23, 2015).

Chesbrough, H., 2006. *Open Business Models*. Watertown, MA: Harvard Business School Press.

Clay, J., 2011. Freeze the footprint of food. *Nature*, **475**, 287–289.

Coley, D., Howard, M., and Winter, M., 2009. Local food, food miles and carbon emissions: a comparison of farm shop and mass distribution approaches. *Food Policy*, **34**, 150–155.

Dijkman, T., Birkved, M., and Hauschild, M., 2012. PestLCI 2.0: a second generation model for estimating emissions of pesticides from arable land in LCA. *The International Journal of Life Cycle Assessment*, **17**, 973–986.

Elhedhli, S. and Merrick, R., 2012. Green supply chain network design to reduce carbon emissions. *Transportation Research Part D: Transport and Environment*, **17**, 370–379.

European Commission, 2012. EU Transport GHG: Routes to 2050 Project, EC, Brussels, http://www.eutransportghg2050.eu/cms/?flush=1 (accessed July 4, 2014).

European Food Sustainable Consumption and Production Round Table, 2013. ENVIFOOD Protocol, Environmental Assessment of Food and Drink Protocol. *Brussels: SCP RT, Working Group* **1**.

European Retail Forum for Sustainability, 2011. Measurement and Reduction of Carbon Footprint of Stores, Issue Paper No. 6, http://ec.europa.eu/environment/industry/retail/pdf/Issue%20Paper%206.pdf (accessed June 23, 2015).

Fitzgerald, W.B., Howitt, O.J., Smith, I.J., and Hume, A., 2011. Energy use of integral refrigerated containers in maritime transportation. *Energy Policy*, **39**, 1885–1896.

Flysjö, A., Cederberg, C., Henriksson, M., and Ledgard, S., 2011. How does co-product handling affect the carbon footprint of milk? Case study of milk production in New Zealand and Sweden. *International Journal of Life Cycle Assessment*, **16**, 420–430.

Foley, J.A., Ramankutty, N., Brauman, K.A., Cassidy, E.S., Gerber, J.S., Johnston, M., Mueller, N.D., O'Connell, C., Ray, D.K., West, P.C., Balzer, C., Bennett, E.M., Carpenter, S.R., Hill, J., Monfreda, C., Polasky, S., Rockström, J., Sheehan, J., Siebert, S., Tilman, D., and Zaks, D.P.M., 2011. Solutions for a cultivated planet. *Nature*, **478**, 337–342.

Food and Agriculture Organization of the United Nations, 2002. *The State of Food Insecurity in the World 2001*. Rome: FAO.

Food and Agriculture Organization of the United Nations, 2014. *Agriculture's Greenhouse Gas Emissions on the Rise*. Rome: FAO, http://www.fao.org/news/story/en/item/216137/icode/ (accessed July 2, 2014).

Garnett, T., 2009. Livestock-related greenhouse gas emissions: impacts and options for policy makers. *Environmental Science and Policy*, **12**, 491–503.

Garnett, T., 2011. Where are the best opportunities for reducing greenhouse gas emissions in the food system (including the food chain)? *Food Policy*, **36**, S23–S32.

General Mills, 2014. Global Responsibility Report, General Mills, Golden Valley, MN, http://www.generalmills.com/en/Responsibility/Overview (accessed July 17, 2015).

Gill, M., Smith, P., and Wilkinson, J.M., 2010. Mitigating climate change: the role of domestic livestock. *Animal*, **4**, 323–333.

Greenhouse Gas Protocol, 2011. Product Life Cycle Accounting and Reporting Standard, World Business Council for Sustainable Development (WBCSD) and World Resources Institute (WRI), http://www.ghgprotocol.org/standards/product-standard (accessed July 4, 2014).

Grimm, J.H., Hofstetter, J.S., and Sarkis, J., 2014. Critical factors for sub-supplier management: a sustainable food supply chains perspective. *International Journal of Production Economics*, **152**, 159–173.

Groupe Casino, 2013. Reducing Greenhouse Gas (GHG) Emissions, Groupe Casino, Saint-Étienne, http://www.groupe-casino.fr/en/environmmentally-proactive-group/reducing-greenhouse-gas-ghg-emissions/ (accessed July 17, 2015).

Guinée, J.B., 2002. Handbook on life cycle assessment operational guide to the ISO standards. *The International Journal of Life Cycle Assessment*, **7**, 311–313.

Gustavsson, J., Cederberg, C., and Sonesson, U., 2011. *Global Food Losses and Food Waste. Extent, Causes and Prevention*. Rome: FAO.

Hendrickson, M.K. and Heffernan, W.D., 2002. Opening spaces through relocalization: locating potential resistance in the weaknesses of the global food system. *Sociologia Ruralis*, **42**, 347–369.

Houghton, R., 2012. Carbon emissions and the drivers of deforestation and forest degradation in the tropics. *Current Opinion in Environmental Sustainability*, **4**, 597–603.

International Dairy Federation, 2010. *A Common Carbon Footprint for Dairy, The IDF Guide to Standard Lifecycle Assessment Methodology for the Dairy Industry*. Brussels: IDF.

International Organization for Standardization ISO/TS 14067:2013/, 2013. *Greenhouse Gases – Carbon Footprint of Products – Requirements and Guidelines for Quantification and Communication*, ISO, Geneva, https://www.iso.org/obp/ui/#iso:std:iso:ts:14067:ed-1:v1:en (accessed 30 June 2014).

International Trade Centre, 2012. Product Carbon Footprinting Standards in the Agri-Food Sector, ITC, Geneva, http://www.intracen.org/Product-Carbon-Footprinting-Standards-in-the-Agri-Food-Sector/ (accessed June 30, 2014).

James, S. and James, C., 2010. The food cold-chain and climate change. *Food Research International*, **43**, 1944–1956.

Jensen, J.K., 2012. Product carbon footprint developments and gaps. *International Journal of Physical Distribution and Logistics Management*, **42**, 338–354.

Johnson, J.M.-F., Franzluebbers, A.J., Weyers, S.L., and Reicosky, D.C., 2007. Agricultural opportunities to mitigate greenhouse gas emissions. *Environmental Pollution*, **150**, 107–124.

Johnson, M.W. and Christensen, C.M., 2008. Reinventing your business model. *Harvard Business Review*, **86**, 50–60.

Kauppinen, T., Pesonen, I., Katajajuuri, J.-M., and Kurppa, S., 2010. Carbon footprint of food-related activities in Finnish households. *Progress in Industrial Ecology – An International Journal*, **7**, 257–267.

Kellogg Company, 2014. Global Sustainability Commitments 2020, Kellog's, Battle Creek, MI, http://www.kelloggcompany.com/content/dam/kelloggcompanyus/corporate_responsibility/pdf/2015/KelloggSustainability2020Commitments02-13-15.pdf (accessed July 17, 2015).

Koh, S.C.L., Genovese, A., Acquaye, A.A., Barratt, P., Rana, N., Kuylenstierna, J., and Gibbs, D., 2012. Decarbonising product supply chains: design and development of an integrated evidence-based decision support system – the supply chain environmental analysis tool (SCEnAT). *International Journal of Production Research*, **51**, 2092–2109.

Lane, R.W., 2008. Avoiding commodity hell. *Research Technology Management*, **51**, 24–29.

Lee, S. and Xu, X., 2004. A simplified life cycle assessment of re-usable and single-use bulk transit packaging. *Packaging Technology and Science*, **17**, 67–83.

Lenzen, M., 2000. Errors in conventional and input–output-based life-cycle inventories. *Journal of Industrial Ecology*, **4**, 127–148.

Levi, M., Cortesi, S., Vezzoli, C., and Salvia, G., 2011. A comparative life cycle assessment of disposable and reusable packaging for the distribution of Italian fruit and vegetables. *Packaging Technology and Science*, **24**, 387–400.

Manfredi, M. and Vignali, G., 2014. Life cycle assessment of a packaged tomato puree: a comparison of environmental impacts produced by different life cycle phases. *Journal of Cleaner Production*, **73**, 275–284.

McKinnon, A.C., 2010. Product-level carbon auditing of supply chains: environmental imperative or wasteful distraction? *International Journal of Physical Distribution and Logistics Management*, **40**, 42–60.

McLaren, S., 2010. Life Cycle Assessment (LCA) of food production and processing: an introduction. In: U. Sonesson, J. Berlin, and F. Ziegler (eds), *Environmental Assessment and Management in the Food Industry*. Cambridge: Woodhead Publishing Limited. pp. 37–58.

Meier, M.S., Schader, C., Berner, A., and Gattinger, A., 2012. Modelling N_2O emissions from organic fertilisers for LCA inventories. In: *8th International Conference on Life Cycle Assessment in the Agri-Food Sector*. Saint-Malo, France, October 1–4, 2012. Rennes: INRA – Research Unit "Soil, Agro and hydroSystem".

Mena, C., Humphries, A., and Choi, T.Y., 2013. Toward a theory of multi-tier supply chain management. *Journal of Supply Chain Management*, **49**, 58–77.

Muñoz, I., Milà i Canals, L., and Clift, R., 2008. A simple model to include human excretion in life cycle assessment of food products. *Journal of Industrial Ecology*, **12**, 521–538.

Nestlé SA, 2012. Nestle in Society: Creating Shared Value and Meeting Commitments, Nestlé, http://www.nestle.com/asset-library/Documents/Library/Documents/Corporate_Social_Responsibility/Nestle-CSV-Full-Report-2012-EN.pdf (accessed August 21, 2014).

Neufeldt, H. and Schäfer, M., 2008. Mitigation strategies for greenhouse gas emissions from agriculture using a regional economic-ecosystem model. *Agriculture Ecosystems and Environment*, **123**, 305–316.

Oglethorpe, D., 2010. Optimising economic, environmental, and social objectives: a goal-programming approach in the food sector. *Environmental and Planning A*, **42**, 1239–1254.

Paarlberg, R., 2006. Are genetically modified (GM) crops a commercial risk for Africa? *International Journal of Technology and Globalisation*, **2**, 81–92.

Parfitt, J., Barthel, M., and Macnaughton, S., 2010. Food waste within food supply chains: quantification and potential for change to 2050. *Philosophical Transactions of the Royal Society*, **365**, 3065–3081.

Pasqualino, J., Meneses, M., and Castells, F., 2011. The carbon footprint and energy consumption of beverage packaging selection and disposal. *Journal of Food Engineering*, **103**, 357–365.

Pelletier, N., Audsley, E., Brodt, S., Garnett, T., Henriksson, P., Kendall, A., Kramer, K.J., Murphy, D., Nemecek, T., and Troell, M., 2011. Energy intensity of agriculture and food systems. *Annual Review of Environment and Resources*, **36**, 223–246.

Ponsioen, T. and Blonk, T., 2012. Calculating land use change in carbon footprints of agricultural products as an impact of current land use. *Journal of Cleaner Production*, **28**, 120–126.

Pretty, J., 2008. Agricultural sustainability: concepts, principles and evidence. *Philosophical Transactions of the Royal Society, B: Biological Sciences*, **363**, 447–465.

Ramankutty, N., Gibbs, H.K., Achard, F., Defriess, R., Foley, J.A., and Houghton, R.A., 2007. Challenges to estimating carbon emissions from tropical deforestation. *Global Change Biology*, **13**, 51–66.

Röös, E., 2013. Analysing the carbon footprint of food: insights for consumer communication. PhD thesis. Swedish University of Agricultural Sciences, Uppsala.

Röös, E., Sundberg, C., and Hansson, P.-A., 2010. Uncertainties in the carbon footprint of food products: a case study on table potatoes. *International Journal of Life Cycle Assessment*, **15**, 478–488.

Röös, E., Sundberg, C., and Hansson, P.-A., 2011. Uncertainties in the carbon footprint of refined wheat products: a case study on Swedish pasta. *International Journal of Life Cycle Assessment*, **16**, 338–350.

Roth, A., Tsay, A., Pullmann, M., and Gray, J., 2008. Unraveling the food supply chain: strategic insights from China and the 2007 recalls. *Journal of Supply Chain Management*, **44**(1) 22–39.

Shelton, T., 2013. *Business Models for the Social Mobile Cloud: Transform Your Business Using Social Media, Mobile Internet, and Cloud Computing*. Hoboken, NJ: John Wiley & Sons, Inc.

Shi, V.G., Koh, S.C.L., Baldwin, J., and Cucchiella, F., 2012. Natural resource based green supply chain management. *Supply Chain Management-An International Journal*, **17**, 54–67.

Siracusa, V., Ingrao, C., Lo Giudice, A., Mbohwa, C., and Dalla Rosa, M., 2014. Environmental assessment of a multilayer polymer bag for food packaging and preservation: an LCA approach. *Food Research International*, **62**, 151–161.

Smith, B., 2008. Developing sustainable food supply chains. *Philosophical Transactions of the Royal Society B*, **363**, 849–861.

Smith, P., Martino, D., Cai, Z., Gwary, D., Janzen, H., Kumar, P., McCarl, B., Ogle, S., O'Mara, F., and Rice, C., 2007. Policy and technological constraints to implementation of greenhouse gas mitigation options in agriculture. *Agriculture, Ecosystems & Environment*, **118**, 6–28.

Soysal, M., Bloemhof, J., Meuwissen, M., and Van der Vorst, J., 2012. A review on quantitative models for sustainable food logistics management. *International Journal of Food System Dynamics*, **3**, 136–155.

Steinfeld, H., Gerber, P., Wassenaar, T., Castel, V., Rosales, M., and de Haan, C., 2006. *Livestock's Long Shadow: Environmental Issues and Options*. Rome: Food and Agriculture Organization of the United Nations.

Stoessel, F., Juraske, R., Pfister, S., and Hellweg, S., 2012. Life cycle inventory and carbon and water foodprint of fruits and vegetables: application to a Swiss retailer. *Environmental Science & Technology*, **46**, 3253–3262.

Tassou, S., De-Lille, G., and Ge, Y., 2009. Food transport refrigeration – Approaches to reduce energy consumption and environmental impacts of road transport. *Applied Thermal Engineering*, **29**, 1467–1477.

Tassou, S., Ge, Y., Hadawey, A., and Marriott, D., 2011. Energy consumption and conservation in food retailing. *Applied Thermal Engineering*, **31**, 147–156.

Tesco PLC, 2013. Tesco and Society Report, Tesco, Cheshunt, http://www.tescoplc.com/files/pdf/reports/tesco_and_society_2013_ipad.pdf (accessedJune 30, 2013).

The CGIAR Research Program on Climate Change, Agriculture and Food Security (CCAFS), 2014. Big Facts. Where Agriculture and Climate Change Meet, CCAFS, Copenhagen, http://ccafs.cgiar.org/bigfacts2014/ (accessed July 1, 2014).

The Economist Intelligence Unit, 2010. Chain Reaction: The Low Carbon Challenge for the Food Industry, UK Trade & Investment, London, http://graphics.eiu.com/PDF/Food%20supply%20chain_final_1802010.pdf (accessed June 23, 2015).

Thompson, J. and Scoones, I., 2009. Addressing the dynamics of agri-food systems: an emerging agenda for social science research. *Environmental Science and Policy*, **12**, 386–397.

Tilman, D., Balzer, C., Hill, J., and Befort, B.L., 2011. Global food demand and the sustainable intensification of agriculture. *Proceedings of the National Academy of Sciences of the United States of America*, **108**, 20260–20264.

Ugaya, C.M.L. and Majijn, B., 2009. *Technology Transfer for Reducing the Carbon Footprint. The Example of CleanerTtechnologies for Food Processing, FinalRreport for UNEP-DTIE*. Paris: United Nations Environment Programme - Division of Technology, Industry and Economics.

Unilever, 2009. Sustainable Development Overview 2009: Creating a Better Future, Unilever, Rotterdam, http://www.unilever.co.kr/Images/sd_UnileverSDReport170310_amended_tcm109-212972.pdf (accessed July 17, 2015).

United Nations, 2012. Doha Amendment to the Kyoto Protocol, UN, New York, NY, treaties. un.org/doc/Publication/CN/2012/CN.718.2012-Eng.pdf (accessed July 4, 2014).

United States Environmental Protection Agency, 2012. *Global Anthropogenic non-CO_2 Greenhouse Gas Emissions: 1990 – 2030. EPA 430-R-12-006.* Washington, DC: US-EPA.

Validi, S., Bhattacharya, A., and Byrne, P., 2014. A case analysis of a sustainable food supply chain distribution system—a multi-objective approach. *International Journal of Production Economics*, **152**, 71–87.

Venkat, K., 2011. The climate change and economic impacts of food waste in the United States. *International Journal of Food System Dynamics*, **2**, 431–446.

Vermeulen, S., Campbell, B., and Ingram, J., 2012. Climate change and food systems. *Annual Review of Environment and Resources*, **37**, 195–222.

van der Vorst, J., Tromp, S., and van der Zee, D., 2009. Simulation modelling for food supply chain redesign, integrated decision making on product quality, sustainability and logistics. *International Journal of Production Research*, **47**, 6611–6631.

Wakeland, W., Cholette, S., and Venkat, K., 2012. Food transportation issues and reducing carbon footprint. In: A. Boyd and Y. Arcand (eds), *Green Technologies in Food Production and Processing*. New York, NY: Springer. pp. 211–236.

Walmart, 2012. Global Responsibility Report, Walmart, Bentonville, AR, www.walmartstores. com/sites/responsibility-report/2012/ (accessed July 4, 2014).

Weber, C.L. and Matthews, H.S., 2008. Food-miles and the relative climate impacts of food choices in the United States. *Environmental Science & Technology*, **42**, 3508–3513.

van der Werf, H., Garnett, T., Corson, M., Hayashi, K., Huisingh, D., and Cederberg, C., 2014. Towards eco-efficient agriculture and food systems: theory, praxis and future challenges. *Journal of Cleaner Production*, **73**, 1–9.

Wiedmann, T. and Minx, J., 2008. A definition of 'carbon footprint'. In: C. Pertsova (ed.), *Ecological Economics Research Trends*. Hauppauge, NY: Nova Science Publishers. pp. 1–11.

Wiedmann, T.O., Suh, S., Feng, K., Lenzen, M., Acquaye, A., Scott, K., and Barrett, J.R., 2011. Application of hybrid life cycle approaches to emerging energy technologies–the case of wind power in the UK. *Environmental Science & Technology*, **45**, 5900–5907.

Williams, H. and Wikström, F., 2011. Environmental impact of packaging and food losses in a life cycle perspective: a comparative analysis of five food items. *Journal of Cleaner Production*, **19**, 43–48.

Wirsenius, S., Hedenus, F., and Mohlin, K., 2011. Greenhouse gas taxes on animal food products: rationale, tax scheme and climate mitigation effects. *Climatic Change*, **108**, 159–184.

(Nel)Wognum, P., Bremmers, H., Trienekens, J., van der Vorst, J., and Bloemhof, J., 2011. Systems for sustainability and transparency of food supply chains – current status and challenges. *Advanced Engineering Informatics*, **25**, 65–76.

Wood, S. and Cowie, A., 2004. *A Review of Greenhouse Gas Emission Factors for Fertiliser Production. For IEA Bioenergy Task 38*. Orange: Research and Development Division, State Forests of New South Wales.

World Bank, 2007. *World Development Report 2008: Agriculture for Development*. Washington, DC: World Bank.

World Trade Organization, 2011. International Trade Statistics, WTO, Geneva, http://www.wto.org/english/res_e/statis_e/its2011_e/its2011_e.pdf (accessed July 2, 2014).

Wu, X., Hu, S., and Mo, S., 2013. Carbon footprint model for evaluating the global warming impact of food transport refrigeration systems. *Journal of Cleaner Production*, **54**, 115–124.

Zampori, L. and Dotelli, G., 2014. Design of a sustainable packaging in the food sector by applying LCA. *The International Journal of Life Cycle Assessment*, **19**, 206–217.

Ziegler, F., Winther, U., Hognes, E., Emanuelsson, A., Sund, V., and Ellingsen, H., 2012. The carbon footprint of Norwegian seafood products on the global seafood market. *Journal of Industrial Ecology*, **17**, 103–116.

9

Quality Management Schemes and Connections to the Concept of Sustainability in the Food Chain

Ioannis Manikas,[1] Karen Hamann,[2] and Anton Sentic[3]

[1] *Department of Systems Management and Strategy, Business School, University of Greenwich, Old Royal Naval College, Park Row, Greenwich, London, UK*
[2] *Institute for Food Studies & Agroindustrial Development (IFAU), Enghave 4, Rungsted Kyst, Denmark*
[3] *ARGE Waste Prevention, Resources Protection and Sustainable Development Ltd, Dreihackengasse 1, Graz, Austria*

9.1   Introduction

Food quality management (FQM) and food quality assurance (FQA) systems in Europe have been developed and refined over the last few decades, motivated both by customer and producer demand for more transparency and a common guideline for ensuring food standards and health safety and by external factors such as recurring food crises, ranging from the bovine spongiform encephalopathy (BSE) and foot-and-mouth disease crises of the 1980s and 1990s to current issues such as the 2013 horse meat scandal. Throughout their development, and following global trends, those systems

Supply Chain Management for Sustainable Food Networks, First Edition. Edited by Eleftherios Iakovou, Dionysis Bochtis, Dimitrios Vlachos and Dimitrios Aidonis.

have been associated with and influenced by ideas of sustainability and sustainable development, some of which have been adopted and subsequently internalised into food quality and food assurance.this chapter will try to identify those ideas and indicators and provide an overview of the connections and interactions of FQM and FQA with sustainability.

In a first step, a general overview will be provided for quality management and assurance schemes in the food supply chain, starting from a debate on the definition of quality in general and, more specific, the quality of food items. This will be followed by a review of quality management scheme concepts and a short discussion on the application of such schemes along the food supply chain. In the course of the discussion, the possible reasons for implementing quality management schemes as well as the structure of the implementation procedure will be reviewed. The third section will provide a general introduction to the ideas of sustainability and sustainable development, starting from a short general history of sustainability and the most important milestones in the history of sustainable development. This will be followed by an outlook on the measurement of sustainability, the ongoing diffusion of sustainability ideas and some comments on the connection of sustainability and the food production sector. In the fourth section, the concepts of food quality (management) and sustainability will be brought together by linking FQM schemes along the food supply chain with ideas of sustainable development, supported with practical examples of sustainable quality management schemes. The focus on the consumer as 'recipient' of the advantages provided by quality management schemes in systems is discussed in the fifth section, where Maslow's needs pyramid is used to provide a possible explanation for the different levels of needs/achieved satisfaction by the consumers through the application of sustainability principles to the different elements of the food supply chain.

9.2 Quality Management Schemes in the Food Supply Chain

9.2.1 Food Quality Definitions

Quality can be defined as any of the features that make something what it is or the degree of excellence or superiority (Kader, 1985). This definition is inadequate as it does not include the concept of consumer satisfaction. Quality is better defined as the ability of a product to accomplish the purpose for which has been produced. Quality is the set of properties and features that satisfy determinate and implicit needs of the consumers (Tzia and Tsiapouris, 1996). Similarly, quality can be also defined as fitness for purpose (Joyce, 2001). The concept of the quality of a product clearly depends on the preferences of the user (Sloof, Tijskens and Wilkinson, 1996). The consumer defines whether a product is of high or low quality. The consumer has the final word to choose a product according to his or her knowledge (e.g. on food safety, nutritive value, etc.), the intended use of the product (e.g. ripe tomatoes for soup, hard tomatoes for salad) and his or her attitude towards the product (Shewfelt, 1999). Quality means different things for each linkage throughout the distribution chain. Consumers, suppliers and producers are concerned about different components of quality (Shewfelt, 1999). Consumers will buy fresh fruits according to their appearance which is an indicator of quality, and are concerned about their nutritive value. Consumers will repeat their purchase only if they are satisfied by the flavour and texture of the product. These two quality components form the edible quality of the product. Moreover, quality attributes such as the flavour and texture are emphasised in a consumer oriented approach of quality, whilst appearance and long shelf life are emphasised in a product oriented approach of quality (Shewfelt, 1999). Producers are mainly concerned with product appearance, safety and extended shelf life. Desirable appearance will ensure product sale, as the consumers largely assess quality based on the external appearance of the produce (Joyce, 2001). The following definition and explanations illustrate the underlying concept of food quality.

9.2.1.1   Definition of Food Quality

The quality management scheme ISO 9000:2000 defines quality as 'the totality of characteristics of an entity (this being a product, service, process, activity, system organisation or person) that bear on its ability to satisfied stated and implied needs' (Will and Guenther, 2007).

Quality today embraces in addition to product quality also:

- The service, organisational, management and in particular process quality.

- The adequacy of its usage.

- The compliance with third-party standards.

- The perception of its excellence at a competitive price.

Quality is associated with all activities related to:

- Standardisation.

- Quality management/assurance as a strategic discipline in company (and supply chain) management.

- Quality control, certification and accreditation.

- Quality schemes and labels.

9.2.2   Quality Management Scheme Concepts

The food industry is the centre point of the food value chain and performs the link between the primary production and the market. Furthermore, the food industry is the most valuable contributing partner in the food chain as the processing and packaging activities transform a primary product to a marketable product. This processing adds a number of features to the product. Basic features are first of all an edible and safe quality; but also a format that increases shelf life and makes the product transportable and ready for trading. Secondary features are those attributes which are additional to the basic features. Here could be mentioned different certifications (e.g. labels such as organic or Fair Trade), a higher level consumer appeal (e.g. more attractive packaging) or that the food product meets a specific purpose. An example of the latter is a food product designed for take-away with the subsequent impact on processing, packaging and maybe also re-heating of the product.

The global trade of agricultural products and processed foods has increased steadily over the last 20 years and at a higher pace than any other global production (Wik, Pingali and Broca, 2008). There are many examples of value chains in the food sector that have developed from local or national value chains to become international value chains. An example is the seafood industry where species are caught in the North Atlantic Sea by European companies, and transported to Asia for processing, then returned to the European Union (EU) market to be sold in local supermarkets. Another example is the collaboration between fruit growers in some African countries as part of a contractual agreement with European producers of organic chocolate or marmalade. The organic products are then traded in the EU market or exported to, for example Japan and China certified as organic food with the EU logo for organic food. This pattern points to a stronger integration of local production into more international value chains.

As the food industry becomes increasingly integrated with its suppliers and customers, and across borders, the importance of a high food safety level stands out even more. This can be illustrated by the food scares that have occurred in the EU including the latest horse meat scandal. This points to the fact that there is a need for a defined regulatory framework and procedures for producing and trading food and agricultural products that can ensure specific standards of the product and its conformity and safety. This is the basis for developing FQM systems.

By its nature a quality management scheme is a set of defined criteria that have to be fulfilled. This can be achieved by implementing routines in practice combined with routines for documentation. The quality management schemes in the food industry are developed to ensure that the food plants produce food products that are safe to eat and of a high quality, thus enabling the food companies [small and medium-sized enterprises (SMEs) and large companies] to trade their products in the market. Some quality management schemes have only local or national scopes, whereas other quality management schemes are internationally recognised such as the International Standard Organisation (ISO) International Standards (Note 9.1).

Companies that export food may have to fulfil quality management schemes imposed by the customers in the overseas markets. An example is the demands from the US Food and Drug Administration to EU pork exporting companies that target the American market. This type of FQM scheme is strictly designed to guarantee to the consumers in the American market that the imported pork is a safe product to eat. Such a scheme lays down very strict procedures for the pork companies about how to

Note 9.1   The ISO International Standards

A standard such as the ISO International Standard is a document that provides requirements, specifications, guidelines or characteristics that can be used consistently to ensure that materials, products, processes and services are fit for their purpose. The ISO International Standards are strategic tools that reduce costs by minimising waste and errors and increasing productivity. These standards help companies to access new markets, level the playing field for developing countries and facilitate global and free trade. The most common ISO standards for the food chain are:

- ISO 9001 with the standards for quality management.

- ISO 22000 with the standards for food safety management including the HACCP principles.

- ISO 14000 with the standards for environmental management.

- ISO 26000 with the standards for social responsibility.

---

www.iso.org.

carry out plant cleaning, the slaughtering and cutting operations, the lab test for food pathogens at the plant, and especially the required documentation needed to trace the pork from the market back to the farmer.

Quality management schemes may be implemented by farmers, food producers, retailers and non-governmental organisations (NGOs), and some schemes target more actors in the value chain. In order to maintain trust from trading partners and consumers in such quality management schemes it is necessary to have an independent body inspecting that the requirements in the particular scheme are followed – all to maintain trust in the product, the logo or the company. The inspection of internationally recognised schemes is performed by an accredited independent organisation such as Bureau Veritas. This organisation will physically inspect the plant or supplier and ensure that all procedures laid down for this scheme are followed, implemented and documented correctly. If not, the producer will lose the certification, thus leading to a reduced competitiveness in the market.

Other schemes are inspected by private bodies as is the case for organic food. For this scheme, each country in the EU or even in the USA and elsewhere has some national pending regulations defining the organic production scheme. The compliance with national regulations by the farmers and producers is inspected by national bodies or private organisations. However, the national bodies and private organisations are then inspected and certified by the International Federation of Organic Agriculture Movements (IFOAM); the international organisation defining 'organic'. This system ensures that consumers across the World can trust the organic label.

9.2.3   Application of Quality Management Schemes to the Supply Chain

Schemes targeting the primary producers may be imposed by companies, retailers, food Authorities or NGOs, thus there are many different stakeholders that show interest in impacting agricultural production to produce a product with a defined profile. Schemes primarily lay down rules for the use of input (medication, feed, pesticides, etc.) and animal welfare conditions, but other schemes are more demanding. Such schemes may target the specific production requirements for a certain crop (e.g. Lammefjord carrots) or focus on the environmental impact of the production (e.g. Swedish Sickle) as in Note 9.2.

Quality management schemes targeting food producers and processors may be internationally recognised ones as exemplified by the ISO International Standards, but they could also be schemes implemented by the EU or national or even local bodies. The EU defined standards of Protected Designation of Origin (PDO); Protected Geographical Indication (PGI) and the Traditional Speciality Guaranteed (TSG) are examples of such quality management schemes (Note 9.3).

From the above it is clear that the EU standards for local products or products with specific connections to a certain area have a significant impact on the local value chains, as these standards influence local agricultural production as well as the processing technologies and applied methods. From this perspective the EU standards mentioned here are contributing to maintaining income and employment in rural areas, and this is in line with the dimension of social sustainability.

Note 9.2   Examples of quality management schemes targeting primary producers

---

Swedish Sickle (*Svenskt Sigill*) is a quality management scheme targeting primary producers in Sweden. The scheme is implemented by the Federation of Swedish Farmers. The scheme targets all subsectors of agriculture and horticulture, and the core idea is to ensure that the agricultural production is undertaken with a minimum application of pesticides and great concern for the environment. For livestock production, the scheme lays down rules that target a very high standard for animal welfare. The latest addition to the scheme's requirements is to include considerations about the climatic impact from the agricultural production.

The PGI scheme has been implemented by the carrot growers in the Lammefjord region in Denmark. The carrots from the Lammefjord region have been certified as a PGI product since 1996. The certification was obtained due to the specific production method (growing, harvesting and storing) for the carrots and the fact that this production method is different from carrot growing elsewhere in Denmark. Danish consumers know the label 'Lammefjord carrots' and connect this name with high quality carrots.

PGI, Protected Geographical Indication.
http://ec.europa.eu/agriculture/quality/schemes/index_en.htm and www.svensktsigill.com.

Note 9.3   Defining the PDO, PGI and TSG products

- *Protected Designation of Origin* (*PDO*): Covers agricultural products and foodstuffs which are produced, processed and prepared in a given geographical area using recognised know-how.

- *Protected Geographical Indication* (*PGI*): Covers agricultural products and foodstuffs closely linked to the geographical area. At least one of the stages of production, processing or preparation takes place in the area.

- *Traditional Speciality Guaranteed* (*TSG*): Highlights traditional character, either in the composition or means of production.

---

http://ec.europa.eu/agriculture/quality/schemes/index_en.htm.

9.2.4   Beneficiaries of Quality Management Schemes along the Supply Chain

The core motivation of implementing a quality management scheme is to remain in business in the food value chain. This is valid for farmers as well as SMEs and large food companies. The quality management scheme may be implemented to differentiate the producer or the product in a competitive market, thus it is necessary that the market (i.e. the consumer) is aware of and understands the scheme. Some challenges arise from this, as consumers are a heterogeneous group with different preferences for food. Organic food is an example to illustrate this issue. Consumers across the EU are familiar with the common EU logo for organic food and in general, European consumers consider organic food as 'healthy' and/or as 'good for the environment'. But there are some regional differences: in the North of Europe, organic food is also linked to higher animal welfare compared with conventional production methods, whereas in the South of Europe purchase of organic food is motivated by taste, health and quality considerations. This shows that despite the fact that the production standards laid down at EU level to define organic production, the European consumers have different perceptions of what organic production actually is.

9.3   Introducing Sustainability and Sustainable Production

Sustainability is, at its core, a modern term describing something that has always been at the heart of any concept of resource use and material production: the need to balance steadily growing production with the fact that some, or all, of the used resources are taken from a finite quantity which, in the best case, replenishes itself at a rate lower than the rate of use or, in the worst case, is not replenished at all. The first applications of sustainability in a form resembling its modern definition were in the forestry sector, from where only a short step leads to agriculture and, more broadly, to food production in general.

In order to introduce the element of sustainability into quality management and quality assurance in the food sector, the term 'sustainability' has to be considered at its base level and the underlying paradigms, as well as the connection of the sustainability to the food and nutrition system in itself.

Sustainability as a term can be defined in a number of ways, most of which are connected to the notion of sustainable development. However, sustainability can be considered to extend to other areas not necessarily directly related to the growth-economy foundations of sustainable development. A number of respected encyclopaedias and dictionaries define sustainable, and by derivation its noun 'sustainability' as 'able to be maintained at a certain rate or level' or 'able to be upheld or defended' (Oxford) or 'of, relating to, or being a method of harvesting or using a resource so that the resource is not depleted or permanently damaged' (Merriam-Webster). From those definitions, it is easy to connect sustainability and sustainable acting as being related to the maintenance (in the broad sense of the word) and use of a resource, which is most often a physical, material resource, but can also have immaterial qualities.

Traditionally, the first uses of the word 'sustainability' are attributed to a German accountant, mining administrator and forestry manager, Hans Carl von Carlowitz, who introduced the first ideas of sustainable management of a resource in his book *Sylvicultura oeconomica, oder haußwirthliche Nachricht und Naturmäßige Anweisung zur wilden Baum-Zucht*, which was published in 1713 (Grober, 1999; Sächsische Hans-Carl-von-Carlowitz-Gesellschaft, 2013). Further development of human civilisation through the industrial revolution and breakthroughs has seen the topic of sustainability come up at different times, but it took until the middle of the twentieth century for sustainability to reappear as a focus of global discussion. The growing and easily observable toll of unchecked and relentless growth and development, and the first signs of using up finite resources, such as the oil crises of the 1970s and 1980s, have further contributed to increased awareness of and orientation towards sustainable ideas, as have more and more common environmental crises and disasters, very often caused by anthropogenic direct action or indirect influence through depletion of resources and alteration of ecosystems. Agendas such as the Club of Rome 'Limits to Growth', spearheaded by Dennis Meadows (Meadows *et al.*, 1972) or the UN-appointed World Commission on Environment and Development (WCED), popularly also known as the 'Brundtland Commission', have delivered probably the most well-known definition of sustainable development in their report 'Our Common Future', published in 1987 and welcomed by the UN General Assembly (UN World Commission on Environment and Development, 1987). Sustainability refers to – as per the definition given by the United Nations Brundtland Commission – in meeting the wants and needs of the present day world without sacrificing the well-being of the future generation. This definition indicates that it is extremely vital to preserve the natural habitat and the environment as well as the resources available such as water, natural gas or rare earth materials (e.g. cadmium) or in fact any sort of resources without sacrificing the profitability of organisations as organisations are not interested in greening their business unless this is combined with economic benefits (Christopher, 2011). Another definition describing the principles of sustainable development also highlights the goal of better living for people. It states for sustainable development, there must be a safe, healthy environment, resources must be used efficiently and environmental issues must be taken into account across various sectors. The principles are then further explained in how they are to be measured. Healthy environment is measured through life expectancies and CO_2 emissions from industries, transportation, households, agriculture and waste. Using resources efficiently is measured through energy and water consumption and waste. This definition is vague in terms of how sustainable development is going to be achieved. It only gives broad areas that are the focus of sustainable development. On the other hand, it gives specific ways in how sustainable development can be measured.

More modern developments in the area of sustainability have included a system of distinction between weak and strong sustainability, which is again connected to the issue of limited resources, with weak sustainability being associated with resources which can be substituted by other, usually man-made resources and strong sustainability being associated with resources where no such substitution is possible. Another area of development addressing the topic of finite resources and non-renewable resources is re-use and the concept of closed-circle economies – here, the use of a certain resource is extended, in the optimal case, to its absolute use limits through returning resources as close as possible to their raw form and re-introducing them into the production stage (the best example being all forms of raw material recycling), re-using resources at the consumption stage if they are still usable according to predetermined indicators or using resources in a different capacity at some stage in the value chain (here, increased use of by-products in the sustainable food industry would come to mind). In general, both current developments and historic facts define the idea of sustainability being a struggle against the issue of limited resources – through reducing their use and at the same time optimising this use as far as possible.

Sustainability in itself is a strongly holistic principle: there is no easy way to exclude any area of modern life and development from sustainability, or find a techno-socio-political dimension that is not affected by the ideas of sustainable development or does not include those ideas in itself. However, to simplify the overview of sustainability and provide an integrated, non-excluding base for the review of quality management and quality assurance schemes, the authors have decided to use the common approach of the three main dimensions of sustainability (UN World Commission on Environment and Development, 1987) Through this approach, sustainability is divided into three main categories: a social dimension, an economic dimension and an ecological dimension. Some documents also include a fourth, political dimension, however, due to the practical orientation of this work this dimension will not be included here (Levy, 1997).

In order to see and create a connection between sustainability and food production/the food industry, an observer does not have to go far beyond the very beginnings of sustainability. Although the first ideas for sustainable management and development came from the forestry sector, food production (especially if one looks at eighteenth-century approaches and standards) also dealt with the need for optimal use of finite resources and the, almost even stronger, need to make sure that these resources, or at least the conditions required to create the resources, remain available for repeated use (own adaptation from UN World Commission on Environment and Development, 1987). Simply said, the very nature and basic characteristics of the food production sector push all successful actors to act in a sustainable way.

In the course of this discussion, it is important to understand that sustainability, as shown above, is a developing, dynamic term and field, with the recognition, aims and goals of sustainability constantly changing depending on the current state of discourse in the public and in the academic and legislative communities, the developments undertaken in the latter usually influencing the ideas adopted by the former. Applied to concrete terms and developments, this means that new developments and the up-take of new ideas in sustainability research, as well as the introduction of new provisions and guidelines by legislative bodies, will reflect on the perception and understanding of sustainability by the general public and, through that, on the way sustainability is interpreted and understood. With regard to the different dimensions of sustainability and their link to fulfilment of different requirements of the public, it can be argued that there are several main factors influencing the ongoing development and the public understanding of sustainability. In the following paragraphs, we will try to discuss some of those factors and, considering the overall topic of this chapter, connect those factors to different parts of the food sector value chain.

9.4 Linking Quality Management Schemes with Sustainability along the Food Chain

Growing environmental, social and ethical concerns as well as increased awareness of effects of food production and consumption on the natural environment have led to increased pressure from consumer organisations, environmental advocacy groups and policy makers to agri-food companies to deal with social and environmental issues related to their supply chains within product life cycles, from 'farm to fork' (Courville, 2003; Weatherell, Tregear and Allinson, 2003; Iilbery and Maye, 2005; Maloni and Brown, 2006; Vachon and Klassen, 2006; Welford and Frost, 2006; Matos and Hall, 2007). Sustainability in the food sector can be applied and approached on a variety of different levels, ranging from using sustainable principles to uphold the capacity of natural resource to produce nutrition products to following strong sustainability paradigms through organic farming methods and avoidance of anthropogenic influences and substitutes throughout food production. While the former application of sustainability, in its roots, comes out of the pure necessity of ensuring a steady flow of a much-needed resource, and can easily be counted as belonging to basic human needs, the use of strong sustainability principles tends towards the upper level of the human need structure as the application of those principles diversifies the products, as well as provides their manufacturers and users with self-gratification (own adaptation from Maslow, 1943, 1954 and Gawron and Theuvsen, 2009). Stakeholders demand corporate responsibility to go beyond product quality and extend to areas of labour standards, health and safety, environmental sustainability, non-financial accounting and reporting, procurement, supplier relations, product life cycles and environmental practices (De Bakker and Nijhof, 2002; Waddock and

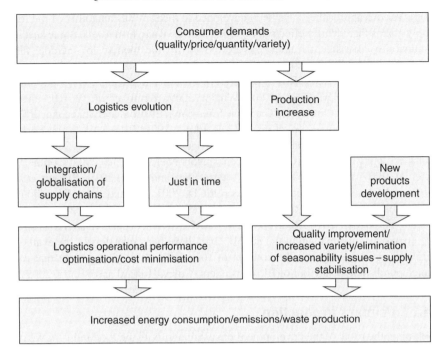

Figure 9.1   Consumer demands and environmental impact in agrifood supply chains.

Bodwell, 2004; Teuscher, Grüninger and Ferdinand, 2006). Sustainable supply chain management expands the concept of sustainability from a company to the supply chain level (Carter and Rogers, 2008) and should provide companies with tools for improving their own and the sector's competitiveness, sustainability and responsibility towards stakeholder expectations (Fritz and Schiefer, 2008). Principles of accountability, transparency and stakeholder engagement are highly relevant to sustainable supply chain management (Waddock and Bodwell, 2004; Teuscher, Grüninger and Ferdinand, 2006; Carter and Rogers, 2008). The impact of consumer demand on the environmental efficiency of the agri-food sector is given in Figure 9.1.

In response to stakeholder pressures for transparency and accountability, agri-food companies need to measure, benchmark and report sustainability performance of their supply chains, whilst policy makers need to measure the performance of sectors within the supply chain context for effective target setting and decision-making.

Although sustainability assessments have traditionally focused on agriculture (McNeeley and Scherr, 2003; Filson, 2004), recently researchers and policy makers have made attempts to develop more holistic approaches by incorporating stages of food processing, food retailing and specifically transportation in the assessment frameworks of food supply chains (Heller and Keoleian, 2003; Green and Foster,

2005). Various approaches have been developed to measure sustainability of the food supply chains that identify effects at regional, industrial and firm levels. Some specific sustainability assessment frameworks developed for the food sector include: farm economic costing (Pretty *et al.*, 2005); life cycle approach to sustainability impacts (Blengini and Busto, 2009; Roy *et al.*, 2009; Heller and Keolean, 2003); food miles (Garnett, 2003; Coley, Howard and Winter, 2009; Kemp *et al.*, 2010); energy accounting in product life cycle (Carlsson-Kanayama, Ekstrom and Shanahan, 2003); mass balance of food sectors (Linstead and Ekins, 2001; Biffaward, 2005; Forkes, 2007; Risku-Norjaa and Mäenpääb, 2007; Lopez, Bunke and Shirai, 2008; Ortiz, 2008); ecological footprint (Gerbens-Leenes, Nonhebel and Ivens, 2002; Collins and Fairchild, 2007; Burton, 2009; Mena, Adenso-Diaz and Yurtc, 2010; Ridoutt *et al.*, 2010); and farm sustainability indicators (OECD, 2001; Fernandes and Woodhouse, 2008; Meul, Nevens and Reheul, 2009; Nickell *et al.*, 2009; Gomez-Limon and Sanchez-Fernandez, 2010; Rodrigues *et al.*, 2010). Some concepts related to sustainable approaches for managing the supply chain have been questioned; for example, there are arguments against the concept of food miles as a driving force that may change purchasing behaviour of EU consumers (Coley, Howard and Winter, 2009).

9.4.1   Primary Production

Looking at the roots of the development of sustainability in the food industry we can observe that the earliest driving factors for food sustainability were related to the production which puts those factors at an early stage in the value chain, being connected to primary and secondary production of food items. The first of those factors can be described as the perception of a certain food item coming from environmentally sustainable production, with the definition of 'environmentally sustainable' being connected to physical indicators describing the individual processes of production as well as the use (or abstinence from using) non-natural supplements during the production process (Note 9.4).

Fish and seafood is an important part of the modern diet, and all figures point to the fact that across the World the consumption of seafood will increase. The increasing demand for fish for industrial purposes (e.g. manufacturing of fish oil and fish meal) and for the manufacturing of food products has led to over-fishing of some species. To restore the natural balance – or to prevent the extinction of a species – the concept of sustainable fisheries has been developed and is now an internationally recognised and applied scheme (Note 9.5).

Some agricultural products are only produced in very few places, but enjoy a global and increasing demand. Vanilla is such a crop. This implies that in order to ensure that natural vanilla will be available in future some measures targeting sustainability have to be implemented (Note 9.6).

9.4.2   Food Manufacturing

In the EU food and drink industry quality management schemes are widely used, and even demanded if the company is supplying to, for example retailers or food service

Note 9.4   Quality management scheme targeting sustainable primary production

The Danish IP scheme targets the horticultural production with criteria defined for the production of open-land and greenhouse cultivations, berries, orchards, and for packers of horticultural products, respectively. Each grower will be inspected by the Danish Plant Directorate to assure compliance with the rules of the certification.

The scope of the certification is to produce high quality fruit and vegetables from an integrated and holistic approach to the production, and care for the environment as well as for human health.

Key criteria in this certification are to optimise the given natural mechanisms for regulating pests; to reduce the application of fertilisers and chemicals to a minimum; controlled and wise watering of the cultivations; and documentation for compliance with these criteria.

IP, integrated production.
www.dansk-ip.dk.

Note 9.5   Marine Stewardship Council (MSC) – sustainable fisheries

The MSC is an independent non-profit organisation which sets a standard for sustainable fishing. Fisheries that wish to demonstrate they are well managed and sustainable compared with the science-based MSC standard are assessed by a team of experts who are independent of both the fishery and the MSC. Seafood products can display the blue MSC eco-label only if that seafood can be traced back through the supply chain to a fishery that has been certified against the MSC standard. More than 13 million European consumers every day enjoy MSC certified fish in more than 7000 McDonald's restaurants across Europe. This certification ensures that the fish is caught in sustainable fishing in the Baltic Sea. The certification is a close collaboration between the MSC (accreditation body), McDonald's and Espersen A/S (the Danish seafood supplier).

operators. The most important quality management schemes are the internationally recognised standards of ISO; the British Retail Consortium standard BRC Global Food Standard and the German originating International Food Standard IFS. Such quality management systems are internationally recognised, focused on the food safety and edible quality of the product, and as such, these schemes function as a 'guarantee of a high quality and safe product'. For example, the IFS standard is used in 96 countries throughout the World, and implemented in more than 11 000 food manufacturing companies (International Trade Center, 2014). The IFS standard is acknowledged for food safety and quality of both the processes and the products. The IFS standard encompasses audits of the firm's quality and safety management systems, the social requirements and the environmental requirements. This illustrates

Note 9.6   Quality management schemes and sustainable production of vanilla (Richard, 2014)

Vanilla is a widely used spice throughout the world, and between 60% and 70% of global ice cream production is made with vanilla flavour. Madagascar is the biggest supplier of natural vanilla flavour which is produced by drying the seed beans from an orchid. Local farmers in Madagascar grow these orchids, and the seed beans are harvested, dried and fermented locally. The vanilla enters into the global market when major international flavour companies buy the vanilla from intermediates. On the one hand, this supply chain puts the local farmers in Madagascar in a vulnerable position and, on the other hand, the major flavour companies are dependent on supplies from a limited geographic area with few suppliers.

In order to establish more sustainable supply chains for natural vanilla several initiatives have been implemented by the major companies. Mane, a French flavour specialist, has incorporated in the CSR strategy measures for ensuring farm income when harvests fail. Such a CSR strategy is in line with the international standard ISO 26000.

Another flavour house, Symrise, has established collaboration with local and international NGOs and local farmers for the purpose of developing local agriculture. To maintain a fixed price on the produce including vanilla has been a target in such programmes, and by this support the sustainability of local farming and local communities. The result was that by 2012 more than 1000 farmers in 29 villages were certified by the Rainforest Alliance; a quality management scheme with a strong focus on environmental sustainability. The Rainforest Alliance is a NGO working to conserve biodiversity and ensure sustainable livelihoods by transforming land-use practices, business practices and consumer behaviour. The Rainforest Alliance's sustainable agriculture programme includes training programmes for and certification of small, medium and large farms that produce tropical crops, including coffee, bananas, cocoa, oranges, cut flowers, ferns and tea. In recent years, the Rainforest Alliance has greatly expanded its work with smallholders, who now account for 75% of the farms (more than 783 000 farmers in all) certified by the organisation. To obtain certification, farms must meet the SAN standard which is designed to conserve ecosystems, protect biodiversity and waterways, conserve forests, reduce agrochemical use and safeguard the well-being of workers' local communities.

CSR, corporate social responsibility; NGO, non-governmental organisation; SAN, Sustainable Agriculture Network.

---

how a globally recognised and widely accepted and applied food standard also touches upon the aspects of sustainability. Concerning the ISO standards, the most important ones are the ISO 9001 which includes the requirements for the HACCP (Hazard Analysis and Critical Control Points) program (emphasising good food safety and hygiene management) and ISO 22000. The latter lays down the requirements for good manufacturing procedures including demands for environmental management.

Note 9.7   Arla Farm – a quality management scheme targeting sustainability and environmental impact

The Arla Farm (*Arlagaarden*) is the quality assurance programme implemented by the major dairy company Arla Foods. The programme was introduced to Danish and Swedish milk suppliers in 2003 and is under implementation among German milk suppliers today. All farmers supplying milk to Arla must participate in this programme. The four cornerstones of the programme are:

• Milk quality: its composition, taste and freshness.

• Food safety and hygiene.

• Animal welfare including conditions about housing, health and feed.

• Environmental considerations including respect for nature, optimisation of the use of resources and nutrients and reduction of the use of hazardous chemicals.

www.arla.com.

Hence, ISO 22000 can be regarded as a quality management scheme for the food industry with requirements for considering the production's impact on the environment, and through this, touching upon environmental sustainability.

Sustainable food manufacturing is connected to a sufficient and sustainable raw material flow (of a specific quality); the optimal use of resources (raw materials, energy and other) and the smallest environmental impact as possible. The quality management scheme Arla Farm is an example of this kind of scheme emphasising quality, sustainability and environmental impact (Note 9.7).

Sustainability may also be related to maintaining a local production and through this have an impact on local income and employment opportunities. Across the EU, locally produced food is attracting much attention from consumers. The number of primary producers and entrepreneurs having ventured into the production of foods and drinks is steadily increasing. By far, such products are produced in small scale facilities thus the market access is quite limited from a geographical perspective. The local products are playing an important role in binding the local communities together, and this pattern is found in all regions of Europe. Furthermore, there are numerous examples of networks being formed by the local producers and this is very often with the purpose of pushing the local food products further into the market. Due to their nature, the local products are dependent on the climatic conditions, production traditions and consumption patterns. There are several examples of quality management schemes targeting local food production. Such schemes typically promote local food chains (Note 9.8).

With a general shift in sustainability research from viewing sustainability issues as being primarily environment-centred towards including and up-valuing economic and social factors, the above-described factor was supplemented by the factor of economically and socially sustainable production. One good example of a quality

Note 9.8   The Waddengroep Foundation in the Netherlands

The Waddengroep Foundation stands for sustainable, environmentally friendly development of the countryside in the Wadden region. The Wadden sea region as a whole includes the Wadden Sea, the Wadden islands and a part of the mainland up to 25 km from the coast. The Waddengroep Foundation stimulates the local economy by promoting:

- Regional production.

- The processing of local raw materials and trade of locally produced goods and foods.

- Developing relations between regional products and nature.

- Stimulating and preserving employment.

- Stimulate internationalisation by, for example the Gastronomy Award.

- Certification of sustainable regional products.

www.waddengoud.nl.

management system reflecting this expansion of perception is the introduction and development of the Fair Trade system. Compared with the initial systems describing and regulating organic food production, where the focus very definitely lays on the production processes and the environmental dimension, the central topic of Fair Trade food products is the regulation and definition of minimum social and economic standards for the primary (and in some cases secondary) producers of the food items. This development clearly demonstrates that the focus of stakeholder attention has shifted from perceiving sustainability as being primarily focused on the product towards sustainability being related to the whole value chain and its participants.

9.4.3   Transportation and Food Retailing

If we move along the value chain towards the consumer, the next important factor related to sustainability is the issue of sustainable transport of the goods. This factor is primarily connected to the environmental dimension of sustainability, but on a secondary level also to social and economic indicators. The connection to the environmental dimension is easily made, as shorter and more efficient transport routes decrease environmental pollution by transport means, while the economic and social dimensions are reflected through lower costs of transport as well as the promotion and encouragement of local and regional products.

Moving even farther along the value chain, the next identified factor is one that has risen in importance in more recent times and where the accompanying indicators are still being developed. For the purposes of this chapter, we will name this factor 'sustainable retail of food items'. This factor is more strictly related to the social and economic dimension, which also helps to explain its relatively late emergence, while

Note 9.9   Danish retailer COOP and the Änglamark certification

COOP is the second biggest retailer in Denmark. COOP has developed a certification with the name of 'Änglamark'; hence this certification is a private label. The certification is used for food, cosmetics and household products that fulfil one or more of the following requirements:
It is a private label organic food product already carrying the EU organic logo.

• Cosmetics and household products are certified according to requirements from the Danish Federation for Asthma and Allergies, or the products are certified according to the criteria in the Nordic environmentally friendly label The Swan.

• All wooden products are made from FSC labelled wood.

• All textiles are made from 100% organic cotton and labelled with the The Swan.

Today, the Änglamark certification is used for more than 600 products ranging from canned tomatoes to detergents, baby food and bed linen.

www.coop.dk.

the environmental dimension plays only a secondary role. Sustainable retail of food items primarily focuses on way the goods are put on the consumer market, some of the underlying trends being a shift from large supermarkets back towards small scale retail, such as smaller organic food stores, farmers' markets and food items being sold directly at their place of production (in the form of farm shops). Other business concepts in this line are supermarkets with a social responsibility profile (e.g. where goods not sold during the day are given away at the close of business), or shops where you can return the packaging for recycling. Sustainability in retailing may also be based on retailers that take an active approach to increase the overall sustainability of the retail chain. In this instance, sustainability may refer to organic products or products that have a reduced environmental impact (Note 9.9).

9.5   Consumers and Quality Management Schemes Addressing Sustainability

This chapter has illustrated how quality management schemes in the food chain can be linked to one or more of the three dimensions of sustainability, and how the importance of the three dimensions vary along the food chain. In the end it is the consumer who decides which product to buy, and here the consumer's personal values (and economic situation) play a crucial role for the choice of food products and priority of certifications. Hence, the added value arising from the purchase of sustainable food is the key concept. The consumer's personal value system can range from merely avoiding hunger, to simple health benefits, over sustainable food being one part of a general sustainable lifestyle, to sustainable food playing the role of a status symbol.

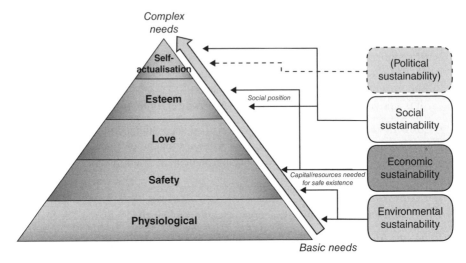

Figure 9.2   The coherence between the consumer's personal value system and the dimensions of sustainability.

The development of sustainability goals in FQM and FQA can be further elaborated if the development of those goals throughout the history of sustainability in food is presented using the Maslow pyramid, which is a graphical approach illustrating Maslow's theory on the hierarchy of human needs. This hierarchy separates human needs into different categories and places them at varying heights on a pyramid, with the most basic, physiological needs forming the bottom layer and the advanced, complex, self-actualisation needs forming the top (adapted from Maslow, 1943, 1954). Figure 9.2 demonstrates a basic graphic concept of Maslow's pyramid and assigns the different dimensions of sustainability, as perceived and applied in FQA and FQM, to different layers of the pyramid.

Figure 9.2 shows that, in FQM and FQA, the dimension of environmental sustainability connects with more basic needs on the physiological and safety level, which are addressed through standards and schemes aimed at sustainable food production and manufacturing. The existence of those schemes and standards serves as a warranty for the consumers that certain basic procedures were followed and certain substances were or, more usually, were not used. In a second step, economic sustainability is addressed through, for example pricing regulations as a part of standardisation procedures. Once those basic needs were satisfied, internalisation of the rulework guaranteeing the fulfilment of those needs took place (Gawron and Theuvsen, 2009) and the 'pushing' needs related to sustainability in FQM and FQA moved up on the pyramid towards esteem and self-actualisation, where they are expressed through strict voluntary systems such as Demeter which assure a very high safety level of the food products, and also add a social, even societal, value to their consumers (Demeter-International e.V., 2013a, b).

Therefore, it can safely be assumed that sustainability in FQM and FQA systems has moved beyond being a side effect of certain requirements of the quality management systems or a purely environment-oriented issue concentrated on primary and secondary production of food items. Sustainability-related topics and drivers in modern FQM and FQA have been extended to cover the entire value chain and to include a multitude of factors that are linked to stakeholder, especially consumer, needs and requirements.

9.6 Conclusion

Paralleling the global diffusion of the sustainability discourse, sustainability issues have entered the field of FQA and FQM systems. While at the beginning sustainability issues were represented only in a niche, the subsequent development has in many ways mirrored the development and diffusion of sustainability topics in a global context, leading to basic sustainability issues connected to the environmental dimension of sustainability being internalised into 'mainstream' FQM and FQA schemes and systems and even more complex sustainability issues from the economic and social dimensions being addressed, with the corresponding standards undergoing a transition from voluntary to mandatory application. This development pattern is evident from the globally recognised quality management systems such as ISO, IFS and BRC.

This chapter has illustrated how quality management systems and quality schemes can by targeting a certain part of the food value chain impact the entire farm-to-fork chain. It is also evident that quality schemes developed specifically for certain dimensions of sustainability can be globally recognised such as the Marine Stewardship Council and Fair Trade, but also be rooted in local supply chains. Hence, some quality management systems are in a position to be implemented in many countries whereas others are linked to local climatic or production-wise conditions. This is a key issue for quality management schemes as a competitive advantage in the local or international food market.

When observed from the customer's point of view, sustainability in FQM and FQA can be divided into its main dimensions, which address different customer needs based on Maslow's needs hierarchy pyramid. This approach can be useful to both assess the different topics and areas addressed by the variety of sustainability-related actions and instruments, and to provide a simple description of the evolution of sustainability within FQM and FQA through the development from addressing simple needs to more complex requirements and issues. However, as the model shown in this chapter is very basic in its scope, depth and level of detail, additional research regarding its refinement and further development is strongly suggested.

References

Biffaward, 2005. Mainstreaming Mass Balance on Farms and in Universities. The Mass Balance Approach for Organisations. Forum for the Future, http://www.adlib.ac.uk/resources/000/236/563/Mainstreaming_mass_balance_on_farms_and_in_universities.pdf (accessed 17 July 2015).

Blengini, G.A. and Busto, M., 2009. The life cycle of rice: LCA of alternative agri-food chain management systems in Vercelli (Italy). *Journal of Environmental Management*, **90**, 1512–1522.

Burton, C.H., 2009. Reconciling the new demands for food protection with environmental needs in the management of livestock wastes. *Bioresource Technology*, **100**, 5399–5405.

Carlsson-Kanayama, A., Ekstrom M.P. and Shanahan, H., 2003. Food and life cycle energy inputs: consequences of diet and ways to increase efficiency. *Ecological Economics*, **44**, 293–307.

Carter, C.R. and Rogers D.S., 2008. A framework of sustainable supply chain management: moving towards new theory. *International Journal of Physical Distribution & Logistics Management*, **38**, 360–387.

Christopher, M., 2011. *Logistics and Supply Chain Management*, 4th edn. London: Financial Times/Prentice Hall.

Coley, D., Howard, M. and Winter, M., 2009. Local food, food miles and carbon emissions: a comparison of farm shop and mass distribution approaches. *Food Policy*, **34**, 150–155.

Collins A. and Fairchild R., 2007. Sustainable food consumption at a sub-national level: an ecological footprint, nutritional and economic analysis. *Journal of Environmental Policy and Planning*, **9**, 5–30.

Courville, S., 2003. Use of indicators to compare supply chains in the coffee industry. *Greener Management International*, **43**, 94–105.

De Bakker, F. and Nijhof, A., 2002. Responsible chain management: a capability assessment framework. *Business Strategy and the Environment*, **11**, 63–75.

Demeter-International e.V., 2013a. *Erzeugung Richtlinien für die Zertifizierung der Demeter-Qualitaet*. Darmstadt: Demeter e.V.

Demeter-International e.V., 2013b. *Verarbeitung Richtlinien für die Zertifizierung "Demeter" und "Biodynamisch"*. Darmstadt: Demeter e.V.

Fernandes L.A.O. and Woodhouse, P.J., 2008. Family farm sustainability in southern Brazil: an application of agri-environmental indicators. *Ecological Economics*, **66**, 243–257.

Filson G.C., 2004. *Intensive Agriculture and Sustainability: A Farming Systems Analysis*. Vancouver: UBC Press.

Forkes, J., 2007. Nitrogen balance for the urban food metabolism of Toronto, Canada. *Resources, Conservation & Recycling*, **52**, 74–94.

Fritz M. and Schiefer, G., 2008. Food chain management for sustainable food system development: a European research agenda. *Agribusiness*, **24**, 440–452.

Garnett, T., 2003. *Wise Moves: Exploring the Relationships Between Food, Transport and Carbon Dioxide*. London: Transport 2000 Trust.

Gawron, J.C. and Theuvsen, L., 2009. Certification schemes in the European agri-food sector: overview and opportunities for Central and Eastern Europe. *Outlook on Agriculture*, **38**, 9–14.

Gerbens-Leenes, P.W., Nonhebel, S. and Ivens, W.P.M.F., 2002. A method to determine land requirements relating to food consumption patterns. *Agriculture Ecosystems and Environment*, **90**, 47–58.

Gomez-Limon, J.A. and Sanchez-Fernandez, G., 2010. Empirical evaluation of agricultural sustainability using composite indicators. *Ecological Economics*, **69**, 1062–1075.

Green, K. and Foster, C., 2005. Give peas a chance: transformations in food consumption and production systems. *Technological Forecasting and Social Change*, **72**, 663–679.

Grober, U. (1999) Der Erfinder der Nachhaltigkeit. Die Zeit (25 November), http://www.zeit. de/ (accessed 12 September 2014).

Heller, M.C. and Keoleian, G.A., 2003. Assessing the sustainability of the US food system: a life cycle perspective. *Agricultural Systems*, **76**, 1007–1041.

Iilbery, B. and Maye, D., 2005. Food supply chains and sustainability: evidence from specialist food producers in the Scottish/English borders. *Land Use Policy*, **22**, 331–344.

International Trade Center, 2014. International Featured Standards – IFS Food, IFS, Berlin, www.standardsmap.org (accessed 24 June 2015).

Joyce, D.C., 2001. The quality cycle. In: R. Dris, R. Niskanen and S.M. Jain (eds), *Crop Management and Postharvest Handling of Horticultural Products*, vol. 1. Ithaca, NY: Science Publishers, pp. 1–11.

Kader, A.A., 1985. Postharvest biology and technology: an overview. In: *Postharvest Technology of Horticultural Crops*. Davis, CA: University of California, pp. 3–7.

Kemp, K., Insch, A., Holdsworth, D.K. and Knight, J.G., 2010. Food miles: do UK consumers actually care? *Food Policy*, **35**, 504–513.

Levy, D.L., 1997. Environmental management as political sustainability. *Organization Environment*, **10**, 126–147.

Linstead, C. and Ekins, P. 2001. Mass balance UK. Mapping UK resource and material flows. Royal Society for Nature Conservation, Newark.

Lopez, D.B., Bunke, M. and Shirai, J.A.B., 2008. Marine aquaculture off Sardinia Island (Italy): ecosystem effects evaluated through a trophic mass-balance model. *Ecological Modelling*, **21**, 292–303.

Maloni, M.J. and Brown, M.E., 2006. Corporate social responsibility in the supply chain: an application in the food industry. *Journal of Business Ethics*, **68**, 35–52.

Maslow, A.H., 1943. A theory of human motivation. *Psychological Review*, **50**, 370-396.

Maslow, A.H., 1954. *Motivation and Personality*, 3rd edn. New York, NY: Addison-Wesley.

Matos, S. and Hall, J., 2007. Integrating sustainable development in the supply chain: the case of life cycle assessment in oil and gas and agricultural biotechnology. *Journal of Operations Management*, **25**, 1083-1102.

McNeeley, J.A. and Scherr, S.L., 2003. *Ecoagriculture: Strategies to Feed the World and Save Biodiversity*. London: Covelo Island Press.

Meadows, D.H., Meadows, D.L., Randers, J. and Behrens, W.W., 1972. *Limits to Growth*. New York, NY: New American Library.

Mena, C., Adenso-Diaz, B. and Yurtc, O., 2010. The causes of food waste in the supplier–retailer interface: evidences from the UK and Spain. *Resources, Conservation and Recycling*, **55**, 648–658.

Meul, M., Nevens, F. and Reheul, D., 2009. Validating sustainability indicators: focus on ecological aspects of Flemish dairy farms. *Ecological Indicators*, **9**, 284–295.

Nickell, T.D., Cromey, C.J., Borja, Á. and Black, K.D., 2009. The benthic impacts of a large cod farm — are there indicators for environmental sustainability? *Aquaculture*, **295**(3-4), 226–237.

OECD, 2001. *Environmental Indicators for Agriculture: Methods and Results, Volume 3. Organisation for Economic Co-Operation and Development*. Paris: OECD Publications Service.

Ortiz, M., 2008. Mass balanced and dynamic simulations of trophic models of kelp ecosystems near the Mejillones Peninsula of northern Chile (SE Pacific): comparative network structure and assessment of harvest strategies. *Ecological Modelling*, **2**, 31–46.

Pretty, J.N., Ball, A.S., Lang, T. and Morison, J.I.L., 2005. Farm costs and food miles: an assessment of the full cost of the UK weekly food basket. *Food Policy*, **30**, 1–19.

Richard, S., 2014. Arômes – La vanilla deviant durable. *La Revue de l'industrie agroalimentaire*, **759**, 38–40.

Ridoutt, B.G., Juliano, P., Sanguansri, P. and Sellahewa, J., 2010.The water footprint of food waste: case study of fresh mango in Australia. *Journal of Cleaner Production*, **18**, 1714–1721.

Risku-Norjaa, H. and Mäenpääb, I., 2007. MFA model to assess economic and environmental consequences of food production and consumption. *Ecological Economics*, **60**, 700–711.

Rodrigues, G., Rodrigues, I., Buschinelli, C. and de Barros, I., 2010. Integrated farm sustainability assessment for the environmental management of rural activities. *Environmental Impact Assessment Review*, **30**, 229-239.

Roy, P., Nei, D., Orikasa, T., Xu, Q., Okadome, H., Nakamura, N. and Shiina, T., 2009. A review of life cycle assessment (LCA) on some food products. *Journal of Food Engineering*, **90**, 1–10.

Sächsische Hans-Carl-Von-Carlowitz-Gesellschaft, 2013. *Die Erfindung der Nachhaltigkeit; Leben, Werk und Wirkung des Hans Carl von Carlowitz*. Munich: oekom verlag.

Shewfelt, R.L., 1999. What is quality? *Postharvest Biology and Technology*, **15**, 197–200.

Sloof, M., Tijskens, L.M.M. and Wilkinson, E.C., 1996. Concepts of modelling the quality of perishable products. *Trends in Food Science and Technology*, **7**, 165–171.

Teuscher, P., Grüninger, B. and Ferdinand, N., 2006. Risk management in sustainable supply chain management (SSCM): lessons learnt from the case of GMO-free soybeans. *Corporate Social Responsibility and Environmental Management*, **13**, 1–10.

Tzia, K. and Tsiapouris, A., 1996. *Hazard Analysis and Critical Control Points (HACCP) for the Food Industry*. Athens: Papasotiriou Publishing.

UN World Commission on Environment and Development, 1987. *Our Common Future*. Oxford: Oxford University Press.

Vachon, S. and Klassen, R.D., 2006. Extending green practices across the supply chain: the impact of upstream and downstream integration. *International Journal of Operations & Production Management*, **26**, 795–821.

Waddock, S. and Bodwell, C., 2004. Managing responsibility: what can be learned from the quality movement? *California Management Review*, **47**, 25–37.

Weatherell C., Tregear, A. and Allinson, J., 2003. In search of the concerned consumer UK public perceptions of food, farming and buying local. *Journal of Rural Studies*, **19**, 233–244.

Welford, R. and Frost, S., 2006. Corporate social responsibility in Asian supply chains. *Corporate Social Responsibility and Environmental Management*, **13**, 166–176.

Wik, M., Pingali, P. and Broca, S., 2008. Global Agricultural Performance: Past Trends and Future Prospects. Background Paper for The World Development Report 2008.

Will, M. and Guenther, D., 2007. *Food Quality and Safety Standards as Required by EU Law and the Private Industry: A Practioner's Handbook*, 2nd edn. Eschborn: Deutsche Gesellschaft für Technische Zusammenarbeit GTZ GmbH.

10

Risk Management for Agri-food Supply Chains

Christos Keramydas, Konstantinos Papapanagiotou, Dimitrios Vlachos, and Eleftherios Iakovou

Laboratory of Quantitative Analysis, Logistics and Supply Chain Management, Department of Mechanical Engineering, Aristotle University of Thessaloniki, Thessaloniki, Greece

10.1 Introduction

Contemporary supply chains experience dynamic evolution and intensive competition due to trends such as globalization, glocalization, and outsourcing, exposing them to several risk sources and thus making them more vulnerable. Increasing product and service complexity have led to the development of complex and constantly evolving supply chain networks, often changing and relocating risk sources, drivers, outcomes, and impacts. The suppliers, facilities, and ship-to-points are now typically dispersed across large geographical regions, possibly involving several countries, and adverse events may be associated directly with their assets/partners, or with the territory over which they are deployed (Klibi, Martel, and Guitouni, 2010). Moreover, the extreme load of asymmetric information that moves through the supply chain threatens the reliability and response of the supply chain's partnership (Yang *et al.*, 2009). Even worse, unpredictable events such as terrorist acts and natural disasters

Supply Chain Management for Sustainable Food Networks, First Edition. Edited by Eleftherios Iakovou, Dionysis Bochtis, Dimitrios Vlachos and Dimitrios Aidonis.
© 2016 John Wiley & Sons, Ltd. Published 2016 by John Wiley & Sons, Ltd.

can cause disruptions and lead to undesirable effects on the supply chain's strategic, tactical, and operational level.

Experience shows that supply chains encounter various kinds of disruptions that may have been prevented and may face severe challenges, such as false supplier management, delays due to a port strike, or product recall initiated by contamination of a carbonated beverage. Especially in today's heightened risk environment, any form of disruption, whether intentional or not, at any point along the supply chain can adversely affect the sustainability of businesses (Pai *et al.*, 2003).

The 2010 eruption of Iceland's Eyjafjallajökull volcano quickly brought about the closure of much of Europe's air space, and the subsequent cancelation of some 95 000 flights. Five days later, the assembly lines at Ford's Flat Rock, Michigan plant ground to a halt, an outage that was to last for 4 days (Wilding, 2013). At 14:46 JST on March 11, 2011, the Tohoku earthquake occurred, causing strong ground motion over large areas and severe Tsunami damage along the 670 km Tohoku Pacific Coastline. Apart from the large number of human casualties, extensive, and unprecedented damage was noted in industrial supply chains, such as the automotive industry (Matsuo, 2014). According to Japan's Ministry of Economy, Trade, and Industry (METI), the total manufacturing production in March 2011 was 15.5% lower than that in the previous month, which was the largest decrease for 58 years since the production index was established. More than half of this decrease was due to a drop in production in the transportation equipment industry, which experienced a 46.7% decline in March 2011 (Kamata, 2011). Months later, Thailand saw the worst flooding for 50 years, which simultaneously hit seven of the country's largest industrial zones, causing factory output to fall by 36% in October 2011 (Wilding, 2013).

To handle these unforeseen conditions, researchers try to develop new analysis and optimization models for demand planning, production scheduling, transportation network design, inventory control, lean initiatives, and other areas along the value chain. This focus on supply chain management (SCM) aspects inevitably turned the spotlight on the extensive study of supply chain risk management (SCRM) and vice versa. Either way, the management of risk in supply chains has now become an established element in the fields of SCM, corporate strategic management, and enterprise risk management (Zsidisin and Ritchie, 2008). It is inevitable to imply that effective SCRM supports and leads to supply chains that can adapt and handle multiple shifts.

Within this SCRM context, contemporary regional agri-food supply chains and global supply networks, as well as the associated risks that threaten their continuity, exhibit a few critical characteristics that render the relevant risk management – at the operational, tactical, and strategic level – a challenging task (Tsolakis *et al.*, 2014). Specifically, there is a set of risks, for example, weather-related risks (e.g., droughts in Kenya in 2007–2008, heat waves in Australia in 2013), natural disasters (e.g., earthquakes in Haiti in 2010, volcanic eruption in Iceland in 2010), biological risks (e.g., avian flu in Southeast Asia in 2004, mad cow disease in Great Britain in 1996), environmental risks (e.g., water scarcity, soil erosion, genetic erosion), or even political risks (e.g., Russian embargo to European Union agricultural products in 2014), economic risks (e.g., food price crisis in 2008), or trade risks (Russian exports ban in 2010), which are either a unique characteristic of, or have an excessive severity on, the agri-food supply chains compared with the typical commercial ones. The particular nature of these risks pose the need for proactively managing them in a specific "ad hoc" manner so as to mitigate their impacts in the aftermath of a disruption.

At the global level, agri-food supply chains have an inherent relationship with the primary level agriculture and food production, as the majority of the supply chain high-end operations depend on the relevant primary sector, for example, livestock breeding, grains farming, and so on. Agri-food supply chains essentially deal, in a direct manner, with – probably – the most fundamental aspect of humankind being, that is, global population nourishment, or what is called, in recent years, food security. Any kind of indifference to risks or ineffective risk management may have critical impacts on both food availability and affordability, in the case of developing countries, while in developed countries temporary short supply disruptions lead to extremely high prices affecting another aspect of food security and food market operations, which is stability.

At the company level, risk types and management practices are, more or less, similar to the ones that a company has to cope with within a typical supply chain. Nonetheless, there are still some risk types that particularly characterize agribusiness, such as traceability and food safety risks, which could potentially threaten a company's own existence or even the reliability of an international supply network as a whole. Moreover, agribusiness is sometimes more closely affected by central governmental action, including changes is agricultural policies (e.g., subsidies, environmental regulations, and food safety regulations), trade policies (e.g., European Union Trade Policy, Common Agricultural Policy), and so on, and thus more vulnerable to the relevant uncertainties.

At the farmer, household, and community level, mainly crop yield uncertainty, and price volatility threaten the stability of farmer incomes and enhance social inequalities. Certainly, these uncertainties are transferred, through production globalization and trade policies, among partners across the entire agri-food supply networks, leading markets to undesirable circumstances of high instability.

The role of public governance and interventions and their effects on the associated risk management balances is enhanced in the case of agri-food supply networks compared with other business sectors. Agri-food risk management is traditionally, either to a great (e.g., Eastern Europe, South America, etc.), or to a limited, extent (e.g., Western Europe, North America, etc.) associated with governmental and/or institutional interference, mostly in terms of regulations, directives, insurance, assistance, subsidies, financing, assurance, and taxation. Moreover, during the history of humankind, agricultural risks – at the farmer, collective, or even at the country level – have influenced critical political decisions several times, while on the other hand, politics – at the local, regional, or international level – have quite often posed extra risks to the involved agri-food supply chain partners.

Consequently, taking all these parameters into account, agri-food SCRM plays a key role in retaining future development within a sustainability context, mainly as regards the social and economic, as well as the governmental aspects of sustainability

(Beske, Land, and Seuring, 2014). To this end, the specific nature and the particular characteristics of agri-food supply chains create the need for additional specialized agri-food-oriented risk management strategies, apart from the implementation of the well-known supply chain-oriented ones, such as dual sourcing, parallel regional and global sourcing, collaboration of suppliers, forward buying, products substitution, and so on (Glendon and Bird, 2013). In this context, farmers, agribusiness companies, and governments employ several strategies in order to mitigate against and/or cope with agri-food risks, including technology development and adoption, enterprise management practices, financial instruments, investments in infrastructure, policy and public programs, and private collective action.

Additionally, given that the relevant final risk management decisions are based, somehow or other, on quantitative inputs, several quantitative tools have been developed thus far in order to support the relevant policy-making, and decision-making processes (Vlachos et al., 2013; Keramydas et al., 2015). In view of the fact that risk, by its very nature, is directly related to randomness and uncertainty, stochastic techniques proved to be more effective in modeling agri-food risk management real-world problems, and thus more acceptable by academicians and practitioners (Iakovou, Vlachos, and Xanthopoulos, 2010; Schmitt and Singh, 2012; Iakovou et al., 2014). In this chapter, we focus on the quantitative management of agri-food supply chain risks in a stochastic environment, with an emphasis on the supply chain-oriented risk mitigation strategies.

The scope of this study is twofold. On the one hand, the manuscript aims to present a set of selected advanced alternative methodologies for modeling real-world SCRM problems of the agri-food sector, capturing a few well-known specific characteristics of agri-food products, customer attributes, and management strategies. On the other hand, this chapter indicatively demonstrates a set of advanced solving methodologies, as well as the associated quantitative tools, that are widely accepted in solving this type of problems. Specifically, the first model demonstrates an analytical mathematical newsvendor-type approach to the issue of revenue management of perishable agri-food products. The second model exhibits a generalized simulation-based methodology to address the issue of emergency supplier contracts in an agri-food supply chain. The third model illustrates a game-theoretic approach to the issue of managing unreliable suppliers within an agri-food supply chain. These models provide a good sense of quantitative agri-food SCRM, while the proposed methodologies could be either directly employed or act as guidelines in effectively building both generic and customized decision-making support tools in the field of agri-food SCRM.

The remainder of the chapter is organized as follows. Section 10.2 provides the general risk management framework for a commercial supply chain, while in Section 10.3 this framework is specialized for the case of the agri-food sector, including risk types and sources, risk mitigation strategies, and quantitative tools for risk analysis. Sections 10.4–10.6 illustrate three cases in order to demonstrate both the theoretical and practical applicability of the aforementioned proposed quantitative agri-food SCRM methodologies. Finally, Section 10.7 closes the chapter by providing the relevant conclusions, as well as some thoughts and ideas for future research.

10.2   Supply Chain Risk Management

In order to better understand the great importance and dynamics of SCRM, it is critical that three key terms should be defined and analyzed under the supply chain framework, namely risk, vulnerability, and resilience. *Risk* has a wide use in many scientific disciplines and it might refer to failures, disruptions, impacts, and/or decisions. One of the most common conceptualizations of risk that includes both qualitative and quantitative components is given by Ellegaard (2008) and incorporates the knowledge of a loss-making event, the probability of a loss-making event and the significance (effect) of the event. Compared with risk, *vulnerability* is wider in scope, even wider than integrated SCM, business continuity planning, commercial corporate risk management, or an amalgamation of all of these disciplines (Peck, 2005). A firm's vulnerability to a disruptive event can be viewed as a combination of the likelihood of a disruption and its potential severity. It is obvious that vulnerability and risk have a strong connection and it is suggested that supply chain vulnerability is the way that the supply chain perceives the existence of risk.

Practice and study of everyday operations and activities show that, regardless of a supply chain's structure, industry type, length, complexity or importance, deviations or disruptions are going to occur. Even the best-managed supply chains will be affected by unexpected turbulences or be impacted by events impossible to forecast, so it is critical that *resilience* is built into them (Christopher, 2005). A supply chain's resilience can be defined as the ability of a system to return to its original (or desired) state after being disturbed (Peck, 2005; Iakovou, Vlachos, and Xanthopoulos, 2007).

It is also crucial to present the important aspects of risk and vulnerability sources and drivers, along with their outcomes and impacts. Sheffi *et al.* (2003) categorized risk sources as the following disruption modes: supply, transportation, facilities, freight breaches, and communications (Iakovou, Vlachos, and Xanthopoulos, 2007). But there are a number of factors (drivers) that have emerged in the last decade or so that are considered to have increased the level of risk (Juttner, Peck, and Christopher, 2003). These include a focus on efficiency rather than effectiveness, the globalization of supply chains, focused factories, and centralized distribution, the trend to outsourcing, and the reduction of the supplier base.

To effectively handle unforeseen events in the supply chain, a clear understanding of both undesirable events that may take place in the supply chain and the associated consequences and impacts from these events should exist (Cheng and Kam, 2008). In general, the effect that risk induces to the strategic, tactical, and operational levels of the supply chain usually determines its degree of vulnerability. Also, according to how the source of risk is perceived and its impact on the supply chain, risk can be divided into various types of losses: financial loss; performance loss; physical loss; psychological loss; social loss; and time loss (Harland, Brenchley, and Walker, 2003).

Most approaches to managing risk follow the generic risk management process that consists of three critical stages (Khan and Burnes, 2007): risk identification; risk analysis; and risk evaluation. In brief, efficient risk management reduces vulnerability by making the supply chain resilient (Bogataj and Bogataj, 2007) and relies on the appropriate decision-making about actions that tend to mitigate risk. These actions can be classified into two categories (Iakovou, Vlachos, and Xanthopoulos, 2007):

- Interventions to reduce vulnerability, in other words to improve security, which encompass initiatives for preventing security breaches, inspections, information protection, compliance to international standards, and so on.

- Interventions to improve supply chain resilience. Mitigating the consequences of a disruption and allowing for a prompt reaction can be achieved by demand-based management, total supply network visibility, balanced inventory management, flexible sourcing, product/process redesign, and so on.

Particularly in the food industry, where the supply chain is extremely regulated and protected, both categories of interventions pose significant implications to supply chain sustainability. For example, in order to adequately plan the operations in the supply chain of fresh products, it is necessary to formulate specific planning models that incorporate issues such as harvesting policies, marketing channels, logistics activities, vertical coordination, and risk management (Ahumada and Villalobos, 2009). Furthermore, changes in food production and consumer habits, as well as increasing trade liberalization and economic globalization, have made food safety an increasingly important global issue (Shen, Liu, and Zhao, 2013).

10.2.1 Improving Security

The major event that initiated both research and regulatory actions for improving security is without doubt the 9/11 attack on the World Trade Center, in 2001. Therefore, when discussing supply chain security, it is reasonable to focus on protection against malicious actions such as terrorist actions. But terrorism is not always the security's main concern, as disruptions such as natural disasters, reliability issues, or even health risks within the food chain also contribute to the vulnerability of contemporary supply chains.

During the last 10 years, several government rules and initiatives have placed security compliance responsibilities on international trade professionals with higher-than-ever penalties for non-compliance. Legal authorities in Europe and the USA have launched acts to improve process security along all stakeholders and channel partners within global supply chains. A summary of major regulatory and certification interventions can be found in Vlachos et al. (2012). However, the implementation of these techniques and regulations has a rather significant cost. For example, The Office of Management and Budget (OMB) has determined that the "10 + 2" regulation will cost industry from US$390 million to US$630 million per year (FedEx, 2008). Additionally, when governmental procedures are concerned, supply chain safety can lead to serious delays, as the following example shows. In July 2004, the US Department of Agriculture received an anonymous mail that containers of Argentine lemons loaded on the New Jersey-bound container ship Rio Puelo were contaminated with a "harmful biological substance." It took over a week to check that there was no contamination and to allow the ship to proceed to its destination, causing all the lemons to rot.

While governments and the private sector are working together to launch new initiatives to create more secure and reliable supply chains, industry is rapidly exploring the potential of new technology solutions to support secure company processes. Radio frequency identification (RFID) tagging and global positioning system (GPS) could lead to the achievement of an overall total product traceability (TPT). This high supply chain visibility can become possible by networking partners and by applying contemporary business models, which link stakeholders and users through specialized security software platforms.

Specifically in the food chain, evidence coming from the study of Fearne, Hornibrook, and Dedman (2001) suggest that quality assurance schemes have the potential to reduce both perceived product category risk – which reflects a person's perception of risk inherent in purchasing any particular product in a specific product category – and product specific risk – which is the perception of risk associated with a particular product within the product class. The former should be the concern of all stakeholders in the industry, including public sector organizations and government, as higher levels of perceived category risk will act as a severe brake on any initiatives to halt the long-term decline in fresh beef consumption. The latter is important for individual firms and supply chains, as it offers a means of creating product differentiation in a commodity sector (Fearne, Hornibrook, and Dedman, 2001).

10.2.2   Facing Disruptions

Even though mitigating strategies adopted by companies are different depending upon whether most of the supply risk is recurrent or results from disruption (Chopra, Reinhardt, and Mohan, 2007), risk mitigating strategies for improving supply chain resilience can be categorized and used by enterprises in case of deviations, disruptions, or disasters (Ellis, Henry, and Shockley, 2010). Snyder *et al.* (2006) give evidence in their study that superior contingency planning can significantly mitigate the effect of a disruption. Strategies that are specific to the category of risk in question are required to cope with low-likelihood high-impact (LLHI) risks, while generic strategies are required to mitigate high-likelihood low-impact (HLLI) types of demand and supply-related risks (Oke and Gopalakrishnan, 2009).

From a single organization view, the mitigating strategies that may be applied to supply chain context can be categorized under four categories: *Avoidance, Control, Cooperation,* and *Flexibility* strategies (Juttner, Peck, and Christopher, 2003). These strategies include interventions to improve resilience, which can be further classified as: (i) Design simulation, Network modeling (*Avoidance*); (ii) Inventory management, Lead time management, Cost minimization, Supplier contracts, Quality management, Redundancy (*Control*); (iii) Forecasting, Social responsibility, Information sharing, Knowledge management, Demand-based coordination (*Cooperation*); and (iv) Flexible sourcing, Infrastructure allocation, Risk pooling, Postponement, Recovery planning (*Flexibility*). These resilience interventions can be applied in the following operations throughout a supply chain: *Supply, Transport, Infrastructure, Communications,* and *Human resources* (Rice and Caniato, 2003).

Focusing on the food supply chain, a variety of risk management measures is available to farmers–producers, the practical use of which depends on a number of factors. Besides the political and market framework, enterprise-inherent factors such as liquidity, the decision-maker's experience with specific risk management measures and the decision-maker's attitude toward risk all influence their application (Heidelbach, 2007). The effectiveness of risk management depends to a great extent on the personal qualifications of the farmer. Especially the use of information sources by the farmers has a close relationship with the survival of the farm business (Backus, Eidman, and Dijkhuizen, 1997). The farmers' economic environment and their perception of risk in this environment are two major issues influencing the adoption of high-yielding technologies (Alocilja and Ritchie, 1990). Farmers often do not face a dichotomous choice whether to invest, but when to invest. Timing of investment is an important risk response. Therefore, information is essential. But the required effort to collect sufficient information can be considerable. More focus by agricultural economists on information as a risk response may contribute to improve farm decision-making under risk and uncertainty (Backus, Eidman, and Dijkhuizen, 1997).

10.3   Risk Management in Agri-food Supply Chains

Agri-food SCRM is vital on several grounds as any lack of concern for, or any failure in proactively managing, risks and mitigating their impacts may have critical consequences, in a straightforward manner, on farmer income, social inequalities, public health, agribusiness profitability, market stability, and food security. Contemporary agri-food supply networks are mainly threatened, among other ways, by crop yield uncertainties at the primary production level, and price volatility at the market level. Additionally, the ongoing climate change exerts extra pressure on agricultural production through extreme weather events, while tight food stocks within the current economic recession environment lead to price surge and price variability. Moreover, water, cropland, and energy scarcity in synergy with the anticipated supply versus demand imbalance due to the rapid population growth and the income increase, will raise new risks in the immediate future.

10.3.1   Risk Types and Sources

The risk factor engaged in contemporary agri-food supply chain networks exhibits a multidisciplinary nature. The main risk types and the associated sources that threaten the continuity of the products, processes, information, and financial flows of these networks are as follows (Jaffee, Siegel, and Andrews, 2010; Schaffnit-Chatterjee, 2010):

- Weather related risks and extreme weather events, for example, *hail storms, extremely low/high temperatures, floods, droughts, hurricanes, and typhoons.*

- Natural disasters, for example, *volcano eruptions and earthquakes.*

- Biological and environmental risks, for example, *crop pests, livestock diseases, and environmental degradation.*

- Production risks, for example, *yield uncertainties.*

- Human resource risks, for example, *seasonal personnel unavailability.*

- Management and operations risks, for example, *forecasting errors, farm and firm equipment failures, and poor decision-making.*

- Logistical, infrastructural, and technological risks, for example, *transportation risks, labor strikes, infrastructure obsolescence, and uncertainty of new technologies adoption.*

- Price and market risks, for example, *price volatility of inputs and outputs due to supply and demand changes, uncertainties of markets' demand in terms of quantity and quality, and loss of customers.*

- Financial risks, for example, *disruptions of farm business financing, and economic recession.*

- Policy, institutional, and regulatory risks, for example, *uncertainties of tax and fiscal policies, change of regulatory policies* (such as *food safety*), *changes trade policies, poor governance, and corruption.*

- Political risks, for example, *political and/or social instability (threats against property and life), bilateral or international disputes among countries, and political trade interventions.*

10.3.2 Risk Mitigation Strategies

There is not a unique universal management strategy in order to cope with the diverse risk types that threaten the continuity of an agri-food supply chain. Risk type and risk severity, personal characteristics of and contrasting interests among supply chain partners, along with the specific contemporaneous natural, institutional, and economic conditions, shape the relevant agri-food risk management landscape and drive the selection of the appropriate form of risk governance, in an "ad hoc" manner (Bachev, 2012). Practically, the majority of agricultural risks are impossible or extremely costly to be completely eliminated; there are always remaining unforeseen risks. Therefore, a rational level of risk forecasting and risk impact mitigation are enough to determine a successful risk management strategy. In general, risk management efficiency in the agri-food sector is always associated with a trade-off balance between risk mitigation benefits (cost savings, etc.) and risk governance costs.

There is a plethora of company-oriented well-established risk mitigating strategies in the field of general SCRM (Simchi-Levi, Kyratzoglou, and Vassiliadis, 2013), that are tailored to and efficiently employed in the agri-food sector, for example, implementation of business continuity plans, dual sourcing, parallel regional and global sourcing, suppliers collaboration, demand collaboration with customers, forward buying/hedging strategy, increased inventory and safety stock levels, employment of distribution centers, near-shoring manufacturing, products substitution, and so on. On the other hand, the specific nature and the particular characteristics of agri-food supply chains create the need for additional specialized agri-food-oriented risk management strategies. In this context, farmers, agribusiness companies, and governments employ several strategies in order to mitigate against and/or cope with agri-food risks, that are further categorized as follows (Jaffee, Siegel, and Andrews, 2010):

- Technology Development and Adoption: *Agricultural research and development of improved varieties and breeds, postharvest technology, software development, information and knowledge technology, and basic and advanced applied education programs.*

- Enterprise Management Practices: *Farm and firm diversification practices, farming systems approaches, just-in-time management, inventory control, improved forecasting capacity, food safety practices, certification of best practices, logistics planning, early warning systems, and so on.*

- Financial Instruments: *Credit and savings (formal and informal), insurance (formal and informal), warehouse financing, price hedging instruments, and other vehicles.*

- Investments in Infrastructure: *Investments in transport and communication infrastructure (including air- and seaports), energy infrastructure, informatics and knowledge transfer infrastructure, storage and handling facilities, marketplaces, processing facilities, weather stations, and other structures.*

- Policy and Public Programs: *Institutional arrangements, regulatory measures, government policies, property and human rights, labor laws, disaster management units, safety nets, and similar programs.*

- Private Collective Action: *Commercial and no-commercial actions taken by farmer groups, cooperatives, industry associations, and other groups, in addition to various types of commercial contractual arrangements and partnerships.*

Finally, three dominant forms of risk governance have been recognized thus far, based on real-world practice, within the agri-food SCRM framework, namely: (i) private modes (individual farmers and cooperatives) that include diverse private initiatives and special contractual and organizational arrangements tailored to the features of risks and agents, such as codes of behavior, diverse (relational, security, future) contracts, cooperatives, associations, business ventures, and so on. (ii) Market modes that include various decentralized initiatives governed by free market price movements and competition, such as risk trading (selling/buying insurance), future contracts and options, production/trade of special (organic, fair-trade, origins) products, and so on. (iii) Public modes that include various forms of third-party public (government, international) intervention in market and private sectors, such as public information, regulations, bans, assistance, funding, assurance, taxation, provision, and so on. According to the aforementioned agri-food risk governance, the associated risk management tools are also categorized into: (i) private; (ii) market-based; and (iii) public.

10.3.3 Quantitative Tools for Agri-food Supply Chain Risk Management

A wide list of well-known classic mathematical and operational research methods have been employed so far to address the issue of SCRM. Risk, by its very nature, renders stochastic techniques more effective in modeling the relevant real-world problems, and thus more received among academicians and practitioners. Indicatively, typical examples of both stochastic and deterministic modeling methods include: simulation, game theory, stochastic processes, mathematical programming, decision tree analysis, scenario analysis and stress testing, fuzzy logic, artificial neural networks, analytical hierarchical process, and multi-agent systems.

In this chapter, three representative agri-food SCRM problems are presented in order to demonstrate, in general, the relevant methodologies implementation and provide an idea of the associated managerial insights. Specifically, the first example deals with a particular characteristic of agri-food supply chains, that is, product perishability. Taking into account consumer preferences, a grocery retailer aims to effectively manage the risk of stock-outs for the limited shelf-life products through an emergency sourcing (ES) risk mitigation strategy in order to maximize the retailer's revenues. A modified newsvendor formulation is employed in the relevant modeling process leading to a closed-form optimal solution. The second example addresses the risk mitigation contracts issue within an emergency dual-sourcing context. A discrete event simulation (DES) methodology is developed in order to determine the optimal premium/contract cost that an agri-food retailer should pay with respect to inventory related costs. Finally, the third example addresses agri-food risk management through a game-theoretic approach, where the supply chain partners could be seen as players in a game defined by multiple goals, constraints and conflicting objectives, sharing communication channels, resources, information, logistic networks, and customer demands, but also facing common risks.

10.4 Case 1: Revenue Management of Perishable Agri-food Products – a Newsvendor-Type Modeling Approach

The rapid increase of the global population and the geographical expansion of modern agri-food distribution channels raise unique risk management challenges for the agri-food supply chain networks, especially in the field of perishable agri-food products. In addition, short life-cycle agricultural commodities enhance the agri-food market segmentation that is usually associated with specific pricing schemas based on the produce quality level (freshness). To that end, complex replenishment policies, involving alternative procurement sources, are a common practice among grocery retailers toward their revenues and profits maximization.

The following modeling example deals with the alternative ordering strategies of a perishable product, where price is related to its freshness. In addition to the regular supply channel, the retailer has the option to employ a more expensive emergency replenishment mode when the regular channel products have deteriorated. Therein, the grocery retailer hedges his or her business against fresh products stock-outs, and therefore manages to handle risk in an affective and profitable manner. A newsvendor-type modeling formulation is employed to determine the optimal quantities of both regular and emergency orders with respect to the retailer's profit maximization. The results of the relevant numerical experimentation indicate that, under certain circumstances, the emergency replenishment option may has a pivotal role as regards the retailer's total revenues.

10.4.1 System Description

We consider a two-echelon agri-food supply chain that deals with a single perishable product. The chain downstream includes a grocery retailer, while upstream involves two grocery producers, either individual farmers, agricultural cooperatives, or grocery wholesalers, trading within a dual-sourcing context. The product shelf-life is assumed to be limited consisting of two consecutive stages, that is, the first stage or "stage of freshness," where the product preserves its original organoleptic characteristics, and the second stage or "stage of deterioration," where a substantial portion of the above characteristics is critically degraded; the product remains marketable, but it is tradable at a lower selling price as the end of this stage coincides with the expiration date of the product (Tajbakhsh, Lee, and Zolfaghari, 2011; Banerjee and Turner, 2012).

The customers, in general, prefer to consume fresh products (first life stage) that are available at a premium price (Ferguson and Ketzenberg, 2006), yet they are willing to purchase the inferior quality products at a lower offer (discount) price when a stock-out of the superior quality ones occurs. The grocery retailer faces the corresponding demand irrespective of the fresh products' availability, which, however, has a critical effect on the grocery retailer's revenues, given the aforementioned consumer preferences and demand patterns. In order to shield the retailer's business continuity against the risk of stock-outs, the retailer implements a procurement strategy with emergency orders within a dual-sourcing framework. Specifically, the retailer contracts a distant regular supplier that is cheaper in price but inflexible in terms of order quantities (higher order volumes are required), as well as a local emergency supplier that is more flexible and willing to deliver orders of lower volumes but at a higher price. The retailer's procurement strategy involves the placement of a regular order, recurrently, at the start of a predefined period (e.g., at the start of each week), which has a length equal to the product's shelf-life. Moreover, during this period the retailer also has the option to place an emergency order so as to minimize fresh product stock-outs, and thus increase his or her revenues through managing risk in a profitable manner.

The grocery retailer has to decide on two critical issues, namely: (i) whether an emergency ordering policy is profitable or not and (ii) which are the optimal regular and emergency order volumes, with respect to the maximization of the expected total inventory-related profit of the system that is the decision-making criterion employed. Specifically, toward achieving higher profitability, the retailer has to increase his or her product sales and concurrently reduce the relevant costs, that is, procurement cost, lost sales cost, and salvage cost of the leftover deteriorated products.

10.4.2 Model Development

A newsvendor-type formulation is employed in modeling the overall agri-food supply chain. In this context, a single period with a length (L) of time units is assumed, that is equal to the shelf-life of a fresh product. Moreover, in order to model the relevant product degradation process, the shelf-life is further divided into two stages (or sub-periods) that represent the product's "freshness" state, and "deterioration" state, respectively. Supplier (R) is the regular supplier, while Supplier (E) is the emergency one. At the beginning of the period, the retailer orders Q_R product units from the regular supplier at a cost of c_R monetary units per product unit. The grocery retailer satisfies the consumer demand for fresh products at the market selling price of p_F monetary units per product unit, while the deteriorated products are less expensive, offered at the discount price of p_D monetary units per product unit ($p_D < p_F$). The leftover expired inventories are disposed of (at no value) at the end of the period (end of the second stage), thus, the associated disposal process does not affect the system revenue's performance. The model also allows for the retailer's fresh products resupply during the selling period. Specifically, the retailer has the option to exploit the emergency supplier and order Q_E product units at the start of the second stage – whereupon the leftover initial inventories decay into their inferior quality state – but at the higher cost of c_E monetary units per product unit $(c_E > c_R)$. Although the lead time of the regular supplier is positive, it is assumed that the transportation process retards the products' decay, so that products are completely fresh when delivered. The emergency supplier's lead time is considerably shorter in comparison with the regular supplier's one, therefore an instant emergency inventory replenishment is assumed.

The aggregate consumers demand for each sub-period (stages 1 and 2) is assumed to be a stochastic continuous variable, that is, X_1 and X_2, respectively, while these variables are further considered to be independent and identically distributed. Unsatisfied demand is assumed to be lost, inducing an extra lost sales cost of b monetary units per lost product sale for the retailer. The customer preferences are modeled as follows: consumers prefer to purchase fresh products, when available, while they are willing to substitute fresh products for the less fresh inferior quality ones at a discount offer price, whenever they do not find the preferable fresh products on the shelf. Indirectly, according to this assumption, the product units that are procured as part of an emergency order are limited to be sold at their fresh state (first stage). This assumption is qualitatively rationalized considering both the facts that the expensive emergency order has a lower product volume, and that these fresh products would be sold in priority during the second sub-period, leading to their sold-out or at least to negligible leftover inventories at the end of the period. This small quantity of leftover inventory, if any, could be probably sold during the successive period, although the relevant probability is rather limited due to the fact that a new regular order of fresh products would have been delivered. Consequently, it seems that the relevant realization probability of the combined event is trivial. Nevertheless, this assumption allows for the newsvendor-type single period modeling of the problem, as well as for its analytical solution, since the products are ordered, delivered, and sold within the same single period; product transfers between periods are not allowed. The overall problem notation is summarized in Table 10.1.

The decision-making process aims to determine the optimal order quantities Q_R and Q_E with respect to the expected total profit $ETP(Q_R, Q_E)$, which is calculated as follows:

$$
\begin{aligned}
ETP\left(Q_R, Q_E\right) = & \int_{-\infty}^{Q_R} p_F x_1 f_1\left(x_1\right) dx_1 + \int_{Q_R}^{\infty}\left[p_F Q_R - b\left(x_1 - Q_R\right)\right] f_1\left(x_1\right) dx_1 \\
& + \int_{Q_R}^{\infty} \int_{-\infty}^{Q_E} p_F x_2 f_1\left(x_1\right) f_2\left(x_2\right) dx_2 dx_1 \\
& + \int_{-\infty}^{Q_R} \int_{-\infty}^{Q_E} p_F x_2 f_1\left(x_1\right) f_2\left(x_2\right) dx_2 dx_1 \\
& + \int_{Q_R}^{\infty} \int_{Q_E}^{\infty}\left[p_F Q_E - b\left(x_2 - Q_E\right)\right] f_1\left(x_1\right) f_2\left(x_2\right) dx_2 dx_1 \\
& + \int_{Q_E}^{\infty} \int_{-\infty}^{Q_R + Q_E - x_2}\left[p_F Q_E + p_D\left(x_2 - Q_E\right)\right] f_1\left(x_1\right) f_2\left(x_2\right) dx_1 dx_2 \\
& + \int_{Q_E}^{\infty} \int_{Q_R + Q_E - x_2}^{Q_R}\left[p_F Q_E + p_D\left(Q_R - x_1\right) - b\left(x_1 + x_1 - Q_E - Q_R\right)\right] \\
& f_1\left(x_1\right) f_2\left(x_2\right) dx_1 dx_2 - c_R Q_R - c_E Q_E.
\end{aligned}
\tag{10.1}
$$

A simplified version of Equation 10.1 is the following:

$$
\begin{aligned}
ETP\left(Q_R, Q_E\right) = & \left(p_F + b - c_R\right) Q_R + \left(p_F + b - c_E\right) Q_E - b\left(\mu_1 + \mu_2\right) \\
& - \left(p_F - p_D\right) Q_R F_1\left(Q_R\right) - \left(p_F + b\right) Q_E F_2\left(Q_E\right) \\
& - \left(p_D + b\right) Q_R F_1\left(Q_R\right) F_2\left(Q_E\right) + \left(p_F - p_D\right) \int_{-\infty}^{Q_R} x_1 f_1\left(x_1\right) dx_1 \\
& + \left(p_D + b\right) \int_{-\infty}^{Q_E} x_2 f_2\left(x_2\right) dx_2 \\
& - \left(p_D Q_R + b Q_R + p_D Q_E + b Q_E\right) \int_{Q_E}^{\infty} f_2\left(x_2\right) f_1\left(Q_R + Q_E - x_2\right) dx_2 \\
& + \left(p_D + b\right) \int_{-\infty}^{Q_R} x_1 f_1\left(x_1\right) F_2\left(Q_R + Q_E - x_1\right) dx_1 \\
& + \left(p_D + b\right) \int_{Q_E}^{\infty} x_2 f_2\left(x_2\right) F_1\left(Q_R + Q_E - x_2\right) dx_2
\end{aligned}
\tag{10.2}
$$

Table 10.1 Notation.

Symbol	Description	Unit
i	Supplier index (1 = regular supplier, 2 = emergency supplier)	—
j	Shelf-life stage (1 = fresh product, 2 = deteriorated product)	—
p_F	Retailer's unit price for fresh products	Monetary units per product unit
p_D	Retailer's unit price for deteriorated products	Monetary units per product unit
c_i	Procurement cost for supplier i	Monetary units per product unit
b	Lost sales cost	Monetary units per lost sale
Q_i	Order quantity for supplier i	Product units
X_j	Stochastic demand during stage j	Product units
$f_j(x_j)$	Probability density function of x_j	—
$F_j(x_j)$	Cumulative probability density function of X_j	Product units

The application of the first-order conditions, in order to determine the optimal solution, leads to two close-form equations, respectively.

$$\frac{\partial}{\partial Q_R} ETP(Q_R, Q_E) = 0$$

$$p_F + b - c_R - (p_F - p_D) F_1(Q_R) - (p_D + b) F_1(Q_R) F_2(Q_E)$$
$$- (p_D + b) \int_{Q_E}^{\infty} f_2(x_2) F_1(Q_R + Q_E - x_2) dx_2$$
$$- (p_D Q_R + b Q_R + p_D Q_E + b Q_E) \int_{Q_E}^{\infty} f_2(x_2) f_1(Q_R + Q_E - x_2) dx_2$$
$$+ (p_D + b) \int_{-\infty}^{Q_R} x_1 f_1(x_1) f_2(Q_R + Q_E - x_1) dx_1$$
$$+ (p_D + b) \int_{Q_E}^{\infty} x_2 f_2(x_2) f_1(Q_R + Q_E - x_2) dx_2 = 0$$

and

$$\frac{\partial}{\partial Q_E} ETP(Q_R, Q_E) = 0$$

$$p_F + b - c_E - (p_F + b) F_2(Q_E) - (p_D + b) \int_{Q_E}^{\infty} f_2(x_2) F_1(Q_R + Q_E - x_2) dx_2$$

$$- (p_D Q_R + b Q_R + p_D Q_E + b Q_E) \int_{Q_E}^{\infty} f_2(x_2) f_1(Q_R + Q_E - x_2) dx_2$$

$$+ (p_D + b) \int_{-\infty}^{Q_R} x_1 f_1(x_1) f_2(Q_R + Q_E - x_1) dx_1$$

$$+ (p_D + b) \int_{Q_E}^{\infty} x_2 f_2(x_2) f_1(Q_R + Q_E - x_2) dx_2 = 0$$

(10.3)

When considering a general distribution for the demand, the proof of the final solution's global optimality is impossible due to the high complexity of the associated Hessian matrix. On the other hand, the above conditions' global optimality could be proven for a set of specific distributions, as illustrated by the following numerical example.

10.4.3   Numerical Example

The role of the numerical example is twofold, as it aims to present the model's applicability, while at the same time it provides valuable managerial insights in a "what-if" analysis context. The basic assumption of this example is that both shelf-life stages are of equal time length, while the stochastic demand for both stages is exponentially distributed with a mean of $1/\lambda_1 = 1/\lambda_2 = 100$ product units per stage (sub-period). The rest of the input parameters (base scenario) are listed below:

$$c_R = 6\left[\frac{€}{unit}\right], c_E = 9\left[\frac{€}{unit}\right], p_F = 16\left[\frac{€}{unit}\right]$$

$$p_D = 9\left[\frac{€}{unit}\right], b = 8\left[\frac{€}{unit}\right]$$

$$f_1(x_1) = \lambda_1 e^{-\lambda_1 x_1}, f_2(x_2) = \lambda_2 e^{-\lambda_2 x_2}, F_1(x_1) = 1 - e^{-\lambda_1 x_1}, F_2(x_2) = 1 - e^{-\lambda_2 x_2}$$

The proof of the expected total profit (ETP) function's concavity with respect to Q_R and Q_E could be rather easily performed given the exponential distribution of the demand. Moreover, first-order conditions provide us with the optimal order quantities $Q_R = 198 [units]$ and $Q_E = 25 [units]$ that correspond to an expected total profit of an $ETP = 557.35 [monetary\ units]$. The close-form solution also allows for an extensive "what-if" analysis of the input parameters' effects on the retailer's profitability performance, and the corresponding order quantities. Indicatively, a "what-if" analysis regarding the procurement cost parameters is presented.

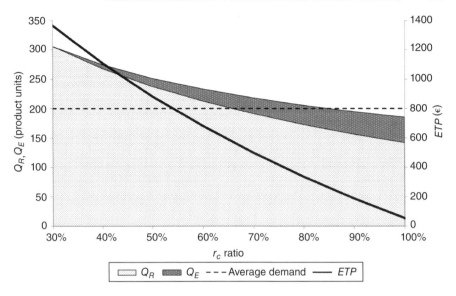

Figure 10.1   Impact of procurement price on the expected total profit and optimal order quantities.

10.4.3.1 "What-If" Analysis

The impact of the procurement cost parameters (c_R, c_E) on the expected total profit, and the optimal order quantities is depicted in Figure 10.1. The procurement cost impact is examined in the ratio form of $r_c = c_R/c_E$, that is, the fraction of the regular supplier procurement cost divided by the emergency supplier one, and is represented by the horizontal axis. Generally, the more economically attractive the emergency supplier procurement cost is, that is, as emergency cost approximates the lower regular supplier procurement cost $(r_c \sim 1)$, the more profitable the emergency orders become. As a result, the optimal sourcing policy includes higher volume emergency orders combined with lower volume regular ones. This is a rational sourcing behavior taking into account that the grocery retailer is able to purchase fresh products that are preferable by the customers at a reasonable price, close to the regular one. The expected total profit is a decreasing function of r_c. Moreover, an interesting managerial insight is the existence of a threshold value $(r_c \sim 0.3 \text{ or } 30\%)$, which highlights the profit inefficiency point of the dual-sourcing policy. Below this threshold, emergency orders are not beneficial anymore for the retailer, who prefers single regular sourcing instead of an emergency dual-sourcing strategy, that is, $Q_R > 0, Q_E = 0$.

10.5   Case 2: Emergency Dual Sourcing Contracts – a Simulation-Based Approach

Emergency supplier contracts within a dual sourcing context is a widely accepted risk management strategy among supply chain experts and practitioners, that is also effective in the agri-food sector, especially at the level of food manufacturers, food processors, wholesalers, and retailers. This second model aims to evaluate emergency dual sourcing in terms of the cost that an agri-food company should pay (e.g., premium to an alternative supplier) so as to ensure its anticipation of and preparedness against supply disruptions. Specifically, a DES methodology is proposed to evaluate the alternative ES strategies and determine the optimal premium that should be paid to the emergency supplier for reserving the necessary "back up" capacity. Moreover, the proposed methodology is able to determine the optimal level of this capacity when the associated reservation fee (premium) offered by the emergency supplier is fixed. The simulation optimization process is performed with respect to the minimization of the associated expected inventory-related costs. The results document the necessity to consider premium cost as a key input to the relevant decision-making process when settling on the appropriate risk mitigation strategy.

10.5.1   System Description

We consider a two-echelon agri-food supply chain for a single Stock-Keeping-Unit (SKU) that includes one food retailer and one supplier (e.g., food processor). The retailer employs a periodic review (s, S) ordering policy to replenish his or her inventory in order to satisfy the stochastic demand. The supplier operates under a make-to-order policy. The relevant production (e.g., processing) capacity is limited, though sufficient to serve the retailer's demand (regular orders).

The supplier's processing and distribution operations are prone to disruptions. A disruption incident renders the food processor completely unable to serve any outstanding orders; the duration of the corresponding operations shutdown is stochastic. The accumulated unfulfilled demand is backordered and satisfied when the required quantities are available.

A risk mitigation option for the retailer is to contract an alternative emergency supplier of the same attributes, that is, lead time and price (SKU), as the main/regular one. This contracted partnership, whereby part of the emergency supplier's capacity is reserved by the retailer at a certain agreed price, allows for the emergency activation of this "stand by" capacity on the part of the retailer when the regular supplier suffers a disruption. The overall agri-food supply chain is presented in Figure 10.2.

In view of the fact that the retailer implements an ES policy she has to decide on the associated reserved capacity level, which is considered as a percentage of the regular supplier's capacity. This percentage varies within the 0–100% range, where the lower bound corresponds to the single sourcing (SS) policy, whereas the upper bound indicates the complete substitution of the regular supplier during the disruption, that is, any disruption incident has no effect on the retailer's sourcing process. The retailer purchases this option, in advance, at a reservation cost (premium) that is a linear proportion of the reserved capacity level. In exchange for this fee, the supplier is bound to release the agreed capacity once the retailer raises the relevant claim.

Furthermore, when the supplier and the retailer transact within an information sharing context that allows for the prompt, without delay, response of the emergency supplier, the switch from the regular supply channel to the emergency one is assumed to be instant. Provided that the suppliers do not collapse at the same time, there is always an available active supplier to serve the retailer's demand. The end of the

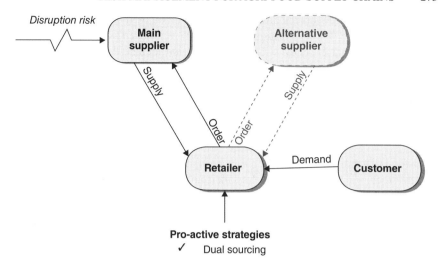

Figure 10.2 Agri-food supply chain diagram.

disruption period highlights the reassignment of the incoming orders to the regular supplier, while the emergency supplier is released upon the delivery of any outstanding orders that she is in charge of.

In this context, the food retailer has to decide on the following issues:

- Is an emergency (dual) sourcing strategy sustainable in terms of cost-efficiency in the long-run?

- What is the optimal emergency capacity level to be reserved when the (contract) premium charged by the supplier is fixed (in a game-theoretic approach, the supplier is the leader and the retailer is the follower) or alternatively what is the optimal premium to seek and negotiate for given that the retailer has already decided on the required emergency capacity level (in a game-theoretic approach, the retailer is the leader and the supplier is the follower)?

The alternative risk mitigation strategies examined herein in order to derive meaningful and practical managerial insights, are the following:

- *Single Sourcing Strategy (SS)*: The retailer undertakes no specific action to mitigate the effects of potential supply disruptions.

- *Emergency Sourcing Strategy (ES)*: The retailer reserves an additional emergency capacity (alternative supplier) to mitigate the effects of potential supply disruptions.

Additionally, the case where no disruption occurs and no risk mitigation acts are performed by the retailer is employed as the reference scenario (Basic Scenario – BS) of the analysis.

The retailer's decision-making process is carried out with respect to the expected inventory-related cost minimization criterion that is employed as the key performance metric. The relevant expected total costs per period include the standard inventory control cost elements, that is, ordering cost, holding cost, and backordering cost, while the premium (capacity reservation cost) is an additional cost element that ramp up total costs when an ES strategy is employed.

10.5.2   Discrete Event Simulation Model

Agri-food risk management problems encapsulate, by their very nature, the effects of several stochasticity sources, therefore their analytical solutions are, in most of the cases, mathematically intractable. Hence, simulation has proved to be a flexible, reliable, and practical operations research tool that is frequently used when addressing SCRM issues within the relevant decision-making processes (Schmitt and Singh, 2009, 2012). Specifically, in this chapter, a generic DES methodology is proposed in order to assess the value of ES contracts, as a risk mitigation strategy, in the agri-food SCRM field.

The proposed DES model emulates the primary supply chain operations, including the typical inventory control processes (e.g., inventory review, order placement, transportation lead time, order delivery, demand fulfillment, stock-outs, etc.), the risk factor (e.g., disruption events, duration of the disruption), and the adopted risk mitigation strategy (e.g., implementation of a single or emergency dual sourcing strategy, decision on the reserved capacity level, determination of the premium cost, etc.). Fine-tuning of the aforementioned model provides the capability to simulate and analyze several alternative scenarios in order to gain meaningful managerial insights.

The proposed DES algorithm and the corresponding modeling parameters are as follows. Consumers arrive at the rate of $\lambda > 0$ individuals per time unit according to a Poisson process, while the demand quantity per individual follows a Poisson distribution with an average of $\mu > 0$ product units. The food retailer reviews his or her inventory every R time units, according to a periodic review (s, S) inventory control policy, where s and S $(0 < s \leq S)$ correspond to the reorder and order-up-to points, respectively. The main supplier's processing capacity $C > 0$ (product units per time unit) is sufficient for serving the retailer's regular orders, which are further delivered according to a triangular distributed lead time with parameters $d > 0, e > 0, f > 0$.

On the other hand, the main supplier's operations are vulnerable to disruptions induced by diverse detrimental events which occur at a probability p per time period. This probability denotes the frequency of a disruption. Additionally, a disruption event renders the regular supplier unable to provide his or her services for a random triangular distributed time period with parameters $l > 0, m > 0, n > 0$. This duration stands as a measure of risk severity. In view of the circumstances and in order to guard against supply disruptions, the retailer reserves extra capacity from an emergency supplier, at a level that is expressed as the reserved capacity RC percentage of the regular capacity C. The emergency supplier charges the retailer a premium or reservation cost of r monetary units per capacity unit that is reserved.

The evaluation of the alternative ES and SS strategies, as well as the relevant decision-making process are performed with respect to the minimization of the expected total cost during the period of interest. This cost is the aggregate result of the following cost elements: the ordering cost k (monetary units per order), the inventory holding cost h (monetary units per time unit and product unit), and the backordering cost b (monetary units per time unit and product unit). The relevant nomenclature is provided in Table 10.2.

The proposed models were subject to a series of practical verification tests so as to ensure their functional adequacy, such as the model elements checking one-by-one, the tracking of the discrete events within each simulation run according to the relevant flow diagram followed by the examination and comparison of the results to the outcomes intuitively expected, and so on. On the other hand, due to the lack of a tangible real-world equivalent system and the absence of the relevant data, few more

Table 10.2 Variables and constants.

Category	Quantity	Type	Value	Unit
Disruption	Probability	Constant	$p = 10$	%
	Duration	Stochastic	Triangular ($a = 25$, $b = 30$, $c = 35$)	Days
Demand	Customer arrivals	Stochastic	Poisson ($\lambda = 10$)	Customers per day
	Order size	Stochastic	Poisson ($\mu = 10$)	Units per customer
Inventory control	Re-order point	Constant	$s = 450$	Units
	Up-to-order point	Constant	$S = 750$	Units
	Review period	Constant	$R = 1$	Days
Suppliers	Production capacity	Constant	$C = 400$	Units per day
	Production lead time	Stochastic	Triangular ($l = 1$, $m = 2$, $n = 3$)	Days
	Reserved capacity	Constant	RC	%
Inventory costs	Ordering cost	Constant	$o = 5$	€ per order
	Holding cost	Constant	$h = 2$	€ per unit and day
	Backordering cost	Constant	$b = 100$	€ per unit and day
	Cost ratio	Constant	$b/h = 50$	—
Contract costs	Premium cost	Constant	r	€ per capacity unit

empirical tests were employed within the models validation context, such as observing, checking, and judging the outcomes based on experience, and the evaluation of the alternative scenarios results on a comparative basis.

Arena™ Simulation Software was employed in developing the models. A complete run of each scenario included 1000 replications. The average simulation run time was 2 minutes on average using an Intel® Pentium® CPU 3.60 GHz. The time-length of a simulation replication corresponds to the time horizon of the analysis, that is, one year.

10.5.3   Numerical Example

A numerical example is provided below in order to illustrate the application of the proposed simulation methodology and provide a brief idea as regards the nature and the type of the managerial insights drawn through the corresponding "what-if" analysis. The overall problem setting input parameters, both constants and stochastic variables, are provided in Table 10.2.

The evaluation of SS and the alternative ES mitigation strategies focuses on the economic assessment of disruption's impacts on the expected system's inventory-related costs, while backorders' clearance time is employed as a secondary performance metric. The supply chain's operation is examined at an annual basis (time horizon), while the corresponding simulation scenarios employ the variables and parameters values presented in Table 10.2. The assumption of just one potential disruption event per year renders the transitional stage, that follows a disruption event, as the focal point of the analysis, and allows for the in depth study of this specific period. Moreover, it is further considered that the time epoch of a potential disruption incident is common among simulation runs, that is, Day 50, so that the involved scenarios could be fairly judged on a meaningful comparative analysis basis; the selection of the aforementioned time epoch allows the system to attain its steady-state operation prior to the disruption incident.

The dynamic nature of the expected total cost is illustrated in Figure 10.3 for both the SS and the ES strategies. All cost functions (curves) are provided in their cumulative form, as the averaging result over 1000 annual operational cycles (simulation replications). A first primary observation is that the slopes of the cost curves exhibit the same constant increase rate (steady-state operation) for the period up to Day 50 (disruption incident), as well as for the period that follows Day 100, that is the time period prior to disruption and post disruption, respectively, when the retailer reverts to the regular supplier sourcing. In the meanwhile, the disruption event at Day 50 induce a rapid increase in the expected total cost, that is graphically indicated by the sharp slope of the corresponding curve, mainly due to the ramp up of the back-ordering cost element. The system's stability is restored after Day 100 (end of transition period), roughly, with a time-to-recover (Simchi-Levi, 2012) of 50 days. Furthermore, a second critical observation is that the expected disruption consequences are substantially reduced when an ES strategy is implemented. Figure 10.3 captures this beneficial effect of the ES strategy on the expected cumulative total cost with respect to the associated reserved capacity level.

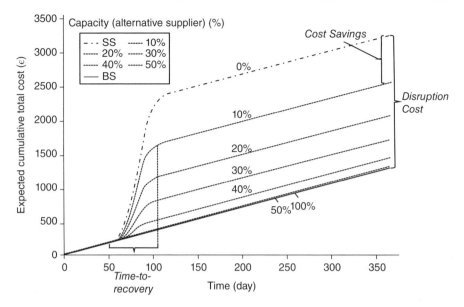

Figure 10.3   Impact of alternative emergency dual sourcing policies on total cumulative cost. Iakovou et al., 2014, figure 4, p. 258. Reproduced with permission of Emerald Group Publishing Ltd.

The upper curve or the highest expected total cost curve corresponds to the SS strategy, that is, $RC = 0\%$. In this case, the retailer suffers the most severe consequences of the associated supply disruptions since, at an earlier stage, he or she has decided on doing nothing to guard his or her sourcing and supply operations against disruptions. On the other hand, the lower curve or the lowest expected total cost curve represents the ES strategy, where the regular supplier is fully substituted by the emergency one, that is, $RC = 100\%$. In this case, the preceding decision of retaining a full "back-up" of the regular capacity safeguards the retailer's business continuity as the associated supply chain experiences no disruption. The expected total cost remains unchanged, disruption effects are counterbalanced, and the overall system behaves identically to the ideal case scenario (BS) of no disruptions. The numerous ES strategies, that is, those that correspond to the incremental levels of the reserved capacity values that fall within the 0–100% range, are represented by the set of intermediate curves.

Furthermore, useful insights could be obtained through the elaborate examination of Figure 10.3, regarding the disruption cost induced in the system, and the associated cost savings that express the economic benefits when an efficient risk mitigation strategy is proactively adopted. The "Disruption Cost" is defined as the expected total cost difference between the SS strategy and the BS, at the end of the period of interest, that is, the end of the year. Consequently, this cost is maximized when the retailer employs no hedging strategy against disruptions, and therefore it incorporates the additional cost that the retailer has to sustain.

In this context, disruption costs could be significantly reduced through the implementation of an efficient risk mitigation strategy. The corresponding cost benefit is considered as the system's "Cost Savings." An indicative example for the ES $RC =$ 10% scenario is illustrated in Figure 10.3. As a matter of fact, these cost savings are not fully retrieved due to the premium paid to the emergency supplier in exchange for the reserved "back-up" capacity; indeed, the retailer buys off these "cost savings" at a certain premium, taking into account the rational assumption that this premium should not exceed the aforementioned savings. According to this consideration, "cost savings" has the role of an upper bound when the retailer decides on the premium price that she could afford to buy the extra "back-up" capacity from the alternative supplier, and justify the adoption of the ES therein. The overall effective cost is determined as:

$$Effective\,Cost = Disruption\,Cost - Cost\,Savings + Premium.$$

Another interesting insight is derived through scrutiny of Figure 10.3. Specifically, the distance between two consecutive curves seems to decrease with the reserved capacity level, and as a result, the curves that correspond to higher reserved capacity values, that is, more than 50%, seem to practically coincide with the BS curve. This leads to the interesting insight, at least for the example under study, that the retailer could regain almost the overall disruption cost by reserving 50% of the regular supply capacity. Actually, a similar upper bound point for the capacity level (lower than 100%) always exists, irrespective of the specific system realization parameters, highlighting the maximum capacity to be reserved from the emergency supplier.

Figure 10.4 illustrates the system performance, when an ES strategy is employed, in terms of the expected backorders' clearance time and the time of recovery, with respect to the contracted capacity level. Specifically, the expected time period (in days) needed to serve any outstanding retailer's orders is presented on the vertical axis, while the time elapsed since the disruption event date is represented on the horizontal one. In the case of the SS strategy or equivalently when there is no emergency supply alternative, expected backorders' clearance time reaches its maximum, exceeding 6 days, while, on the other hand, clearance time is critically lower when an alternative emergency supplier has been proactively contracted. Moreover, Figure 10.4 further allows for the identification of the relevant time to recovery, that is, the time needed for the system to return to the regular response time to an order (negligible expected backorders' clearance time).

The results indicate a considerable decrease of the overall sourcing costs when efficiently applying an emergency dual sourcing strategy, through counterbalancing disruption costs and increasing the corresponding cost savings. These improvements seem to be feasible at a reasonable premium. Moreover, the upper limit proposed regarding the cost effectiveness of the sourcing operations could act both as a threshold when contracting a supplier in the long term, as well as a supplier selection criterion within the relevant decision-making process, or even a milestone for shifting to a lower cost risk mitigation strategy. The benefits of emergency dual sourcing goes further to the intuitively sound reduction of the expected backorders' clearance time, and a significant reduction of the time needed to recover in the aftermath of a disruption.

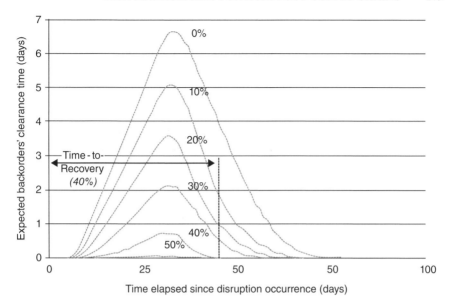

Figure 10.4   Impact of alternative emergency dual sourcing policy on response time and time to recovery. Iakovou et al., 2014, figure 5, p. 259. Reproduced with permission of Emerald Group Publishing Ltd.

10.6   Case 3: Managing Agri-food Supply Chain Disruption Risks – a Game-Theoretic Approach

Risk and vulnerability sources, drivers, outcomes, and impacts have a strong link and call for mitigating strategies, in order to face the challenges that stem from them. Consequently, by responding to risk, each independent "actor" attempts to optimize its individual objectives. But each decision taken affects the performance of the other parties in the supply chain. Furthermore, an interruption occurring at a supply chain partner may have impact on others and the influence may propagate across the chain (More and Babu, 2011). Thus, the necessity of action alignment and coordination emerge and a game that involves players, decisions, policies, and results commences. This complexity in interactions is enhanced when attempting to handle unforeseen events.

However, it can be successfully confronted by combining risk management with game theory: the study of situations of cooperation or conflict among heterogeneous actors, which is ideally suited to deal with this kind of interaction. In game theory, a game usually consists of three basic elements: the set of players, the strategy space, and the payoff functions (Fudenberg and Tirole, 1991). Whatever the kind of game, the pursuit of balance and stability of the supply chain is usually the ultimate objective; hence equilibrium is formed and studied. For example, Nash equilibrium is a profile of strategies such that each player's strategy is an optimal response to the other players' strategies. Stackelberg equilibrium is applied when there is an asymmetry in power or in moves of the players.

Although game theory is broadly known by its application to economic and political science problems, recently it has been applied extensively to various scientific disciplines, including that of SCRM. From our perspective, the supply chain partners could be seen as players in a game defined by multiple goals, constraints and conflicting objectives, sharing communication channels, resources, information, logistic networks, and customer demands, but also facing common risks. For the purpose of this book, we focus on discrete-time, random exogenous demand, single period (newsvendor problem) games. In particular, we focus on a stocking game with vertical competition, random exogenous demand, where a local food retailer decides on the quantity to purchase in order to replenish his or her stock, while a farmer sets the wholesale price to sell some of his or her stock.

10.6.1   Model and Problem Description

We apply game theory to a supply chain consisting of a farmer and a local food retailer with stochastic exogenous demand and risk sharing policies. We consider a single period inventory system where a single ordering decision is to be made before the sales period begins so as to maximize the expected total profit (classic newsvendor problem). If we take into account a farmer that independently maximizes his or her profit and thus impacts the retailer's optimal solution by choosing a wholesale price, the newsvendor problem is reduced to a game. In order to better portray the system, Table 10.3 provides a summary of the model's basic parameters.

The supply chain faces through the retailer a stochastic demand that is assumed to follow a specific pattern in time with a demand function of $D(X,t) = Xk(t)$, where X is a positive stochastic random variable with probability density function $f(\cdot)$ and cumulative distribution function $F(\cdot)$ and (t), $0 \leq t \leq T$ is a positive function of t with $\int_0^T k(t)dt = 1$, which determines the demand pattern in time. The retailer orders quantity Q depending on the wholesale price set by the supplier. We consider v as the unit purchase cost paid to the farmer and c as the unit production cost of the farmer. The

Table 10.3   Model parameters.

Market demand function faced by food retailer	$D(X,t) = Xk(t)$
Demand pattern in time	$k(t)$
Stochastic demand variable	X
Stochastic demand – probability density function	$f(\cdot)$
Stochastic demand – cumulative distribution function	$F(\cdot)$
Quantity ordered by food retailer	Q
Unit purchase cost paid to the farmer	v
Farmer's unit production cost	c
Unit selling price to market	P
Salvage value	g
Lost sales cost	B

unit-selling price is denoted by P and the surplus stock that remains unsold after the end of the period can be sold in a secondary market at a unit salvage value g. The salvage value is assumed to be less than v. In addition, B indicates the lost sales cost. It should be noted that any proof of propositions and equations that follow are depicted in detail in Papapanagiotou and Vlachos (2012).

10.6.2   Study of Game Interactions

To study the effect of interactions between the players of the game, we consider various different scenarios, altering cooperation, and disruption conditions.

10.6.2.1   Decentralized Solution without Disruption (DC,ND)

Considering a decentralized supply chain with no disruption and assuming that both players make their decisions simultaneously (Nash equilibrium, vertical competition), a game where each player maximizes his or her own profit without taking under consideration the decision of the other is formed. The expected profit of the retailer $\pi_0^r(Q)$ is given by:

$$\pi_0^r(Q) = \int_0^Q \left[Px - vQ + g(Q-x) \right] f(x) dx + \int_Q^\infty \left[PQ - vQ - B(x-Q) \right] f(x) dx \quad (10.4)$$

while the expected profit of the farmer $\pi_0^s(Q)$ is given by:

$$\pi_0^s(Q) = (v-c)Q \quad (10.5)$$

By applying the first-order optimality conditions for the retailer's problem (10.4), we have that the best retailer's response is determined by:

$$F(Q^*) = \frac{P+B-v}{P+B-g} \quad (10.6)$$

10.6.2.2   Centralized Solution without Disruption (C,ND)

In order to find a system-wide optimal solution to the game, we identify that the centralized problem is to maximize the sum of both profits, Equations 10.4 and 10.5. So, the corresponding centralized problem when no disruption occurs is:

$$\max \left\{ \pi_0^r(Q) + \pi_0^s(Q) \right\} \quad (10.7)$$

By applying the first-order optimality conditions with respect to Q, the system-wide optimal solution provides after simple manipulations the optimal order quantity Q^* that satisfies:

$$F(Q^*) = \frac{P+B-c}{P+B-g} \quad (10.8)$$

From Equations 10.7 and 10.8 we can conclude that since $v > c$, when there is vertical competition rather than a centralized solution, the retailer's order quantity is lower, and so is the customer service level. Furthermore, from Equation 10.6 we conclude that there is a maximum wholesale price v, $v_{max} = P + B$. Regarding the farmer, from his or her objective function (Equation 10.5) we conclude that he or she would set the wholesale price as high as possible under the Nash strategy, that is, $v = v_{max}$. In such a case, the retailer makes no profit and orders nothing. As a result, there is no business between the farmer and the retailer.

Similar analysis applies when the supply chain under examination faces a product delivery disruption from the farmer to the retailer during the trade-off period. We denote i the probability of this single delivery disruption. Moreover, we denote q the percentage of the total demand that could potentially be satisfied (if Q is adequate) because of a disruption. Thus, q captures the "severity" of the impact of the disruption. We distinguish two cases of delivery disruption, depending on the effect of the disruption:

- *"No Risk Sharing" Policy*: In the first case the risk is not shared among the partners of the supply chain and thus the disruption affects only the retailer.

- *"Risk Sharing" Policy*: In the second case the disruption risk is shared between the players and thus the disruption has an effect on both the farmer and the retailer.

It should be noted that in both cases the delivery disruption denotes the quantity delivered to the retailer without affecting in any way the crop yield of the farmer. Transportation or conservation failures are typical examples of such a disruption.

10.6.2.3   "No Risk Sharing" Policy – Decentralized Solution with Disruption (DC,D)

Under this policy the players agree that if a delivery disruption occurs, it affects only the retailer, by altering the retailer's profit as follows:

$$\pi_1^r(Q) = \int_0^{Q/q} \left[Pqx - vQ + g(Q - qx) - B(x - qx) \right] f(x) dx$$

$$+ \int_{Q/q}^{\infty} \left[PQ - vQ - B(x - Q) \right] f(x) dx \tag{10.9}$$

while the expected profit of the farmer remains

$$\pi_1^s(Q) = (v - c)Q \tag{10.10}$$

Should we consider a decentralized supply chain with disruption and assume that both players make their decisions simultaneously (Nash equilibrium, vertical competition), each player maximizes his or her own profit, without taking under consideration the decision of the other. By applying the first-order optimality conditions for the retailer's problem (10.9), we have that the best retailer's response is determined by:

$$F\left(\frac{Q^*}{q}\right) = \frac{P+B-v}{P+B-g} \qquad (10.11)$$

10.6.2.4 "No Risk Sharing" Policy – Centralized Solution with Disruption (C,D)

In order to find a system-wide optimal solution under this policy and type of game, we identify that the corresponding centralized problem is to maximize the sum of both profits (Equations 10.9 and 10.10):

$$\max\left\{\pi_1^r(Q)+\pi_1^s(Q)\right\} \qquad (10.12)$$

which after applying the first-order optimality conditions with respect to Q provides

$$F\left(\frac{Q^*}{q}\right) = \frac{P+B-c}{P+B-g} \qquad (10.13)$$

From Equations 10.11 and 10.13 we can conclude that since $v > c$, when there is vertical competition rather than a centralized solution, the retailer's order quantity under disruption is lower, and so is the customer service level.

10.6.2.5 "No Risk Sharing" Policy – Centralized Solution with Disruption Probability i (C,Di)

When the sole disruption occurs with a probability i, the expected total profit of the centralized problem is:

$$\pi_{d1}^t(Q) = (1-i)\left[\pi_0^r(Q)+\pi_0^s(Q)\right] + i\left[\pi_1^r(Q)+\pi_1^s(Q)\right] \qquad (10.14)$$

Applying the first-order conditions with respect to Q, we obtain the optimal order quantity Q_d^*:

$$(1-i)F\left(Q_d^*\right) + iF\left(\frac{Q_d^*}{q}\right) = \frac{P+B-c}{P+B-g} \qquad (10.15)$$

10.6.2.6   "Risk Sharing" Policy – Centralized Solution with Disruption

Under this policy, the players agree that in the case of a delivery disruption, the retailer will get a discount from the farmer, expressed by a factor a, with $a < 1$, meaning that the retailer will buy the units sold by the farmer at a wholesale price of av rather than v.

The expected profit of the retailer now turns into:

$$\pi_2^r(Q) = \int_0^{Q/q} \left[Pqx - avQ + g(Q - qx) - B(x - qx) \right] f(x)\,dx$$
$$+ \int_{Q/q}^{\infty} \left[PQ - avQ - B(x - Q) \right] f(x)\,dx \tag{10.16}$$

while the expected profit of the farmer is given by:

$$\pi_2^S(Q) = (av - c)Q \tag{10.17}$$

The corresponding centralized problem formulates as:

$$\max \left\{ \pi_2^r(Q) + \pi_2^S(Q) \right\} \tag{10.18}$$

which after applying the first-order optimality conditions with respect to Q provides:

$$F\left(\frac{Q^*}{q} \right) = \frac{P + B - c}{P + B - g} \tag{10.19}$$

10.6.2.7   "Risk Sharing" Policy – Decentralized Solution with Disruption

We consider now a decentralized supply chain with delivery disruption and assume that both players make their decisions simultaneously (Nash equilibrium, vertical competition), each player maximizes his or her own profit, without taking under consideration the decision of the other. By applying the first-order optimality conditions for the retailer's problem (10.16), the best retailer's response is determined by:

$$F\left(\frac{Q^*}{q} \right) = \frac{P + B - av}{P + B - g} \tag{10.20}$$

It is easy to derive that a "risk sharing" game that faces delivery disruption and shares the disruption risk through a farmer discount expressed by a factor a, provides a centralized solution with better trade-off conditions than vertical competition when:

$$1 > a > c / v \tag{10.21}$$

The "risk sharing" game that satisfies Equation 10.21 proposes a centralized solution – as opposed to the vertical competition – that coordinates the supply chain, raises custumer service level, shares the risk of delivery disruption, and provides profitable conditions for the whole trade-off.

10.6.3   Numerical Example

In order not only to validate the model's propositions, but also to visualize and explain its managerial implications, we employ MathCAD™ software. We consider the values shown in Table 10.4 for P, v, g, B, q, c, and i and assume that probability density function $f(x)$ equals that of a normal distribution with values of mean and standard deviation as shown in the same table.

Under these assumptions, Figure 10.5 can represent the profit functions that satisfy Equation 10.3 [Decentralized, No Disruption (DC,ND)], Equation 10.5 [Centralized, No Disruption (C,ND)], Equation 10.8 [Decentralized, Disruption (DC,D)], Equation 10.10 [Centralized, Disruption (C,D)], and Equation 10.15 [Centralized, Disruption with probability i(C,Di)], in relation to the order quantity.

It is straightforward that the profit functions maximize at the optimal quantity and they vary depending on the quantity and game theory policy. Also, we observe that both profit and order quantity are higher in the coordinated supply chain than in the case of vertical competition between the farmer and the retailer, even when we make a comparison between a non-disrupted decentralized supply chain and a disrupted centralized supply chain. Additional findings can be extracted from the numerical example, as shown in Table 10.5. The supply chain maximizes its profit and optimal order quantity when the players are in a centralized game with no disruptions, whereas a significant drop in value of these variables is observed in each other alternative. The optimal order quantity decreases from 1.68 to 24.44%, while the respective profit decreases from 3.22 to 54.02%.

Table 10.4   Parameter values and assumptions (numerical example).

Parameter	Value
P	US$45
v	US$24
g	US$10
B	US$15
q	0.9
$f(x)$ mean deviation	400
$f(x)$ standard deviation	130
c	US$15
i	0.2
$1-i$	0.8

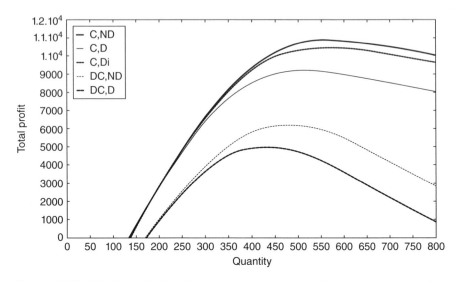

Figure 10.5   Total profit functions with respect to order quantity (numerical example).

Table 10.5   Optimal quantity (Q^*) and maximum (max) values of total profit functions [π^t (Q)].

Supply chain game	Q^*	π^t (Q)max	Q^* change (%)	π^t (Q) change (%)
Centralized, No Disruption (C,ND)	566.82	10 870	0	0
Centralized, Disruption (C,D)	510.14	9 182	−10.00	−15.53
Centralized, Disruption with probability i (C,Di)	557.28	10 520	−1.68	−3.22
Decentralized, No Disruption (DC,ND)	475.88	6 220	−16.04	−42.78
Decentralized, Disruption (DC,D)	428.29	4 998	−24.44	−54.02

This model proves that vertical competition decreases the efficiency of the supply chain and the greater the wholesale price, the lower the quantity that the retailer orders. Furthermore, in case of potential disruptions during the trade, the model shows that it is necessary to follow certain rules, in order for the whole supply chain to gain from the exchange. Obviously, the selection of game policy largely affects the efficiency, profit, and response of the supply chain and at the same time it influences the managerial decisions regarding cooperation, risk mitigating strategies, and marketing policies. For example, a "risk sharing" game that faces delivery disruption and shares the disruption risk through a farmer discount, assuming the same values of Table 10.5, provides a centralized solution with better trade-off conditions than vertical competition if the retailer could get a discount greater than 62.5%. Common practice also shows that large discounts could cause a risk-averse food retailer to proceed in contract with a farmer not hesitant to risk. In our case, a value of a near c/v provides the retailer with a "safety net" to buy at a price near the cost c of the farmer, in the case of a delivery disruption.

10.7   Conclusions

Modern agri-food supply chains experience a wide variety of natural, environmental, technological, operational, institutional, and financial risks, such as natural disasters, unfavorable weather conditions, biological incidents, market instability, logistical and infrastructural disruptions, public policy interventions, institutional reforms, and so on (Jaffee, Siegel, and Andrews, 2010). These risks may inhibit normal operations of agri-food supply chains and could provoke deviations, disruptions, or shutdowns to the supply chain's fundamental products, processes, information, and financial flows. To a further extent, they may have a dramatic impact on cost, efficiency, and reliability of the included activities and operations. There is not a unique universal management strategy in order to cope with the diverse risk types that threaten the continuity of an agri-food supply chain. Risk type and risk severity, personal characteristics of and contrasting interests among supply chain partners, along with specific contemporaneous natural, institutional, and economic conditions, shape the relevant agri-food risk management landscape and drive the selection of the appropriate form of risk governance in an "ad hoc" manner.

In this context, three representative cases were analyzed in order to provide critical guidelines regarding risk management in agri-food supply chains. Specifically, based on a modified single-period newsvendor modeling approach, the first case addressed the issue of perishability risks within a revenue management context, in terms of the corresponding network design and the involved order quantities, taking also into account customer preferences for fresh products. Secondly, a simulation-based approach was employed in determining the optimal risk mitigation strategy, that is, single or dual sourcing, for an agri-food manufacturer, food processor, wholesaler, or retailer, as well as the optimal contracted reserved capacity of the emergency (dual) supplier, in terms of the relevant cost-to-recovery, or equivalently the optimal contract premium to be paid to the emergency supplier. Finally, the third case employed game theory and a single-period formulation in a stocking game with vertical competition, including the risk sharing alternative, where a local food retailer decides on the quantity to purchase in order to replenish his or her stock, while a farmer sets the wholesale price to sell some of his or her stock level.

Well-established proactive risk management strategies documented through robust quantitative tools could prove to be beneficial in real-world risk management situations. They could moderate or even eliminate stock-outs, protect a retailer's service level or a company's market share, increase the economic efficiency of the implemented policies, reduce the costs to recover from a disruption, increase the company's or the supply chain's competitiveness, protect a company's brand name against the loss-of-goodwill, and so on. Indicative high-value managerial insights drawn from the three selected cases that are herein analyzed are presented below. According to the analysis of the first case, an agri-food supply chain could be both effectively and efficiently designed in order to handle perishable products taking into account consumer preferences. A proper mix of regular and emergency suppliers allows the company to hedge its business against stock-outs, and handle risk in a profitable manner. According to the analysis of the second case, it seems that the emergency supply mode within a dual sourcing context could lead to a considerable decrease of the overall sourcing costs, at a reasonable premium, through the increase of the corresponding cost savings. Moreover, there is a premium threshold that a company should consider either as an upper price limit when negotiating with back-up suppliers or as a milestone when deciding on shifting from the single to a dual sourcing strategy. Furthermore, the relevant benefits go further to the significant reduction of the expected backorders' clearance time, as well as of the time needed to recover in the aftermath of a disruption. The analysis of the third case highlights that vertical competition decreases the efficiency of the supply chain, while in the case of potential disruptions during the trade, it seems that it is necessary for the involved partners to follow certain rules, in order for the whole supply chain to gain from the exchange. Of course, the selection of game policy largely affects the efficiency, profit, and response of the supply chain, and at the same time it influences the managerial decisions regarding cooperation, risk mitigating strategies and marketing policies. Finally, common practice also shows that large discounts could cause a risk-averse food retailer to proceed in contract with a farmer not hesitant to risk.

The nature of the overall decision-making process is twofold, definitely stochastic and purely dynamic, as it unfolds in real time within an uncertain environment that changes continuously, providing new challenges, and opportunities. Consequently, the decisions along with the associated implemented strategies should be continuously evaluated and reconsidered in order to ensure the entire long-term agri-food supply chain efficiency and sustainability. In this context, a wide list of well-known classic mathematical and operational research methods have been employed to address the issue of agri-food SCRM. Taking into account the very nature of risk, stochastic techniques seem to be more effective in modeling the relevant real-world problems, and thus more acceptable among academicians and practitioners.

In light of the new quantitative analysis advances, agri-food SCRM in the near future should focus on effectively analyzing and managing the rising climate change risks that threaten agriculture and global agri-food networks. In parallel, traditional agribusiness risks such as price volatility and crop yield uncertainties should be more elaborately addressed, in terms of forecasting precision and proactive risk management. Simulation techniques, as well as game theory, could play a pivotal role in addressing the relevant real-world problems. Given the intrinsic presence of the "random" factor as regards risks, along with the complexity of agri-food chains and the contrasting interests among partners, the aforementioned techniques seem to be the basis for future analytical steps.

Acknowledgments

This research has received funding from the European Union's Seventh Framework Programme (FP7-REGPOT-2012-2013-1) under Grant Agreement No. 316167, Project Acronym: GREEN-AgriChains, Project Full Title: "Innovation Capacity Building by Strengthening Expertise and Research in the Design, Planning, and Operations of Green Agri-food Supply Chains," Project Duration: 2012–2016. This chapter reflects only the authors' views; the European Union is not liable for any use that may be made of the information contained herein.

References

Ahumada, O. and Villalobos, J.R., 2009. Application of planning models in the agri-food supply chain: a review. *European Journal of Operational Research*, **196**, 1–20.

Alocilja, E.C. and Ritchie, J.T., 1990. The application of SIMOPT2: RICE to evaluate profit and yield-risk in upland-rice production. *Agricultural Systems*, **33**, 315–326.

Bachev, H., 2012. Risk management in the agri-food sector. *Contemporary Economics*, **7**, 45–62.

Backus, G., Eidman, V., and Dijkhuizen, A., 1997. Farm decision making under risk and uncertainty. *NJAS Wageningen Journal of Life Sciences*, **45**, 307–328.

Banerjee, P.K. and Turner, T.R., 2012. A flexible model for the pricing of perishable assets. *Omega*, **40**, 533–544.

Beske, P., Land, A., and Seuring, S., 2014. Sustainable supply chain management practices and dynamic capabilities in the food industry: a critical analysis of the literature. *International Journal of Production Economics*, **152**, 131–143.

Bogataj, D. and Bogataj, M., 2007. Measuring the supply chain risk and vulnerability in frequency space. *International Journal of Production Economics*, **108**, 291–301.

Cheng, S.K. and Kam, B.H., 2008. A conceptual framework for analyzing risk in supply networks. *Journal of Enterprise Information Management*, **21**, 345–360.

Chopra, S., Reinhardt, G., and Mohan, U., 2007. The importance of decoupling recurrent and disruption risks in a supply chain. *Naval Research Logistics*, **54**(5), 544–555.

Christopher, M., 2005. Managing risk in the supply chain. In: M. Christopher, ed. *Logistics and Supply Chain Management: Creating Value-Adding Networks*. Harlow: Financial Times – Prentice Hall, Pearson Education, pp. 231–258.

Ellegaard, C., 2008. Supply risk management in a small company perspective. *Supply Chain Management: An International Journal*, **13**, 425–434.

Ellis, S.C., Henry, R.M., and Shockley, J., 2010. Buyer perceptions of supply disruption risk: a behavioral view and empirical assessment. *Journal of Operations Management*, **28**, 34–46.

Fearne, A., Hornibrook, S., and Dedman, S., 2001. The management of perceived risk in the food supply chain: a comparative study of retailer-led beef quality assurance schemes in

Germany and Italy. *The International Food and Agribusiness Management Review*, **4**, 19–36.

FedEx, 2008. FedEx Trade Networks, FedEx Corporation, http://ftn.fedex.com/news/NewsBulletinDisplay.jsp?url=010308b&lang=en (accessed August 1, 2010).

Ferguson, M. and Ketzenberg, M.E., 2006. Information sharing to improve retail product freshness of perishables. *Production and Operations Management*, **15**, 57–73.

Fudenberg, D. and Tirole, J., 1991. *Game Theory*. London: MIT Press.

Glendon, L. and Bird, L., 2013. *Supply Chain Resilience 2013: An International Survey to Consider the Origin, Causes and Consequences of Supply Chain Disruption*. Caversham: Business Continuity Institute, http://www.zurich.com/internet/main/SiteCollectionDocuments/insight/supply-chain-resilience-2013-en.pdf (accessed September 10, 2014).

Harland, C., Brenchley, R., and Walker, H., 2003. Risk in supply networks. *Journal of Purchasing and Supply Management*, **9**, 51–62.

Heidelbach, O., 2007. Efficiency of selected risk management instruments: an empirical analysis of risk reduction in Kazakhstani crop production. *Studies on the Agricultural and Food Sector in Central and Eastern Europe*, **40**, 1–174.

Iakovou, E., Vlachos, D., Keramydas C., and Partch D., 2014. Dual sourcing for mitigating humanitarian supply chain disruptions. *Journal of Humanitarian Logistics and Supply Chain Management*, **4**, 246–264.

Iakovou, E., Vlachos, D., and Xanthopoulos, A., 2007. An analytical methodological framework for the optimal design of resilient supply chains. *International Journal of Logistics Economics and Globalization*, **1**, 1–20.

Iakovou, E., Vlachos, D., and Xanthopoulos, A., 2010. A stochastic inventory management model for a dual-sourcing supply chain with disruptions. *International Journal of Systems Science*, **41**, 315–324.

Jaffee, S., Siegel, P., and Andrews, C., 2010. Rapid Agricultural Supply Chain Risk Assessment: *A Conceptual Framework*. Agriculture and Rural Development Discussion, Paper 47. Washington, DC: The International Bank for Reconstruction and Development/The World Bank.

Juttner, U., Peck, H., and Christopher, M., 2003. Supply chain risk management: outlining an agenda for future research. *International Journal of Logistics Research and Applications*, **6**, 197–210.

Kamata, I., 2011. The Great East Japan Earthquake: A View on Its Implication for Japan's Economy, La Follette Policy Report 21(1), Robert M. La Follette School of Public Affairs, University of Wisconsin–Madison, pp. 12–16.

Keramydas, C., Tsiolias, D., Vlachos, D., and Iakovou, E., 2015. A simulation methodology for evaluating emergency sourcing strategies of a discrete part manufacturer. *International Journal of Data Analysis Techniques and Strategies*, **7**, 141–155.

Khan, O. and Burnes, B., 2007. Risk and supply chain management: creating a research agenda. *International Journal of Logistics Management*, **18**, 197–216.

Klibi, W., Martel, A., and Guitouni, A., 2010. The design of robust value-creating supply chain networks: a critical review. *European Journal of Operational Research*, **203**, 283–293.

Matsuo, H., 2014. Implications of the Tohoku earthquake for Toyota's coordination mechanism: supply chain disruption of automotive semiconductors. *International Journal of Production Economics*, **161**, 217–227.

More, D.S. and Babu, A.S., 2011. Supply chain flexibility: a risk management approach. *International Journal of Business Innovation and Research*, **5**, 255–279.

Oke, A. and Gopalakrishnan, M., 2009. Managing disruptions in supply chains: a case study of a retail supply chain. *International Journal of Production Economics*, **118**, 168–174.

Pai, R.R., Kallepalli, V.R., Caudill, R.J., and Zhou, M.C., 2003. Methods toward supply chain risk analysis. Conference on Systems, Man and Cybernetics, IEEE (Institute of Electrical and Electronics Engineers), Washington, DC, October 5–8, 2003, vol. **5**, pp. 4560–4565.

Papapanagiotou, K. and Vlachos, D., 2012. Managing supply chain disruption risks: a game-theoretical approach. *International Journal of Mathematics in Operational Research*, **4**, 515–530.

Peck, H., 2005. Drivers of supply chain vulnerability: an integrated framework. *International Journal of Physical Distribution and Logistics Management*, **35**, 210–232.

Rice, J. and Caniato, F., 2003. Building a secure and resilient supply network. *Supply Chain Management Review*, **7**, 22–33.

Schaffnit-Chatterjee, C., 2010. *Risk Management in Agriculture: Towards Market Solutions in the EU*. Frankfurt: Deutsche Bank.

Schmitt, A.J. and Singh, M. (2009) Quantifying supply chain disruption risk using Monte Carlo and discrete-event simulation. 2009 Winter Simulation Conference, Austin, TX, December 13–16, 2009, pp. 1237–1248.

Schmitt, A. J., and Singh. M., 2012. A quantitative analysis of disruption risk in a multi-echelon supply chain. *International Journal of Production Economics*, **139**, 22–32.

Sheffi, Y., Rice, Jr, J.B., Caniato, F., Fleck, J., Disraelly, D., Lowtan, D., Lensig, R., and Pickett, C., 2003. *Supply Chain Response to Terrorism: Creating Resilient and Secure Supply Chains, Supply Chain Response to Terrorism Project – Interim Report*. Cambridge, MA: MIT Center for Transportation and Logistics.

Shen, X., Liu, Q., and Zhao, D., 2013. Application of neural networks model to assess agricultural products safety risks. *Journal of Applied Sciences*, **13**, 3049–3054.

Simchi-Levi, D., 2012. SCDIGEST VideoCast: Dr David Simchi-Levi`s Risk Exposure Index 3.0. http://www.supplychainvideocasts.com/videocasts/risk-exposure-index-3.0-the-ford-story/login.php (accessed September 10, 2014).

Simchi-Levi, D., Kyratzoglou, I.M., and Vassiliadis, C. G., 2013. *Supply Chain and Risk Management: Making the Right Decisions to Strengthen Operations Performance*. London: PwC and MIT Forum for Supply Chain Innovation.

Snyder, L.V., Scaparra, M.P., Daskin, M.S., and Church, R.L., 2006. Planning for disruptions in supply chain networks. *Tutorials in Operations Research*. Baltimore, MD: INFORMS, pp. 234–257.

Tajbakhsh, M.M., Lee, C., and Zolfaghari, S., 2011. An inventory model with random discount offerings. *Omega*, **39**, 710–718.

Tsolakis, N., Keramydas, C., Toka, A., Aidonis, D., and E. Iakovou, 2014. Agrifood supply chain management: a comprehensive hierarchical decision-making framework and a critical taxonomy. *Biosystems Engineering*, **120**, 47–64.

Vlachos, D., Iakovou, E., Papapanagiotou, K., and Partsch, D., 2012. Building robust supply chains by reducing vulnerability and improving resilience. *International Journal of Agile Systems and Management*, **5**, 59–81.

Vlachos, D., Tsolakis, N., Keramydas, C., and Iakovou, E., 2013. Design of agrifood supply chains with dual sourcing and emergency replenishments. Commission Internationale de l'Organisation Scientifique du Travail en Agriculture (CIOSTA), XXXV Conference on Intelligent Agriculture and Forestry, Billund, Denmark, July 3–5, 2013.

Wilding, R., 2013. Supply chain temple of resilience. *Logistics and Transport Focus*, **15**, 54–59.

Yang, Z., Aydın, G., Babich, V., and Beil, D.R., 2009. Supply disruptions, asymmetric information, and a backup production option. *Management Science*, **55**, 192–209.

Zsidisin, G.A. and Ritchie, B., 2008. *Supply Chain Risk: A Handbook of Assessment, Management and Performance*. New York: Springer Science + Business Media, LLC.

11

Regulatory Policies and Trends

Sirpa Kurppa

Natural Resources Institute Finland, Myllytie 1, Jokioinen, Finland

11.1   Regulations as Tools

In the previous chapters, different phenomena, requirements, and essential components of sustainable food supply chains have been discussed. In all of these contexts, certain regulations set limits, constrain rights, create or limit duties, or allocate responsibilities for different stakeholders of particular supply chains or networks. Practical expressions of regulations that can be faced are controlled market entries, prices that might include certain taxes, predetermined wages, predefined payments, controlled rewards, predetermined subsidies, approvals for specific development actions, controlled limits for resource use or pollutant releases, or obligation for pollution impacts, standards of production for certain goods and services, or employment for certain experts in particular actions and industries (Aruoma, 2006). The economics of imposing or removing regulations related to single stakeholders, supply chains, and networks is strongly linked with various trade-offs in the markets.

Supply Chain Management for Sustainable Food Networks, First Edition. Edited by Eleftherios Iakovou,
Dionysis Bochtis, Dimitrios Vlachos and Dimitrios Aidonis.
© 2016 John Wiley & Sons, Ltd. Published 2016 by John Wiley & Sons, Ltd.

Expressed forms of regulations are based on and rely on specific justifications: legal restrictions authorized by a government, contractual obligations between collaborating parties, contractual obligations between food chain stakeholders and shareholders of the particular industry, between industry and insurance or a funding body or as self-regulation by the industry itself. Self-regulation can be managed with the help of a trade association, following social norms, co-regulation between partners of supply chains (business to business), third-party regulation, certification or accreditation processes, or market regulation. A well-established approach is Corporate Responsibility that has been defined at three levels: (i) individual responsibility; (ii) corporate responsibility; and (iii) the responsibility of the societal system as a whole (Murphy, Öberseder, and Laczniak, 2013). At the level of individual responsibility, the focus is on the values by which self-interest is balanced in terms of fairness and the good of others, both within and outside the company. At the level of corporate responsibility, the focus is on the acknowledged or unacknowledged obligations that every company has as it drives for its economic objectives. At the level of the societal system, business ethic faces the pattern of cultural, political, and economic driving forces and values that define democratic capitalism in a global environment. Corporate social responsibility, with its regulations, is a one-way approach. Recently, co-creation processes between a supply chain and consumer groups have been gradually introducing mutually agreed practices, regulations, and norms between business and consumers (business to consumer). This is an interactive and inclusive process, either business to business or business to consumer.

Supporting justified policy regulations is a policy approach, or when concerning sustainability, statements from different policies. When emphasizing sustainability, three or four different dimensions should be included: economic, environmental, social, and cultural. Normally each of these has been, unfortunately, approached and managed separately by different institutions within a society. Up to now, economic sustainability has led, and environmental and social sustainability followed as subordinate through their specific paths. Mostly development of social and environmental sustainability has been seen as controversial and conflicting with economic success. Cultural sustainability is, often, the last to get attention. Regionally or locally, an adapted balance between each dimension of sustainability would be the optimal basis for a regulatory process, but this is an optimistic goal.

A mainstream activity for implementing self-regulation for responsibility in an area of environmental or more extensive overall sustainability is sustainability reporting pursued by bigger companies and global corporations (Hess, 2014). Several European countries mandate companies and corporations to produce a sustainability report, preferably following standards such as of the Global Reporting Initiative. The Initiative practically defines the types of information that corporations should publicly disclose. However, sustainability reports are not working as an effective policy mechanism unless they are placed in a context of governance that can effectively exploit the information and motivates or obliges corporations to change their strategies and practices in accordance with modern requirements and value change.

11.2   Environmental Externalities and Savings as Drivers of Regulations

The economy has become a driver of the food supply chain, and with globalization economic connectedness through supply chains has further strengthened. This has further enhanced the decisive role of the economy. Food supply chains are highly regulated and introduction of new regulations always reorients the relations of stakeholder groups. These issues are always discussed by an economic terminology, as questions of fairness of allocation of costs or sharing of benefits through production chains or networks. Economic benefit normally peaks at the top of the material hierarchy at the interface between trading company and consumer. Ecological values peak at the interface between basic production and nature. This is a dichotomy that has affected regulative approaches by differentiating regulative actions at an ecological level totally from those at an economic level.

Supply chain connectedness, in terms of environmental measures and drivers, has grown during the last 10–15 years. During this time, the life cycle assessment (LCA) approach has been developed and become established. Even now, supply chain environmental connectedness is, however, not yet adequately facilitated. LCA methods for gauging climate change (carbon footprint) and eutrophication impacts are available. From resource use impacts, methods for water use (water footprint), land use, and energy use impact measurements have become available. Assessment methods for ecotoxic (and human toxic) impacts are also under development. Methods for assessing nutrient use efficacy (nutrient footprint) are also on the way. Biodiversity is regarded as an important but very complex environmental phenomenon, indeed, being a driver of many other secondary impacts in the environment. Methods for assessing chain-based impacts of biodiversity have been very difficult to develop. Impact on landscapes can be assessed qualitatively or semi-quantitatively. A number of these impacts from a food supply chain have been addressed in Figure 11.1.

The described impacts of a production chain are termed externalities to the production environment. Externality is typically used in regulatory speech and refers to a cost or benefit that affects a third party who did not choose to incur that cost or benefit. Normally cost and benefit to society are defined as the sum of the imputed monetary value of benefits and costs to all parties involved. If this process is ignored and unregulated, markets in goods or services with significant externalities generate prices that do not reflect the full social cost or benefit of their transactions. Therefore it is a typical process that, in societies, is targeted by regulations in order correct for obvious unfairness.

Environmental impacts are typical externalities that appear in the form of climate change impacts, eutrophication impacts, or ecotoxic impacts. These externalities normally appear in the form of spoilt or destroyed areas of the environment for which restoration costs remain to be paid by society. In practice, these impacts are diminished by application of a set of regulations directed at farmers and other food chain stakeholders and waste managers. A "pollution pays" principle is,

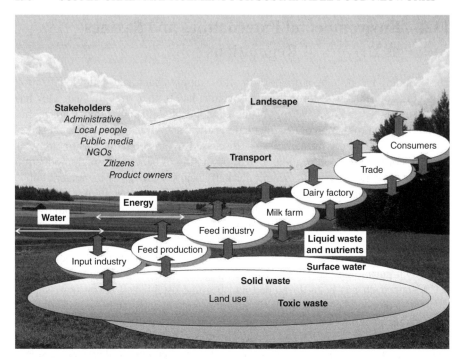

Figure 11.1   Description of some of the linkages between the food supply chain and the environment. NGOs, non-governmental organizations.

however, not normally applied to farming. Thus regulations do not normally push the food business to transfer external costs to consumers through increased prices. Food system externalities remain to a large extent net losses or societal costs. This is an area where discussion has been frequently raised, normally arguing that the current regulatory process is not covering all necessary areas or that the process is not efficient enough.

Environmental management is, however, approaching a new era. Many of the environmental impact measures assess resource efficiency. These include nutrient footprint, (blue) water footprint, measures of energy use and land use, and also emergy analysis, which describes the use efficacy of sun-based energy as a resource measure. These measures are not linked to such externalities as defined in Figure 11.2. The externality-based impact measures, such as carbon footprint, might gradually be compensated for by use of energy efficiency (or energy assessment) nutrient footprint and land use, especially if land use is complemented by carbon balance. Resource efficiency requirement is correlated with saving of expenses. Saving resources and money could be expected to be a strong extra motivation for enhancing environmental processes. Quite an obvious future trend is a preference for resource efficacy indicators. Thus, gradually more regulations will likely be launched to restrict use of resources that simultaneously confer savings to the targeted stakeholder and reduce extra costs to the consumer or society.

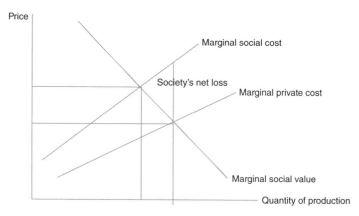

Figure 11.2   Impact of negative externalities on societal net loss. (You can learn about his at http://welkerswikinomics.wikifoundry.com/page/Externalities)

11.3   Diversity as a Driver for Informal Regulations and Trends

Setting up requirements and launching regulations within the framework of food systems should be based on principles of food and nutrition security anywhere in the world. Needs for energy, proteins, minerals, vitamins, and other vital and health benefitting minor components should be met at any time.

At the global level, diets have been continuously unified and simplified in terms of components. Driving forces leading to this tendency are to be found in the global development of regulatory systems. These are governed by big multinational companies, and simplifying regulatory processes has been seen as a pathway toward efficiency in the global supply chain. In terms of sustainability, this is not necessarily beneficial for nutritional security or ecological sustainability. In addition, local and regional cultural values are easily ignored when diets are unified and globalized. This change is very strong at present in emerging societies of Asian and Africa. There a typical plant-based diet is changing quickly to a mixed diet and further to an animal-based diet. The quick change is problematic in terms of the phenomenon of epigenetics. Living in a certain type of food culture has modified the reaction of the human body to the particular typical well-adapted and appropriate diet. This adaptation has been shown to be effective up to the second generation, and therefore quick changes cause functional imbalance and have been shown to lead to an obesity problem, for instance. In two or three generations, society reaches again a state of better adaptation. Even though major problems are caused, the situations are very difficult if not impossible to target by any regulatory process. The only potential lies in regulations to build awareness among food chain stakeholders and consumers.

Unifying and simplifying food ingredient consumption is not necessarily beneficial for ecological sustainability because it is a driver to simplified and less diverse production systems. Already in the Millennium Assessment a profitable scenario was seen to include the adaptive mosaic, which emphasized adaptive means to utilize local and regional resources as sustainably as possible. The results of ignoring of this principle can be seen in a number of examples of monoculture in many parts of the world. These appear in the form of eroded, salt-filled, nutrition-enriched, or chemically polluted land areas. Partially as a result of that, for example, 50% of the fertile growing area of Mediterranean countries is expected to be lost between 1960 and 2020. The problems are case specific and often result from ignoring regulations introduced to avoid such problems. In organic agriculture, global normative regulations prohibit the detrimental impact of monoculture, and as a result there is convincing and widespread evidence that in terms of biodiversity organic agriculture surpasses conventional. Cultivation method will become a potential target of official and informal regulation in striving for food sustainability.

A very challenging aim for regulation of food supply is that the demand chain should be integrated into the diversity requirement for a food consumption system and a food shed (Berger et al., 2013). A food shed refers to a production area from where food ingredients for a certain local population are generated. Linked to physical region or locality this approach equates with a local food system. But theoretically the two integrated areas may be distant from each other but nonetheless be linked through historical, cultural, or socioeconomic interconnections. The basis of this kind of connectedness can be built on collaborative and co-creative processes. This could be a strategy adopted by a location, village or municipality, or even a larger area. How all-inclusive the collaboration can be in meeting food demand depends on climate, diversity, and fertility of the production area, and scale and diversity of requirements of the particular group of consumers (Remans et al., 2014). At present social media supports people to build up collaborative production and consumption "islands," agreeing with their own normative regulations for supply–demand chain practices. These cases have appeared when people have to live in a neighborhood where they have not been able to fulfill their culture-based, ethical, safety or other requirements of their food. This encourages them to build an informal external supply chain. Sometimes a group of people spontaneously build such a network. These types of separate collaborative supply–demand islands can become a trend similarly as for local food systems. Both are potential platforms for a diverse set of informal regulatory processes in the future.

11.4   Nutrition and Environmental Issues Regulated at the Food Plate Level

Most consumers receive a food portion on their plate every day at least once. A food portion is the potential crystallization of a number of regulations on dietary issues, food safety, and nutrition. Regulated food portions in public catering are an efficient way to implement national dietary recommendations. A food portion is also a rational target of many other dimensions of sustainability to be presented in the form of recommendations or regulations. Subsidized pricing of a portion is an effective way to motivate people to consume a proper lunch during a working day and simultaneously take care of adequate equity in food and nutrition security at a national level. Socio-cultural linkages to food portions can also be guarded. This happens often through cultural promotions and by providing services that follow specific regulations regarding religious, ethical, or even mythical issues, and so on. Regulations come into play when equal rights of different cultural groups express their values or visualize and communicate their culture through food.

The most common additional dimension linked to the food portion is, however, the environmental impact value. LCA has provided a method to confer life-cycle-based environmental impacts in a food portion context. The number of ingredients in one food portion may be high, but if the impacts of 90% of those are assessed, the figure is sufficiently complete. There are numerous ways to communicate environmental impacts for a food portion to consumers. A demonstrative group of plates of a low carbon footprint can be designed and served as an alternative climate-safe food that contrasts with conventional offerings. Actual results of carbon footprints, eutrophication or other indicators per portion may be presented to people on Web pages, by mobile phone systems or by labeling the actual food portions (see indicatively Figure 11.3). Gradually these procedures may lead to informal regulations developed through co-creational actions between food service providers and consumers. These regulations will be expressions of shared responsibility on management of environmental sustainability in a future food system.

11.5   Citizens–Consumers Facing Regulations at the Market

In management of a food supply chain, the role of food consumers has recently been raised. In a market-oriented system food consumers are the final arbiters when decisions are made about what is to be produced and sometimes, also, how food is to be produced. There is, however, a strong inconsistency in human behavior when performing as a citizen or a consumer. As a citizen people are ready to require altruistic moves and actions in markets and humankind on behalf of ecological sustainability. But as consumers people easily turn out to be interest-chasers, following their egoistic values and attitudes that run counter to holistic sustainability. The motivations to follow different regulations depend on the role that a person at a particular case takes and pursues.

How confident citizens or consumers are with normative regulations in a food context depends, at least, on awareness, education, age, gender, income, profession, residence (rural/urban), nationality, religion, and ideology (Rentingô, Marsden, and Banks, 2003). People are positive about incremental development of regulations as far as no major behavioral change in everyday life is required. Plastic bags can be quite easily changed to paper bags by regulating; people accept eco-packages and happily accept withdrawal of household chemicals or detergents harmful to the environment. In contrast, it has appeared extremely difficult to formulate regulations that would combine aims of beneficial dietary value and improved dietary impact on the environment. Taxing of consumer products that contain an additional or excessive amount of sugar

Figure 11.3   A food plate model, in this case including baked pike as a main course with boiled potatoes, peas, bread, a glass of milk, and a dessert. Wild fish from a lake have a beneficial impact by removing nutrients from surface waters. Therefore, the eutrophication impact of a fish is counted to be positive and mitigates the impacts of the whole portion. (www.foodweb.ut.ee)

and fat are good examples of unsuccessful attempts. Consumer behavior is most unpredictable: people might refocus their preferences on products that are problematic in some other aspect. Open markets provide numerous alternatives to most products.

Another way to integrate dietary aims of human consumption with the aims of environmentally appreciated food production is to impose taxes on the production chain of disadvantageous food ingredients. These regulations are aimed at reorienting stakeholders to choose processes or inputs that are not damaging in any way. A very successful example of this is a pesticide tax that has diminished the use of many pesticides. This has been an additional impact to the total withdrawal of the most hazardous pesticide and has led to a very positive tendency toward improved human and environmental safety (Trienekens and Zuurbier, 2008). In some countries this has appeared as an additional driver for proliferation of organic cultivation. The tendency to reregulate use of synthetic chemicals in agriculture and food processing will be most obviously strengthened in the coming years. National action plans have been launched for implementation of an integrated plant protection policy. In food processing natural aseptic, aromatic, and other improved compounds are gradually replacing synthetic chemicals. Anxiety and doubts linked to overall chemicalization are drivers of this regulatory trend.

11.6   Food Production as a Component of a Future Bioeconomy

Sustainable food production has been promoted within the European Union (EU) by the Common Agricultural Policy (CAP) and environmental legislation. These regulations, primarily agreed at the EU level and implemented through national regulations, have outlawed the most unsustainable practices from agriculture and horticulture. In recent years, a clear focus has been set on reduction of greenhouse gases. This is, of course, linked to a larger set of policies outside agriculture. The Renewable Energy Directive (RED) and the Fuel Quality Directive have set obligatory requirements for compliance with sustainability criteria. Sustainability criteria in these directives cover land use and greenhouse gas reduction, which have not been the focus in the CAP environmental approach.

In agricultural and most possibly also in other biomass production, regulation of indirect land use change (ILUC) will be particularly problematic. In farming, this has been a critical issue concerning beef production in South America. Some calculations have been published where the carbon footprint appears to be two to three times higher than from intensive North American and European beef farming. The results are based on exploiting natural pastures as increased soya production has pushed animal farming into new areas where even land ownership is not totally under control. In the expanding bioeconomy, the discussion on impact of ILUCs has been recognized in Europe in the form of discussions centered on land use for food production or for biomass production. This discussion, and also the affinity for a regulatory approach, will be strengthened with the potential shortage of biomass in Europe and also because of the cascading principle. Cascading refers to prioritizing the high value ingredients in processing of biomass. Many food ingredients are of high value, but a high value ingredient may appear in the biochemical or even cosmetic sector. Up to now, the use of food crops for biofuels has, because of this discussion, been limited by regulations. The larger regulatory process linked to cascading and land is yet to be resolved.

In the current food production sector there are a number of private schemes that focus on more sustainable production processes, for instance on sugar, palm oil, coffee, and fish. GLOBALGAP is the most widespread non-governmental organization that sets voluntary standards for the certification of agricultural products around the globe. The certification systems are fairly complex and cover both primary producers and consumers. However, these "sustainably produced products" have had difficulties entering consumer markets partly because of price differences. Consumers' willingness to pay has not taken off, but niche markets representing these products have appeared. As informal regulatory procedures, these seem to be potentially viable.

The modern bioeconomy requires that agriculture, the main source of food, modernizes its strategies and procedures in many ways. From the aspect of the bioeconomy, there is potential to increase yield by applying new techniques, or choosing the most efficient crops (for biofuels these are often food crops), using unused land, restoring degraded lands and soil quality, and (re)introducing integrated animal–plant systems. The circular economy, as an additional approach to the bioeconomy, will be a strategic challenge for future agriculture. Farming during recent decades has developed toward a direction of the linear processing plant receiving most external inputs and processing an accelerating quantity of output with improved cost-efficacy. The circular economy claims a systemic approach and necessities horizontal and vertical networking of food chain stakeholders. The numbers of private, bilateral, or multilateral contracts will inevitably grow, and with those informal regulations will increase considerably.

Forestry has been and will continue to be, a specific source of food ingredients. Traditional forest products are berries, mushrooms, and game animals. Highly variable regulation, even in Europe, governs this area. In the Nordic countries, the right of public access complicates the business activities associated with wild berries and mushrooms. Various types of semi-farmed systems and social contracts with informal regulations have been planned for this purpose. Wood material itself can be a source of food ingredients; xylitol being one of the established examples. Sustainable use of future forests, and in the framework of the cascading requirement, could be improved by introducing more productive species into fertile areas and a more diverse set of species into less fertile forests. Some of the high tech ingredients, such as resin, require specific vegetation and management in a particular manner.

Expectations, as well as anxiety, for highly regulated land use and cascading regulation may turn out to be a fairly temporary issue. The three key questions for sustainability of the future bioeconomy would be (European Commission, 2014):

• How do we build up synergy between extensive and intensive use of land areas?

• How do we build up synergy between different production sectors, business models, and visions?

• How do we build up synergy between production or service systems with different boundaries, some of them more restricted locally and some more global?

What does this mean for regulatory processes? Before long, this will mean a total shake-up of the present regulations and will lead to a new structural combination between official, institutional regulations, and private informal regulatory processes. The creation of policy to support development of the bioeconomy is clearly a complex operation. The policy framework, between different directorates and national ministries needs to be coherent, holistic and supportive, evaluating risks and benefits in collaboration with all relevant policy sectors, academia, industry, and civil society.

11.7 Future Regionalism Related to Regulation

A regional approach that promotes and ensures sustainable practices for a food system may be a workable and applicable alternative for future food production. Integrating the circular economy into the bioeconomy represents a regional approach that takes into account divergent natural and social circumstances, material and human capital. Motivation of primary producers can be raised at regional levels and commitment strengthened. According to a few preliminary examples, regional scale investments in the bioeconomy have been particularly successful. Regional and national support, in terms of policy harmonization, regulatory processes, and funding, has played a critical enabling role in those cases.

The previous approach was challenging to the European Commission and, also, to national regulatory institutions. Long-term political commitment and a supportive, predictable, and science-based regulatory environment should be provided and applied to regional needs to find solutions to the challenges in their specific cases. Some rethinking on food safety and security measures and regulations may be needed when emphasizing a regional context (Gorris, 2005). But there is also a clear need to develop and implement a coherent communications strategy to raise consumer awareness to the regional circular economy for food systems and the opportunities for and barriers to its development. This should be done in the context of the grand challenges facing future generations.

Effective cascading inevitably requires centralizing in relation to regional emphasis. In the food sector this refers largely to functional food components and their specific processing. Setting up of specific interregional programs, directed toward regional cooperation, funding, and development of a joint strategic policy for specified high tech production, will be important. This will become a platform to integrate principles of regulatory processes that were originally developed for the diverse source areas from where the raw materials are gathered for the cascading process.

An EU-wide public procurement program has played an important role in boosting awareness of sustainability in the food sector. Up to now, it has somewhat worked against regional and local food systems, but new regulation represents steps toward more regional and locality-supportive schemes. Furthermore, a regional approach could provide stronger legislation for ecologically sensitive habitats, soil and water protection, and collection and recycling of biomass. New, more regional approaches could be taken up for management of residues, wastes, and other bio-based products.

Labeling enables consumers and businesses to identify sustainable products. Labeling has been strongly integrated into regions and localities as the origin of the food product has been regarded as a kind of certification of quality and sustainability (Clifton, 2014). Inspections for an area have been backing the label. Traditions of regional- or local- based labeling are alive, applicable, and regulated. The Product Environmental Footprint (PEF) by DG Environment, at present organizing a pilot test process, plays an important role in gathering information on the labeling process for environmentally sustainable products and coordinating development of generic regulations for sustainability assessments in the food and drink sector. At the same time, at national level, many countries are looking for potential means to support sustainability assessments and potential ways to link sustainability indicators with food labeling targeted at end users or food service companies. In terms of sustainability, all these regulatory processes have been informal business to business or business to consumer linkages. The key issue, to be solved in future, is how regional labeling and product-based labeling are to be integrated. For smaller businesses, it would be helpful to be engaged with the local labeling process that would be composed of a number of key sustainability indicators applied to the particular locality. For larger companies that need to combine food ingredients from different sources, the PEF type of approach, might be more applicable. Interpretations of results from both systems and sustainability communication should be possible. No doubt, both of these labeling processes will gain in importance and publicity, and lead to various regulatory processes in the future.

11.8 What Is Needed for Regulatory Policy Development

Regulatory processes for sustainability (bioeconomy or circular economy) cannot be managed under a single policy, but integration of all necessary policies should be the aim. Policy coherence in the topic of sustainable food production is required at all levels: at the EU, national, regional, and local levels. In practice, this refers to relationships, collaboration, and conflicts between different policy sectors and different lines of ministries, different economic sectors, and also different business and process owners (Beulens *et al.*, 2005). In addition, effective mechanisms and institutional models (e.g., examples, good practices, and inspiring examples) of horizontal and vertical policy integration in the topic of sustainable food are needed, specifically those taking into account reliability issues and cooperation with stakeholders.

Re-evaluation of currently applied systems and policies is needed: success factors, best practices and bottlenecks should be used for learning in the policy cycle.

The key issue to understand is that sustainability is not a static linear development but a cyclic dynamic process during the course of which it is essential to know which state of the process is ongoing. The cyclic process of the dimensions of sustainability was described by Holling (2001).

A new research approach and a new type of connectivity are also needed for future regulation of sustainability. The best ways are needed to communicate research results about conflicts and trade-offs between the environmental, social, and economic aspects of sustainable food production and consumption. Innovative ways of how to link research to policy-making and related success factors must be created. Knowledge management about future developments – using scenarios and visioning processes is needed. But most importantly, support to stakeholders is needed to identify success factors and take up challenges of how to include contextual/framework-supporting resources into business. When striving for a sustainable bioeconomy, regulatory processes should necessarily be facilitative and compatible in a wider, multifunctional context.

References

Aruoma, O.I., 2006. The impact of food regulation on the food supply chain. *Toxicology*, **221**, 119–127.

Berger, G., Pisano, U., Szlezak, J., and Csobod, E., 2013. The CORPUS Research Agenda for Sustainable Food Consumption in Europe. The SCP-Knowledge Hub, http://www.scp-knowledge.eu/ (accessed June 23, 2015).

Beulens, A.J.M., Broens, D.-F., Folstar, P., and Hofstede, G.J., 2005. Food safety and transparency in food chains and networks relationships and challenges. *Food Control*, **16**, 481–486.

Clifton N, 2014. Towards a holistic understanding of county of origin effects? Branding of the region, branding from the region. *Journal of Destination Marketing & Management*, **3**, 122–132.

European Commission, 2014. Where next for the European bioeconomy? The latest thinking from the European Bioeconomy Panel and the Standing Committee on Agricultural Research Strategic Working Group (SCAR). EC, Brussels, Directorate-General for Research and Innovation, Directorate F – Bioeconomy Unit F1 – Strategy, http://ec.europa.eu/research/bioeconomy/pdf/where-next-for-european-bioeconomy-report-0809102014_en.pdf (accessed June 23, 2015).

Gorris, L.G.M., 2005. Food safety objective: an integral part of food chain management. *Food Control*, **16**, 801–809.

Hess, D., 2014. The future of sustainability reporting as a regulatory mechanism. In: D.R. Cahoy and J.E. Colburn, eds. *Law and the Transition to Business Sustainability. Perspectives on Sustainable Growth*. New York: Springer International Publishing, pp. 125–139.

Holling, C.S., 2001. Understanding the complexity of economic, ecological, and social systems. *Ecosystems*, **4**, 390–405.

Murphy, P.E., Öberseder, M., and Laczniak, G.R., 2013. Corporate societal responsibility in marketing: normatively broadening the concept. *Academy of Marketing Science Review*, **3**, 86–102.

Remans, R., Wood, S.A., Saha, N., Anderman, T.L., and DeFries, R.S., 2014. Measuring nutritional diversity of national food supplies. *Global Food Security*, **3**, 174–182.

Rentingô, H., Marsden, T.K., and Banks, J., 2003. Understanding alternative food networks: exploring the role of short food supply chains in rural development. *Environment and Planning*, **35**, 393-411.

Trienekens, J. and Zuurbier, P., 2008. Quality and safety standards in the food industry, developments and challenges. *International Journal of Production Economics*, **113**, 107–122.

Index

Supply Chain Management for Sustainable Food Networks, First Edition. Edited by Eleftherios Iakovou,
Dionysis Bochtis, Dimitrios Vlachos and Dimitrios Aidonis.
© 2016 John Wiley & Sons, Ltd. Published 2016 by John Wiley & Sons, Ltd.